COMPREHENSIVE

Section 1. THE PRACTICE AND THEORY OF KINETICS

Volume 1 The Practice of Kinetics

Volume 2 The Theory of Kinetics

Volume 3 The Formation and Decay of Excited Species

Section 2. HOMOGENEOUS DECOMPOSITION AND ISOMERISATION REACTIONS

Volume 4 Decomposition of Inorganic and Organometallic Compounds

Volume 5 Decomposition and Isomerisation of Organic Compounds

Section 3. INORGANIC REACTIONS

Volume 6 Reactions of Non-metallic Inorganic Compounds

Volume 7 Reactions of Metallic Salts and Complexes, and Organometallic Compounds

Section 4. ORGANIC REACTIONS (6 volumes)

Volume 9 Addition and Elimination Reactions of Aliphatic Compounds

Volume 10 Ester Formation and Hydrolysis and Related Reactions

Volume 13 Reactions of Aromatic Compounds

Section 5. POLYMERISATION REACTIONS (2 volumes)

Section 6. OXIDATION AND COMBUSTION REACTIONS (2 volumes)

Section 7. SELECTED ELEMENTARY REACTIONS (2 volumes)

Additional Sections

HETEROGENEOUS REACTIONS

SOLID STATE REACTIONS

KINETICS AND TECHNOLOGICAL PROCESSES

COMPREHENSIVE CHEMICAL KINETICS

CHEMICAL KINETICS

EDITED BY

C. H. BAMFORD
M.A., Ph.D., Sc.D. (Cantab.), F.R.I.C., F.R.S.
Campbell-Brown Professor of Industrial Chemistry,
University of Liverpool

AND

C. F. H. TIPPER
Ph.D. (Bristol), D.Sc. (Edinburgh)
Senior Lecturer in Physical Chemistry,
University of Liverpool

VOLUME 10

ESTER FORMATION AND HYDROLYSIS AND RELATED REACTIONS

ELSEVIER PUBLISHING COMPANY
AMSTERDAM – LONDON – NEW YORK
1972

ELSEVIER PUBLISHING COMPANY
335 JAN VAN GALENSTRAAT
P. O. BOX 211, AMSTERDAM, THE NETHERLANDS

AMERICAN ELSEVIER PUBLISHING COMPANY, INC.
52 VANDERBILT AVENUE
NEW YORK, NEW YORK 10017

LIBRARY OF CONGRESS CARD NUMBER 74-168908
ISBN 0-444-40957-2

WITH 3 ILLUSTRATIONS AND 74 TABLES

PRINTED IN THE NETHERLANDS

COMPREHENSIVE CHEMICAL KINETICS

Contributors to Volume 10

S. J. Benkovic	Department of Chemistry, College of Science, The Pennsylvania State University, University Park, Pa., U.S.A.
A. J. Kirby	University Chemical Laboratory, Cambridge, England
R. J. E. Talbot	Roche Products Limited, Welwyn Garden City, Herts., England

Preface

Section 4 deals almost exclusively with reactions recognised as organic in a traditional sense, but excluding (unless very relevant) those already considered in Sections 2 and 3 and biochemical systems. Also oxidations, *e.g.* of hydrocarbons, by molecular oxygen, polymerization reactions and fully heterogeneous processes are considered later. The relationships of mechanism and kinetics, *e.g.* effects of structure of reactants and solvent, isotope effects, are fully discussed. Rate parameters of individual elementary steps, as well as of overall processes, are given if available. We have endeavoured, in conformity with our earlier policy, to organise this section according to the types of chemical transformation and with the minimum of recourse to mechanistic classification. Nevertheless it seemed desirable to divide up certain general processes on the basis of their nucleophilic or electrophilic character.

Volume 10 is devoted to formation and solvolysis of esters and related reactions, with discussion of the effect of neighbouring groups and biological implications, *e.g.* enzyme action, where appropriate. The first chapter deals mainly with esters of the inorganic acids of phosphorus and sulphur, Chapter 2 with the formation and solvolysis of esters of organic acids and the final chapter with the solvolysis of related derivatives of carboxylic acids, *e.g.* halides, amides, anhydrides, cyanides, carbamic acid derivatives.

The editors are very grateful for the continuing support and advice from members of the editorial board, and also for much invaluable assistance from Dr. D. Bethell of the Department of Organic Chemistry.

C. H. Bamford
C. F. H. Tipper

Liverpool
April, 1972

Contents

Chapter 1

Hydrolytic Reactions of Inorganic Esters

S. J. BENKOVIC

1. Introduction

The intent of this chapter is to review recent developments that extend our understanding of the mechanism of hydrolysis of various inorganic esters. Owing to the fact that a number of excellent critical summaries on particular aspects of this topic have been recently written[1–6], the author will concentrate on those experimental discoveries and interpretations which, in his opinion, have and will significantly influence the construction of hypotheses and experiments in this field. As a consequence, this chapter will be mainly concerned with the chemistry of the esters derived from phosphorus and sulfur (not with esters of nitrogen acids) and in particular with systems and findings that are biochemically relevant. No attempt will be made to provide a thorough historical review of this area except in those cases where a brief summary enables the reader to appreciate more fully the topic being discussed. The relevance of the chemistry to certain biological systems will be mentioned on occasion, in the hope that the reader might utilize this information to initiate his own comparisons.

2. Phosphate esters

2.1 METAPHOSPHATE MECHANISMS FOR MONOESTERS

Converging lines of evidence have led to a general acceptance of the monomeric metaphosphate mechanism for the hydrolysis of phosphate monoester monoanions. The pH rate profile for aryl and alkyl phosphate monoester hydrolyses commonly exhibits a rate maximum near pH 4, where the concentration of the monoanion is at a maximum. The proposed mechanism is based on these principal points of evidence: (*a*) a general observation of P–O bond cleavage; (*b*) the entropies of activation for a series of monoester monoanions are all close to zero, which is consistent with a unimolecular rather than a bimolecular solvolysis where entropies of activation are usually more negative by 20 eu[7]; (*c*) the molar product composition (methyl phosphate : inorganic phosphate) arising from the solvolysis of the monoester monoanion in a mixed methanol–water solvent usually approximates the molar ratio of methanol:

References pp. 51–56

water, which suggests the presence of a highly reactive electrophilic species and (*d*) the existence of a linear free-energy relationship between the logarithmic rate of hydrolysis of the monoanions and the dissociation constant of the corresponding phenol or alcohol (Fig. 1). The observation cited under heading

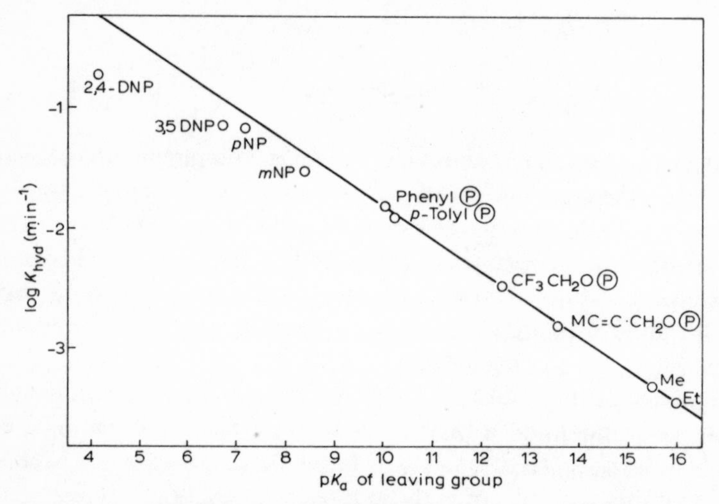

Fig. 1. Hydrolysis rates for the monoanions of phosphate esters at 100°C as a function of the pK_a of the conjugate acid of the leaving group. D, p and mNP are di-, *p*- and *m*-nitrophenyl phosphates.

(*c*) is not universal and we will return to this point in later discussion. The small slope observed for the structure–reactivity correlation (0.3) suggests that the mechanism involves loss of alcohol or phenol rather than alkoxide or phenoxide and, therefore involves proton transfer to the oxygen atom in a preequilibrium step[8,9], *viz.*

$$\text{RO-}\overset{\text{O}}{\overset{\|}{\underset{\text{O}^-}{\underset{|}{\text{P}}}}}\text{-OH} \rightleftharpoons \underset{\text{H}}{\overset{+}{\text{RO}}}\text{-}\overset{\text{O}^-}{\overset{\vdots}{\underset{\text{O}^-}{\underset{\vdots}{\text{P}}}}}\text{=O} \longrightarrow \text{ROH} + [\text{PO}_3^-] \qquad (1)$$

Proton transfer may proceed directly or *via* a six-membered cyclic transition state involving a molecule of water. A calculation of the intermediate zwitterionic concentration for the hydrolysis of methyl phosphate monoanion, based on the pK_a values for methanol and methyl phosphate dianion, predicts the first-order rate coefficient for zwitterion decomposition to be *ca.* 10^6 sec^{-1} at 100°C. This value is in good agreement with the observed rate of hydrolysis and, considering the assumptions involved, with the rate of P–O bond fission of the presumed zwitterionic intermediate (2) formed in the Hg(II) catalyzed solvolysis of phosphoenolpyruvic acid, a model reaction for pyruvate kinase[10].

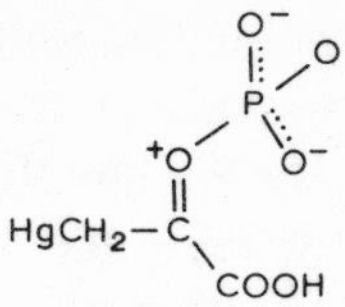

$$HgCH_2-C(COOH)=O^{+}-PO_2(O^{-})_2 \qquad (2)$$

Negative deviations from the linear free-energy relationship are encountered when the leaving group corresponds to a $pK_a < 7.0$, implying that the rate of diffusion apart of the zwitterion fragments is now limited by proton transfer. Indeed a deuterium solvent isotope effect $k_{H_2O}/k_{D_2O} = 1.45$ is found for the hydrolysis of the monoanion of 2,4-dinitrophenylphosphate. In contrast, the value for methyl phosphate monoanion hydrolysis is 0.87. The above mechanism reasonably accommodates the available results; other pathways involving pentacovalent phosphorus as an intermediate species are made less likely by the lack of isotopic evidence for the required addition–elimination reaction and the low susceptibility of the monoanion to nucleophilic attack.

A similar mechanism has been proposed for the hydrolysis of S-aryl phosphorothioates based on analogous chemical results[11]. Internal comparison of the rate coefficients for hydrolysis of pNO$_2$, pCl and p-H substituted S-aryl phosphorothioate monoanions reveals almost a complete lack of sensitivity to the pK_a of the leaving group, again emphasizing that protonation of the leaving group is important in the transition state. Moreover, the great facility with which thiophosphate monoanions hydrolyze is not a consequence of leaving group pK_a. This follows from the sixty-fold ratio of rate coefficients for monoanion hydrolysis of the p-NO$_2$ S-aryl relative to 2,4-dinitro O-aryl phosphate (normalizing the pK_a). This factor probably partially reflects the considerable difference (50 kcal·mole^{-1}) in P–S bond energy relative to P–O. It should be noted that a deuterium solvent isotope effect ranging from 1.8 for pNO$_2$ to 1.4 for p-H was detected which may signify partial rate-determining protonation. However, in the absence of other evidence, *i.e.* deviation from a structure–reactivity correlation, this interpretation is not unequivocal since solvent isotope effects may arise from secondary changes in solvation during the progress of the reaction[12].

The study of the monoanions of a third class of phosphorus-containing monoesters, the phosphoramidates, has been restricted for the present to p- and o-substituted aryl phosphoramidates, and phosphoramidate and its O-methyl derivatives[13–16]. Two striking differences from the behavior of O- and S-phosphate monoester monoanions are noted in the experimental criteria compiled in order to postulate mechanism; (*a*) solvolysis in mixed organic solvents, particularly 50% v/v dioxan–water, results in a 50% decrease in the rate of hydrolysis of o- and p-substituted aryl phosphoramidate monoanions and all ionic forms of phosphoramidate and its O-methyl derivatives excepting the parent monoanion, and (*b*) partitioning of the aryl phosphoramidates and phosphoramidate in mixed methanol–water solvents leads to a preponderence of methyl phosphate over inorganic phosphate with a selectivity ratio for methanol approaching 10 to 1. The kinetically indistinguishable reactions involving attack of water on the phosphoramidate monoanion, or hydroxide ion on the neutral species of phosphoramidate have been resolved in favor of the former by demonstrating that the reaction of the neutral species of O-methyl phosphoramidate with hydroxide ion is insignificant. Granting the equivalence of a methyl and hydrogen in bimolecular substitution reactions on phosphorus, this result further suggests that hydrolysis of the monoanion form of phosphoramidate proceeds because of zwitterion formation. The proposed mechanism for *p*- and *o*-substituted aryl phosphoramidate and phosphoramidate monoanions, therefore, is a facsimile of equation (1) with the additional proviso that the generated metaphosphate is not a free species, in order to accommodate the methanol selectivity. Some support for this view is gained by (*a*) the decrease in methyl phosphate formation as the leaving group is improved through reaction with hypochlorous acid[17] and (*b*) the tendency for methyl phosphate formation in excess of the molar ratio of methanol to water in the mixed solvent to diminish as the phosphoramidate is successively protonated[15,16]. The progressive decrease in the unimolecular character of the transition state for phosphoramidate hydrolysis upon protonation can be viewed as resulting from the lesser resonance stabilization of a protonated metaphosphate to the parent ion. Alternatively the increasing participation by water may be ascribed to an increase in the electropositive character of the central phosphorus atom owing to a decrease in π bonding upon the protonation of the oxygen atoms[18]. The effect of dioxan on the rate of hydrolysis of the various species has been rationalized in terms of such a model.

An obvious difficulty arises with this rather elaborate rationale when phosphoramidate and aryl phosphoramidate monoanions are compared; for example, the dissimilarity of the dioxan effect yet the identity of product distribution observed in methanol–water competition experiments. Preliminary studies in the author's laboratory have revealed striking differences in the hydrolytic behavior between a series of phosphoramidates derived from primary aliphatic amines and the above aryl systems. No linear structure–reactivity relationship between the logarithmic rate of hydrolysis of the monoanion species and the pK_a of the amine is observed[19]. Moreover, the rate of hydrolysis of phosphoramidate monoanions derived from aliphatic amines is at least 10^4 times slower than those formed from aryl amines. In contrast, only a thirty-fold decrease in rate is observed for the corresponding ΔpK_a in the O-phosphate monoester series. The suspicion that mechanism (1), even with the above proposed modification, is not an accurate description of phosphoramidate monoanion hydrolysis derives some further support from the observation that the monoanion is subject to nucleophilic attack by substituted pyridines although the sensitivity to basicity is low (β, 0.2)[20]. A consequence of this susceptibility to nucleophilic attack is that N-methylimidazolium phosphoramidate (3), an analogue of the presumed labile zwitterionic intermediate, can be successfully employed in the efficient synthesis of a series of aryl and aliphatic phosphoramidate monoesters in aqueous solution[21].

$$CH_3-\overset{+}{N}\text{(imidazolium)}N-P(=O)(O^-)_2 \quad Na^+ \qquad (3)$$

Whether this unsatisfactory state of affairs is a consequence of a "borderline" situation as encountered in displacements on carbon, or the result of a pentacovalent addition intermediate, remains to be demonstrated.

With few exceptions the dianions of monoalkyl and monoaryl phosphates are unreactive but with a good leaving group, *e.g.* carboxylate, dianions undergo hydrolysis. The same reasoning applies to a dinitrophenoxide anion and the pH–rate profile for the hydrolyses of 2,4- and 2,6-dinitrophenyl phosphates differs from those of other aryl phosphates in that the dianion is more reactive than the monoanion species (Fig. 2)[22,23]. This reactivity is at-

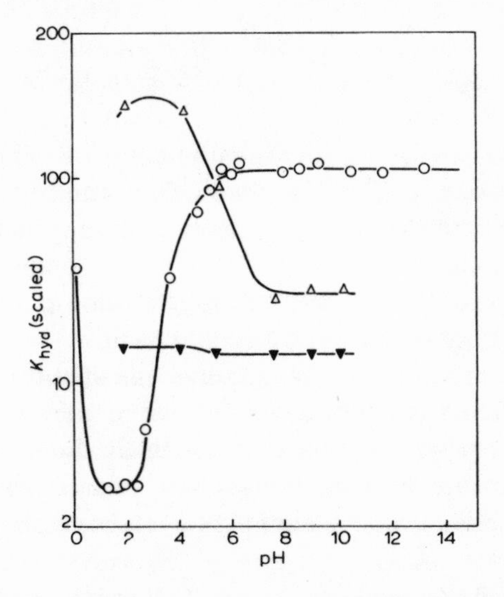

Fig. 2. pH–rate profiles for three representative esters at 39°C, ionic strength 1.0, on adjusted scales. △, $k_{hyd} \times 10^7$ min^{-1} for 2,4,6-trichlorophenyl phosphate; ▼, $k_{hyd} \times 10^6$ min^{-1} for 2-chloro-4-nitrophenyl phosphate; ○, $k_{hyd} \times 10^4$ min^{-1} for 2,4-dinitrophenyl phosphate.

tributed to spontaneous hydrolysis of the dianion and not the kinetically indistinguishable attack of hydroxide ion upon the monoanion owing to (*a*) the improbably large rate coefficient associated with the latter mechanism and (*b*) the results of solvolysis in aqueous alcohols which are consistent with the elim-

References pp. 51–56

ination of some form of metaphosphate which is captured preferentially by alcohol. For 2,4-dinitrophenyl phosphate dianion the approximate relative reactivities toward metaphosphate are methanol : water = 6 : 1. The absence of a correlation between the acceleratory effects of mixed methanol–water solvents on the rate of solvolysis and product composition suggests that the rate-limiting and product-forming steps are different. It is of interest to note that the relative reactivity of alcohols toward the dianion of 2,4-dinitrophenyl phosphate is in the order methanol > ethanol > water > 2,2,2 trifluoroethanol > 2-propanol. Whether these observations can be employed as an accurate probe to detect the extent of unimolecular P–O or, as in the above case, unimolecular P–N bond scission is problematic. There is the possibility that "free" metaphosphate may itself exhibit selectivity in aqueous solution. Insofar as the analogy between isoelectronic sulfur trioxide and metaphosphate is deemed valid, the above order is observed in the solvolysis of monomeric sulfur trioxide generated from thermal decomposition of a sulfur trioxide polymer[24]. Nor can one ignore the possibility of specific solvation effects especially in the considered cases which involve *o*-substitution.

Additional compelling evidence for the unimolecular elimination mechanism (4)

$$Ar{-}O{-}P(=O)(O^-)_2 \longrightarrow [PO_3^-] + ArO^- \quad (4)$$

however, is the high sensitivity of the logarithmic rate of hydrolysis to the pK_a of the leaving group (slope, −1.23), the small positive entropy of activation, the insignificant deuterium solvent isotope effects, and the small sensitivity (β = 0.1) of dianions to nucleophilic attack by amines. A mechanism involving unimolecular decomposition of S-(4-nitrophenyl)phosphorothioate dianion to metaphosphate has been deduced on a similar basis.

The validity of the assumption that random phosphorylation of the components in mixed solvents constitutes experimental evidence for monomeric metaphosphate can often be questioned. A relatively direct demonstration for the existence of an analogue of metaphosphate results from a study of the alkaline solvolysis of optically active methyl N-cyclohexylphosphoramidothioic chloride[25], *viz.*

$$(CH_3O)(C_6H_{11}NH)P(=S)Cl \quad (5)$$

Previous investigations by Traylor and Westheimer[26] and Crunden and Hudson[27] had strongly supported the suggestion that the rapid alkaline reaction of phosphorodiamidic fluorides or chlorides and pyrophosphoramides which contain at least one amide hydrogen was by way of an anion derived from the amidate, *viz.*

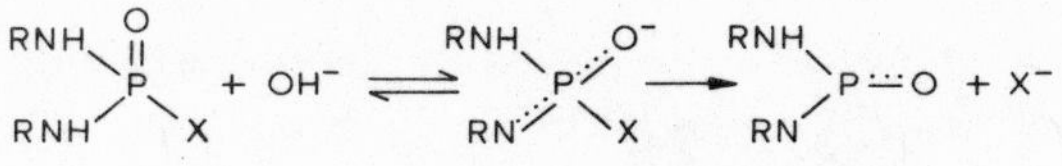

$$(RNH)_2P(=O)X + OH^- \rightleftharpoons (RNH)(RN)P(O^-)X \longrightarrow (RNH)(RN)P{=}O + X^- \quad (6)$$

In the present case, alkaline solvolysis forms almost completely racemic product in contrast with that from the neutral hydrolysis which is stereospecific. It is clear that the alkaline solvolysis involves a symmetrical intermediate, a metaphosphorimidothioate, which reasonably may be expected to be planar (isoelectronic with SO_3). The close similarity in reactions of thiophosphoryl and phosphoryl centers is well established for bimolecular displacements on phosphorus and the resemblance apparently extends to the metaphosphate eliminations.

Although the metaphosphate mechanism for hydrolysis is well documented, such a pathway remains to be demonstrated in a biological system. Our present knowledge of many enzymic reactions allows, at best, the formulation of a preliminary mechanism, *i.e.* the chemical identity of substrates and enzymic intermediates and the minimal kinetic scheme. For example, much recent attention has been focused on the remarkable stability of the covalent phosphoryl-enzyme (an O-phosphoryl serine) derived from *E. coli* alkaline phosphatase[28] and inorganic phosphate, and on a systematic kinetic study of the enzyme's substrate specificity (O-, N- and S-monoesters)[29]. Dephosphorylation of the enzyme, however, does not appear to be *via* a metaphosphate mechanism[30].

The problem of oxidative phosphorylation has been approached through model studies that utilize the phosphorylating potential of metaphosphate. In the mitrochondrial process inorganic phosphate and adenosine diphosphate are converted to adenosine triphosphate. Wieland, in a series of papers (*e.g.* ref. 31), has shown that a variety of thiolactones can activate inorganic phosphate in the presence of bromine for transfer to adenosine diphosphate (ADP). The intermediate may be an acyl phosphate or a sulfonium salt similar to that postulated by Higuchi and Gensch[32] and by Lambeth and Lardy[33], *viz.*

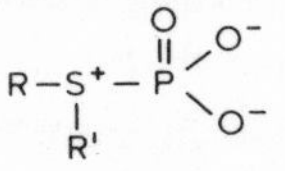

$$R{-}\overset{+}{S}(R'){-}P(=O)(O^-)_2 \quad (7)$$

A particularly fascinating model is the air oxidation of ferrohemochrome in the presence of imidazole, inorganic phosphate and adenosine diphosphate which generates ATP. The active phosphorylating agent apparently is an imidazolyl phosphoramidate which has been detected chromatographically. The intriguing question is the formation of the phosphoramidate which requires the dehydration of inorganic phosphate. Wang *et al.*[34] presently favor a free-radical mechanism which features a nucleophilic imidazolyl radical formed by two-electron oxidation of the ferrohemochrome, *viz.*

References pp. 51–56

$$\text{Imidazole} + H_2PO_4^- \xrightarrow{+e^-} \text{Im-N-}PO_3^{2-} + H_2O \qquad (8)$$

However, owing to the complexity of the system and the difficulty of nucleophilic attack on nonesterified phosphate, other mechanisms appear possible, *i.e.* activation of phosphate through a ferrohemochrome–phosphate complex followed by nucleophilic attack of imidazole.

2.2 NUCLEOPHILIC REACTIONS AT ACYCLIC PHOSPHORUS ESTERS

Previous studies on the reactions of nucleophiles at the phosphorus atom of $-PO_3^{2-}$ include the reaction of tertiary amines and fluoride anion with acetyl phosphate[35], and the measurement of the reactivity of a series of amines towards *p*-nitrophenyl phosphate dianion[36] and the zwitterion derived from the phosphoramidate monoanion. A striking and encompassing feature of these reactions is the small sensitivity of their rates to the basicity of the nucleophile as reflected in the Bronsted β of 0.22 and 0.13 for phosphoramidate and *p*-nitrophenyl phosphate respectively. These have been interpreted as displacement reactions in which only a small amount of bond formation to the nucleophile has occurred in the transition state and have been used to reinforce the above arguments for the predominant unimolecular or metaphosphate character of these reactions. Kirby and Varvoglis[37] have examined the reactions of triethylenediamine, hydrazine, and various pyridines with a series of 2-nitro-4-substituted-phenyl phosphate dianions. Logarithmic plots of the second-order rate coefficients for amine attack against pK_a of the leaving group reveal that the amine reaction is influenced by the nature of the leaving group (average slope *ca.* 1.0). Moreover, the displacement reaction is responsive to the chemical identity of the amine with the sensitivity to the leaving group being significantly greater for the less basic nucleophile. Hydrazine apparently does not exhibit an α-effect. Since there is no doubt that a bimolecular displacement reaction on phosphorus is involved—trapping of the anticipated phosphoramidate products, lack of a deuterium solvent isotope effect, $\Delta S^{\ddagger} \cong -20$ eu, etc.—it is remarkable that the rates of the second-order substitution reactions become completely independent of the basicity of the nucleophile with the dianion of 2,4-dinitrophenyl phosphate. The zero Bronsted coefficient β (Fig. 3) for the reactions with this dianion indicates that bond formation to the nucleophile in the transition state has not proceeded even to the small extent observed with *p*-nitrophenyl phosphate dianion. These amine reactions, although typically second-order, may represent a bimolecular substitution or more probably, especially with hindered tertiary amines such as 2-picoline

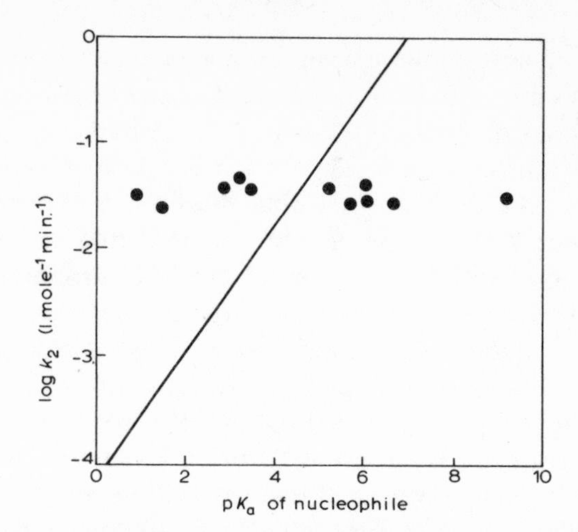

Fig. 3. Bronsted plot for the reactions of substituted pyridines with the dianion of 2,4-dinitrophenyl phosphate. The line, which is taken from the corresponding plot for the monoanions, is included for comparison.

or 2,6-lutidine, a specific molecular interaction more akin to solvation of the developing electrophilic center rather than the mechanism of bimolecular displacement encountered with sp^3 hybridized carbon. Reactions of the same series of pyridines with the monoanion of 2,4-dinitrophenyl phosphate, however, correlate with a Bronsted coefficient of 0.3–0.6. Furthermore, the logarithmic rate of reaction of nicotinamide with a series of 2-nitro-4-substituted phenyl phosphate monoanions as a function of the pK_a of the leaving group is linear with slope of *ca.* −1. Consequently, as for the above case of the dianion species, nucleophilic attack by amines on the phosphorus of a monoanion is sensitive to the pK_a of the leaving group and rapidly becomes less significant as the pK_a increases. However, unlike the dianion species, the strong dependence of reaction rate on the nature of both the nucleophile and leaving group is consistent with a classical S_N2 mechanism with bond making and bond breaking both well advanced in the transition state. A similar mechanism may be written for the nucleophilic attack of anions including the attack of fluoride ion on the monoanions of acetyl phosphate[35] and 2,4-dinitrophenyl phosphate[37], of alkoxide and hydroxide on the dianions of 2,4- and 2,6-dinitrophenyl phosphate[38] and the intramolecular nucleophilic attack by alkoxide ion in the hydrolysis of diester monoanions of 2-hydroxypropyl aryl or alkyl phosphates[39]. It is intriguing that these reactions occur despite the electrostatic repulsions. However, these are cases involving either nucleophiles of maximal basicity (alkoxide) or exceptional nucleophilicity (fluoride) towards phosphorus.

Despite the biological importance of simple acyclic phosphate diesters, which include the nucleic acids, little detailed work had been done on their hydrolytic reactions although monoester monoanions serve as preliminary models. Kirby and Younas[40] have now established a structure–reactivity correlation for the spontaneous hydrolysis of a series of diaryl phosphate diester monoanions which exhibits a high sensitivity to the nature of the leaving group (slope -1.16, 39°C). The experimental difficulties establishing the relationship become apparent with the prediction that diphenyl phosphate would have a half-life of nearly 180 years in water at 100°C. The proposed mechanism based on the usual experimental criteria – structure–reactivity correlation, entropy of activation, deuterium solvent isotope effects – involves a simple bimolecular attack by water on phosphorus, perhaps subject to general base catalysis. There is no evidence for elimination of a metaphosphate ester in agreement with the results of Bunton and Farber[41]. Extending the study to include a range of nucleophiles[42] revealed that (*a*) the rate of attack by nucleophilic anions is retarded on average by a factor of 100 owing to electrostatic repulsions and (*b*) separate plots for the attack by a series of pyridines or oxyanions (acetate, formate, and *p*-substituted phenoxides) on 2,4-dinitrophenyl methyl phosphate are characterized by Bronsted coefficients of 0.31. As expected, the Bronsted β for nucleophilic attack by amine decreases as the pK_a of the leaving group decreases. The relatively low sensitivity to the basicity of the nucleophile as measured by β defines the phosphorus center of diesters as a hard electrophile, comparable to the aromatic carbon of the 2,4-dinitrophenyl system[42].

The remaining class of simple esters, the dialkyl aryl triesters, also hydrolyze with a spontaneous rate which is of intermediate sensitivity to the nature of the leaving group (slope -0.99, 39°C). By comparing the various linear free-energy relationships at 39°C, it is seen that the order of reactivity is diaryl diester anions $<$ dialkyl aryl triesters $<$ monoester monoanions (Fig. 4). Monoester dianions are more reactive than triesters and monoester monoanions, when the leaving group of the former has a $pK_a < 10$ and < 5 respectively. Comparable reactivity for all ester types excepting the monoester dianion is encountered with a leaving group of $pK_a \approx 2$. The hydrolysis of the triester, therefore, is less sensitive to the basicity of the leaving group than that of any other species with the exception of the monoester monoanions. The proposed mechanism is similar to that above for diester hydrolysis. With a series of nucleophiles and various six-membered cyclic triesters, a number of structure–reactivity correlations also have been established[43]. The six-membered ring confers no special properties on phosphorus. For example, logarithmic rates for the reaction of hydroxide with a series of 2-(substituted-phenyl)-1,3,2-dioxaphosphorinane-2-oxides (9) are a linear

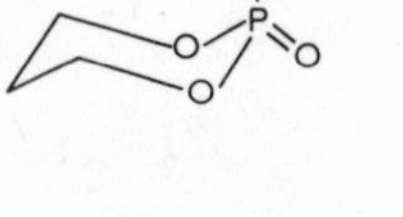

(9)

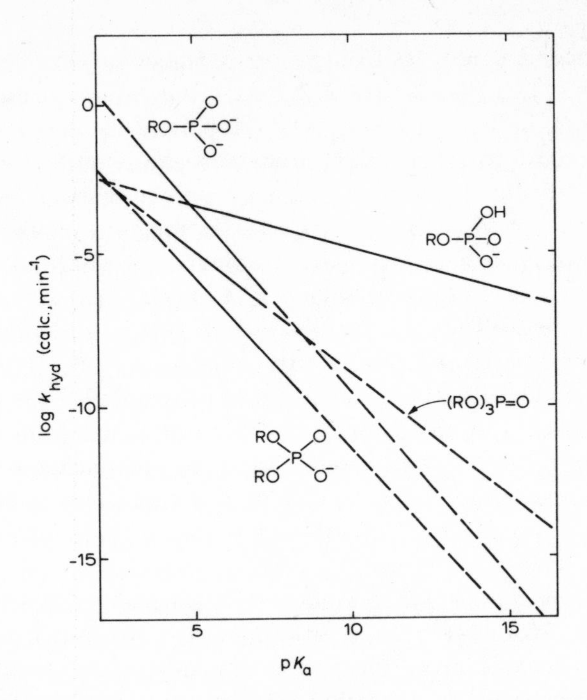

Fig. 4. Plot of the hydrolytic rate coefficients as a function of pK_a of the leaving group for various ester types.

function of the pK_a of the leaving group with slope of -0.41. A similar slope is found for hydroperoxide (exhibits an α-effect) and trifluoroethoxide but changes rapidly to -0.65, -0.88 and -0.99 for phosphate dianion, acetate and water, respectively. This observation is consistent, but does not uniquely require a mechanism for the displacement reaction *via* a pentacovalent addition compound. One can argue either in terms of changes in the rate-determining step, or in the position of the transition state relative to the reaction coordinate as a function of the basicity of the nucleophile. Since the observed reactions involve predominantly nucleophilic catalysis, the variations in slope are not a result of the incursion of general base catalysis which dominates for the poorer leaving groups (X = H). For a series of pyridines, attack on the 2,4-dinitrophenyl ester of (9) correlates with a β of 0.61 reflecting the anticipated increased degree of bond formation in the transition state as dissociation of the phosphoryl oxygens is prevented. Consequently in terms of β there is a progressive increase in bond formation between nucleophile and ester in the transition state in the order triester > diester monoanion > monoester mono- and dianion. Since the dependency of rate coefficients for a given nucleophile on the variations in the leaving group remains relatively constant ($\beta = -1.0$ to -1.1) regardless of the degree of substitution of the ester, the transition state appears to progress from predominantly unimolecular (metaphosphate-like) to a coupled S_N2 displacement process from mono- to triesters. The fact that β for nucleophilic attack is less than β for the leaving group argues against a pentacovalent intermediate in the monoester mono- and dianions and diester monoanion cases[19]. In regard to biochemical processes one point is unmistakably clear, that is the difficulty in carrying out nucleophilic displacements in O-phosphate esters except in cases with superior leaving groups ($pK_a < 7$). It should also be pointed out that progressive protonation or esterification may in some cases lead to mechanisms involving C–O bond cleavage. Prominent examples are the hydrolyses of D-ribose- and D-glucose-1-phosphates[44], isopropyl phosphate, and tri- and mono-*t*-butyl phosphates[45–47].

Stereochemical evidence for the mechanism of nucleophilic substitution at a tetracovalent phosphorus atom has been mainly forthcoming from the experiments of Michalski *et al.*[48–50] with esters containing thiophosphoryl centers. Until recently, the inaccessibility of suitable optically active phosphorus compounds precluded stereochemical investigations on the mechanisms of their hydrolysis.

Optically active phosphoryl compounds of the general formula RR'P(O)X (where X = Cl, OR, SR or NR_2) are hydrolyzed to the optically inactive acids RR'POOH. The preparation of the stable optically active O-ethyl ethylphosphonochloridothionate, Et (EtO)P(S)Cl, has presented the opportunity for the synthesis and stereochemical study of the features of nucleophilic displacement reactions for a wide range of optically active organo phosphorus derivatives[48]. The alkaline hydrolysis of optically active pyrophosphonothionates and the parent chloride[49,50] proceed *via* inversion of configuration, confirming the initial studies of Green and Hudson[51]. The most likely transition state appears to be a trigonal bipyramid where the two axial bonds may be viewed as *pd* hybrids and the three basal bonds as sp^2 hybrids, with the entering nucleophile and the leaving group occupying axial positions[52]. Other bonding schemes also suffice[66]. An alternative bipyramidal transition state resulting in inversion, but with the nucleophile and the leaving group occupying basal positions, is plausible. An argument against this latter case had been constructed by Hamer[53] based on the reactions of the monoesterified phosphoramidic acids depicted below (10), *viz.*

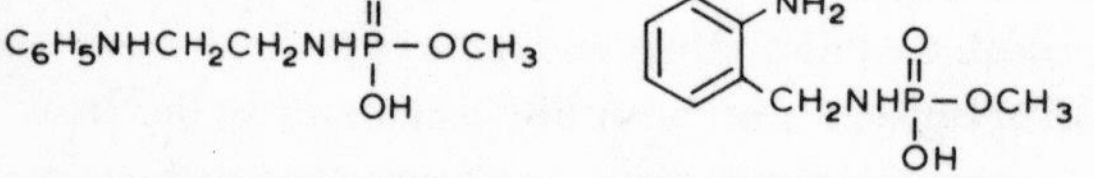

$$C_6H_5NHCH_2CH_2NHP(=O)(OH)-OCH_3 \qquad o\text{-}(H_2N)C_6H_4CH_2NHP(=O)(OH)-OCH_3 \quad (10)$$

Several considerations recommended their employment; (*a*) aromatic amines readily displace the more strongly basic aliphatic amines from phosphoramidic monoesters[54]; (*b*) the transition state for displacement would require the accommodation of phosphorus and the entering and leaving groups in a five-

and six-membered ring. No evidence was found for an intramolecular phosphoryl migration under conditions where intermolecular phosphoryl transfer to 1-naphthylamine by the above compounds was demonstrated. Thus, a linear arrangement would appear to be necessary, which in conjunction with the overall stereochemistry of the displacement, supports an apical–apical mechanism, and not one involving equatorial bonds, (11).

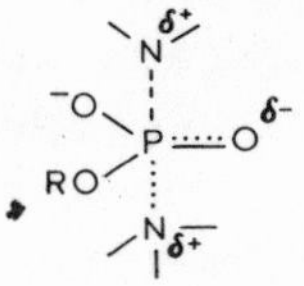

(11)

The question then naturally arises as to whether a metastable pentacovalent intermediate exists in these nucleophilic displacement reactions. In the case of the thiophosphoryl center one might have expected (*a*) racemization as a result of pseudorotation of the trigonal bipyramid and (*b*) oxygen exchange into the starting material as a consequence of an addition–elimination mechanism. These processes, however, would only have been observed if pseudorotation were competitive with the rate of leaving group expulsion from the intermediate. In fact the generality of the phosphoramidate result may be questioned on similar grounds since the predominant zwitterionic ground state may locate the electronegative leaving group ($R{-}NH_2^{\delta+}\cdots$) only in an apical position. With the reference to the term pseudorotation and the partial development of a discussion about trigonal bipyramidal transition states, it is necessary to turn to the chemistry of cyclic phosphates where the significance of these concepts to the chemistry of inorganic esters may be explored more thoroughly.

2.3 CYCLIC PHOSPHATE ESTERS

The hydrolysis in acid or base of five-membered cyclic esters of phosphoric acid proceed some 10^6 times faster than their acyclic analogs or the six- and seven-membered cyclic phosphates[1,55,56]. Unlike that of dimethyl phosphate, the hydrolysis of hydrogen ethylene phosphate is accompanied by rapid oxygen exchange into the unreacted substrate[57], *viz.*

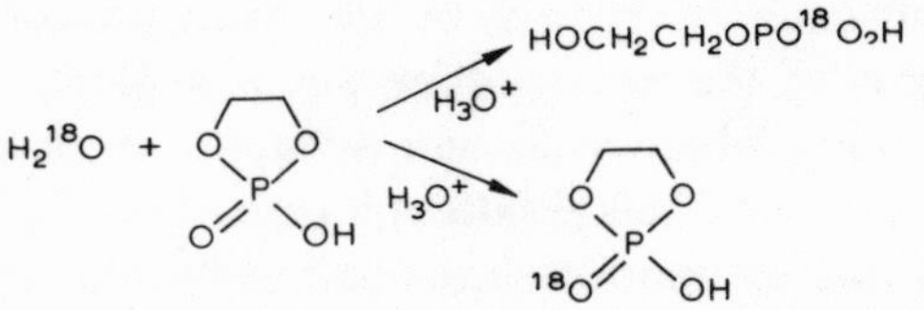

(12)

References pp. 51–56

This fact, in conjunction with the discovery that the hydronium-ion-catalyzed hydrolysis of methyl ethylene phosphate proceeds with at least 30% cleavage of the exocyclic methoxyl group preserving the ring structure, but at a rate 10^6 times greater than trimethyl phosphate, poses a crucial question. What feature of ring structure can affect a rapid acceleration of hydrolysis of the exocyclic group? The answer, suggested independently by Hamer[53] and Dennis and Westheimer[58] but its implications confirmed mainly through the efforts of this latter group, invokes the concept of pseudo-rotation between trigonal bipyramidal intermediates. Since this postulation has far-reaching implications for the hydrolyses of a variety of cyclic and possibly acyclic inorganic esters, it will be examined in some detail.

The mechanism of pseudorotation was first proposed by Berry[59] to account for the NMR spectrum of the pentacovalent fluorophosphorane, PF_5. This compound, as well as gaseous PCl_5, has been shown by a variety of physical measurements to be a trigonal bipyramid with the two apical bonds longer than the three equatorial bonds[60–63]. However, only one type of fluorine can be detected in its simple NMR spectrum which is a doublet ($J_{P-F} = 1010$ c/sec). Berry explained this result by proposing that the groups in apical and equatorial positions of a trigonal bipyramid can exchange places readily *via* an intramolecular process of pseudorotation.

To effect one pseudorotation in compounds containing pentacovalent elements, one generally selects an equatorial group as a pivotal point. The two axial groups are pushed backwards closing the axial–central atom–axial group bond angle from 180 to 120°. Simultaneously, the remaining two equatorial groups are pulled forward opening the equatorial–central atom–equatorial group bond angle from 120 to 180°. The result is a new trigonal bipyramid in which the pivoted group has remained in an equatorial position but the other four groups have exchanged their positions. The above process is illustrated below (the central atom is not shown).

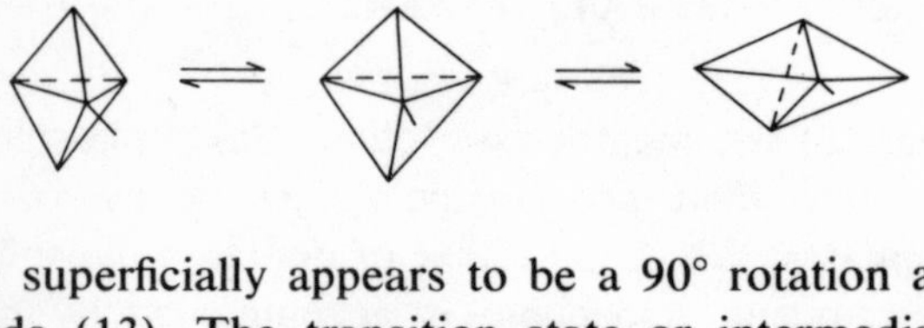

(13)

and superficially appears to be a 90° rotation about one of the interatomic bonds (13). The transition state or intermediate structure in the process apparently is a tetragonal pyramid[64].

This initial hypothesis was subsequently refined by the deduction of two preference rules for the structures of the intermediate pentacovalent phosphorus derivatives, the phosphoranes. Muetterties and Schunn[65], in order to rationalize the NMR spectra of a number of alkyl fluorophosphoranes $(CH_3)_n$-PF_{5-n}, formulated the hypothesis that the more electronegative fluorine atoms preferentially occupy the axial positions, whereas the less electronegative alkyl groups prefer the equatorial positions. Theoretical arguments based on extended Huckel theory assuming a trigonal bipyramidal geometry reveal that the phosphorus 3*s* character is greater in equatorial than axial bonds[66]. Since atomic *s* character is concentrated in orbitals directed toward electron-donating substituents (if a choice is possible), it follows that electron-donating substituents will occupy the equatorial positions. Moreover, additional energy lowering for the trigonal bypyrimidal structure occurs if the more electronegative groups occupy the apical positions owing to the greater *p* character of these bonds[67]. It is not necessary to employ *d* orbital participation to obtain the above results although there is probably *d* character in both sets of bonds.

The second preference rule is derived from X-ray crystallographic determination of the stereochemistry of oxyphosphoranes, a series of pentacovalent phosphorus-containing compounds whose synthesis has been largely developed by Ramirez (and his coworkers)[68]. The trigonal bipyramidal structure of the phenanthrenequinone–triisopropyl phosphite adduct, (14) below, has the following pertinent features; (*a*) an O–P–O ring bond angle of nearly 90°; (*b*) a five-membered ring which spans one apical and one equatorial position and (*c*) extensive repulsive steric interactions which occur owing to several short non-bonded distances, *i.e.* the apical oxygens are within 2.7 A of the first carbon of the equatorial isopropyl groups. The main argument in favor of the pentacovalency of the central phosphorus atom is the large positive value of the ^{31}P NMR shift which is *ca.* 40–50 ppm relative to H_3PO_4. The spanning of one apical and equatorial position by the five-membered ring is probably a general phenomenon and constitutes the second rule. The minimization of steric repulsion by constraining an apical and equatorial group into a five-membered ring may account, to a large extent, for the greater stability of the cyclic *versus* acyclic oxyphosphorenes.

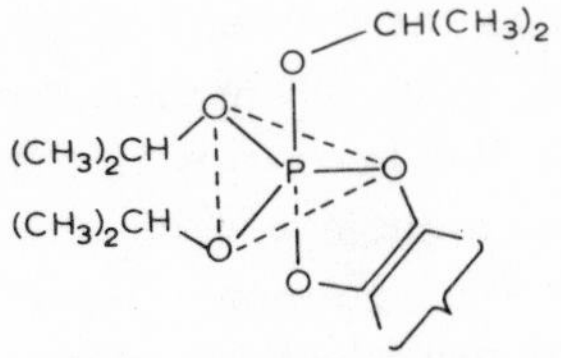

CH(CH3)2
(CH3)2CH
(CH3)2CH
(14)

As in the above case for the alkyl fluorophosphoranes, the H^1 NMR spectrum of the oxyphosphoranes possesses features which are only readily interpretable in terms of a pseudorotation mechanism[68,69]. The H^1 spectrum of the cyclic tetraoxyalkylphosphorane synthesized from trimethylphosphate and 3-benzylidene-2,4-pentadione shows two kinds of methoxyl groups at low temperature (−65°C), a single methoxyl signal at room temperature, and a

References pp. 51–56

coalescene at 100°C of the NMR signals arising from the acetyl and the methyl group attached to the phospholene ring. These observations are consistent with an oxyphosphorene (15) which has a barrier to pseudorotation, and as a

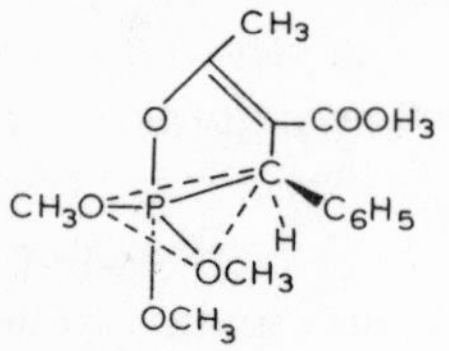

(15) 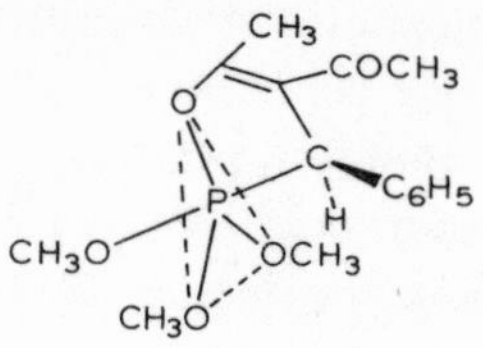 (16)

consequence, equilibration of the methoxyl groups at low temperature is hindered. In order for (15) to pseudorotate, it is necessary to place an alkyl group in an apical position (equatorial methoxyl groups as pivots) or to distort the O–P–O ring bond angle to 120° by forcing the ring into a diequatorial position (ring carbon as pivot). The latter should be accompanied with an increase in ring strain, as discussed later, and the former process, although formally not allowed, is probably the means of equilibration (16). At higher temperatures the cyclic structure presumably is in equilibrium with the acyclic dipolar form (17)[70].

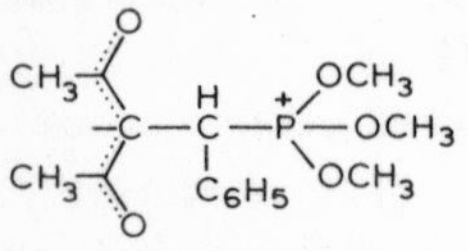

(17)

The concept of pseudorotation *per se* does not explain the rapid rate of acid and alkaline hydrolysis of the ring of ethylene phosphate which exceeds that of dimethylphosphate by *ca.* 10^8. Initially this kinetic acceleration of attack on the phosphorus atom was attributed to strain in the five-membered ring based on evidence provided by measurements of the heats of saponification of the acyclic and five-membered cyclic compounds[71,72]. These showed that the cyclic compound has a heat of saponification 5–6 kcal·mole^{-1} higher than the acyclic analog. Similarly ethylene sulfate liberates 5–6 kcal·mole^{-1} more heat on hydrolysis than does the acyclic standard[72]. In contrast the heats of hydrolysis of 5-membered cyclic alkyl and aryl sulfites are normal[73a,b], consistent with the observations that these compounds hydrolyze at rates comparable to their respective acyclic standards[74]. An apparent exception is *o*-phenylene sulfite whose spontaneous rate of hydrolysis is 10^3 faster than diphenyl sulfite[75,76]. We will return to this point later.

Calculation of the bond angles and conformation of the cyclic phosphates by minimization of ring and eclipsing strain leads to bond angles closely in accord with the X-ray crystallographic data. Although absolute strain values may not be computed in this manner, similar calculations indicate that the relief of strain (3–6 kcal·mole^{-1}) in the phosphate ring on passing from the ground state with a ring angle at phosphorus of 99° to a trigonal bipyramidal transition state (ring angle, 90°) contrasts sharply with a strain energy of 5–7 kcal·mole^{-1} introduced by proceeding to a transition state with the ring angle at 120°[77]. In the former the ring spans apical–equatorial positions; in the latter equatorial–equatorial positions. Although this approach attributes the observed relative rate acceleration primarily to a difference in σ energy associated with the angle distortions in the cyclic structure, an alternate calculation based on Huckel theory emphasizes the difference in π energy between a cyclic and acyclic ester. The origin of π bonding is overlap between the $2p$ orbitals on the oxygens, not involved in the σ system, and the five $3d$ orbitals on the phosphorus atom[18]. The conformational restraints imposed by the ring structure increase the net positive charge on the phosphorus atom by interrupting π resonance relative to its acyclic reference and presumably account for the former's increased reactivity[78]. A corollary manifestation of the change in π bonding and the deshielding of the central phosphorus atom may be reflected in a shift of *ca.* -18 ppm in the ^{31}P NMR spectra of the five-membered cyclic phosphate triesters relative to their acyclic forms[79]. Since the difference in the free energy of activation between the alkaline hydrolysis of methyl ethylene phosphate and trimethyl phosphate is approximately 8.5 kcal·mole^{-1}, it is apparent that factors in addition to the thermochemical difference in energy are involved, *i.e.* possible changes in transition state π bonding. As alluded to above, the degree of $p\pi$–$d\pi$ bonding in phosphorus-containing substrates remains an unsettled question. Although such π bonding presumably is geometrically feasible, the answer to the problem of mixing the energetically and radically dissimilar $2p$ and $3d$ orbitals to generate strong bonding has proved elusive[80]. Thus, the influence of π bonding on the hydrolytic rates of phosphate esters is at present an unknown quantity. A third factor, termed the "entropy-strain" principle, which apparently may also apply in part to these systems[81], will be treated in the section dealing with cyclic sulfur-containing esters.

Before returning to the ethylene phosphate problem, let us review several additional observations derived from the hydrolysis of five-membered phostonates which are pertinent to the application of the combined concepts of ring strain and pseudorotation. The rate of hydrolysis of the five-membered propylphostonate acid (18),

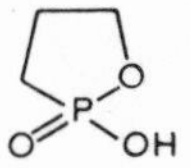

(18)

is 8×10^4 times greater in acid and 6×10^5 in alkali than that of hydrolysis of ethyl ethylphosphonate, and thus parallels the extraordinary reactivity of the cyclic phosphates[82]. More striking, however, is the experimental fact that the hydrolysis of the methyl ester of propylphostonic acid, (19) below, occurs almost exclusively with ring opening[81,83]. On the other hand, the rates of hydrolysis of esters from five-membered cyclic phosphinates (20) are normal

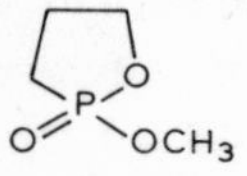

(19) 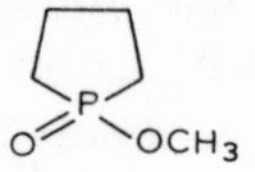 (20)

relative to their reference standards[83]. By utilizing arguments based on ring strain and pseudorotation, Westheimer *et al.* have shown that these diverse observations have a common rationale.

The acid hydrolysis or oxygen exchange of ethylene hydrogen phosphate presumably proceeds through a trigonal bipyramidal species with the ring occupying one apical and equatorial position and the attacking water at an apical position, *viz.*

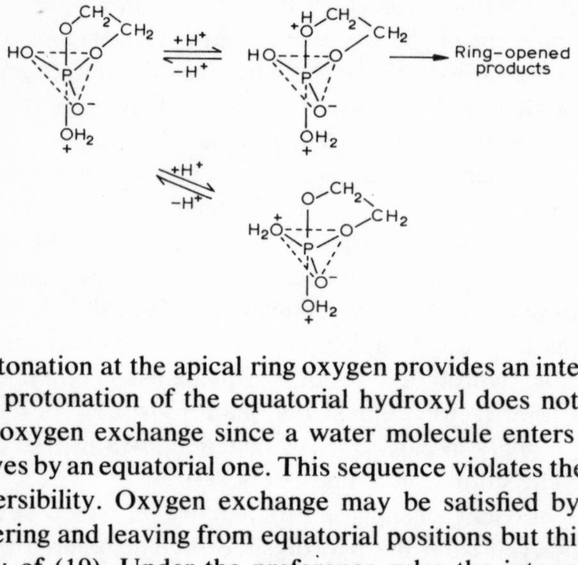

Ring-opened products

(21)

Protonation at the apical ring oxygen provides an intermediate for ring opening but protonation of the equatorial hydroxyl does not provide an intermediate for oxygen exchange since a water molecule enters at an apical position but leaves by an equatorial one. This sequence violates the principle of microscopic reversibility. Oxygen exchange may be satisfied by having water molecules entering and leaving from equatorial positions but this is contrary to the chemistry of (19). Under the preference rules the intermediate derived from (19) may be depicted as (22) where the methylene group of the ring is equatorial

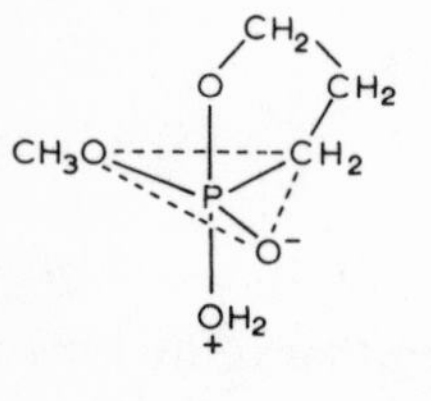

(22)

and the oxygen atom is apical. If both entering and leaving groups were equatorial, then ring opening would not occur which is contrary to experiment. The fact that loss of the methoxyl group is negligible is rationalized in terms of the preference rules which dictate that the trigonal bipyramidal intermediate cannot undergo pseudo-rotation since (*a*) pseudorotation about the equatorial carbon atom would expand the ring angle from 109° to 120° and (*b*) pseudorotation about either of the other two equatorial substituents as pivot would force the methylene into an apical position. Oxygen exchange in ethylene phosphate hydrolysis, or acid catalyzed loss of the methanol from methyl ethylene phosphate, therefore, arises from allowed pseudorotations which transform the initial equatorial hydroxyls to apical ones. Similarly, the rate of hydrolysis of the cyclic phosphinates should be low, since in order to expel the external methoxyl group energetically unfavorable processes identical to the operations described in (*a*) and (*b*) above must be performed. The reduction in strain owing to the formation of the 90° ring angle of the trigonal bipyramidal intermediate coupled with the mechanism of pseudorotation thus furnish an elegant explanation of how ring strain can induce rapid reaction external to the ring. To the list of preference rules should now be added that groups enter at and leave from apical positions only.

An in-depth examination of the pH–product profile (Fig. 5) for the hydrolysis of methyl ethylenephosphate has revealed additional facts that are

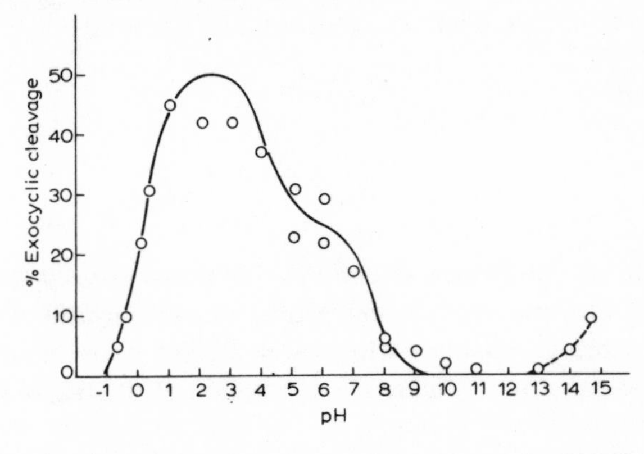

Fig. 5. pH–product profile for the hydrolysis of methyl ethylene phosphate in water at 25°C.

reasonably correlated with the above preference rules. Recall that the loss of the exocyclic methoxyl group takes place only after pseudorotation about the central phosphorus atom. In the pH–product profile the % of exocyclic cleavage markedly decreases from 50 to 0% as pH decreases from 4 to strong acid. The anticipated trigonal bipyramidal intermediate formed initially by apical attack of a water molecule and equatorial positioning of the phosphoryl oxygen can only yield ring-cleavage products; pseudorotation about the phosphoryl oxygen as pivot is necessary to give the required trigonal bipyramidal intermediate for exocyclic cleavage.

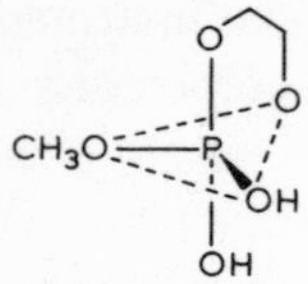

(23) 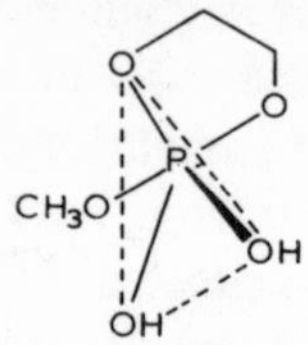 (24)

Since the rate of reaction in this pH region is directly proportional to hydrogen-ion concentration it appears that an acidity has been attained, where the rate of protonation and decomposition of the cationic intermediate derived from (23) to yield ring cleavage products exceeds the rate of pseudorotation of (23) to (24). This change in rate-determining step would account for the above decrease in exocyclic cleavage. The notion that pseudorotation occurs *via* the neutral unprotonated intermediates, rather than the conjugate acids follows from the preference rule which places electron-withdrawing groups in the apical position. Protonation of the apical ring oxygen militates against pseudorotation, owing to its conversion to an equatorial protonated oxygen by this process. Rate-limiting pseudorotation also appears in the acidic region of the pH-rate profile for the hydrolysis of the ester of bicyclic phosphinic acid (25) as a negative rate deviation. The rate coefficient for alkaline and acid hydrolysis

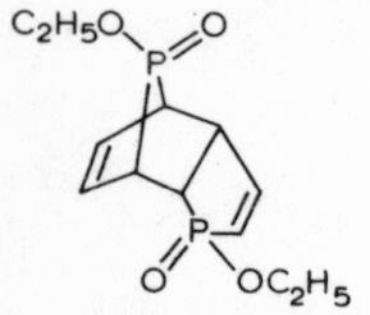

(25)

of the ester group on the bridge is *ca.* 10^4–10^5 times greater than the monocyclic analogue[84]. The enormous acceleration can be correlated with the high strain energy at a bridge position which is sufficient to overcome the barrier to placing an alkyl substituent in an apical position of the trigonal bipyramidal intermediate[85].

Species (23) and (24) function as the intermediates for the water-catalyzed hydrolysis of methyl ethylene phosphate which gives 25% exocyclic cleavage. Ionization to generate (26) followed by pseudorotation to (27), *viz.*

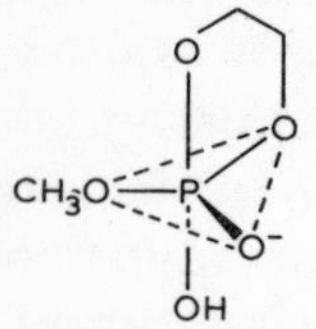

(26) 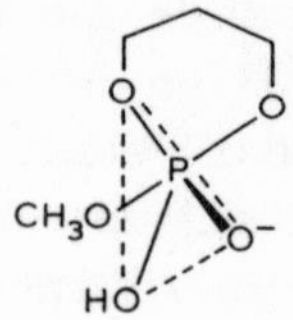 (27)

places the electron-donating O^- in an equatorial position in both species, as required by the preference rules. As the fraction of exocyclic cleavage approaches zero in the pH region where these species dominate, it must be assumed that the rate of decomposition of (26) greatly exceeds the rate at which (26) pseudorotates to (27). Two details should be noted: (*a*) that an expulsion of alkoxide ion is suggested by the principle of microscopic reversibility and (*b*) that neither the kinetics nor product distribution require a discrete pentacovalent intermediate.

In strong alkali (10 *M* NaOH) up to 15% exocyclic cleavage is again found. The anticipated intermediates in alkali are the dianions (28) and (29), *viz.*

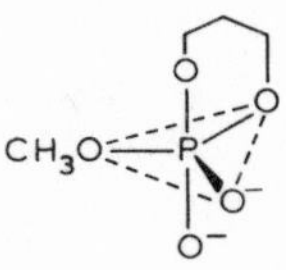

(28) 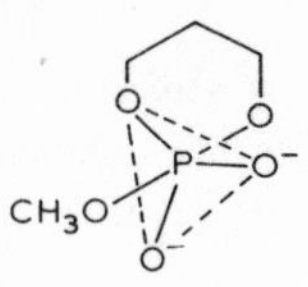 (29)

Since (28) possesses a negative charge at an apical position, it should be unstable relative to (29). Rapid pseudorotation to (29) should then compete with decomposition of (28) to yield products arising from exocyclic cleavage. This explanation requires, however, a pentacovalent intermediate and that the rate of pseudorotation of (28) be increased by a minimal 10^3 relative to (26) to accommodate the experimental findings. By implication then the above base-catalyzed hydrolysis also is thought to involve the pentacovalent species.

A fundamental question posed, although not answered by the above arguments, is the actual rate of pseudorotation. Calculations based on the temperature-dependent NMR spectrum of a cyclic oxyphosphorene, (15) p. 16, in which pseudorotation is inhibited sets a free energy of activation for the exchange process at roughly 12 kcal·mole^{-1}. A figure of 5–6 kcal·mole^{-1} corresponding to a rate coefficient of 10^8–10^9 sec^{-1} at 25°C is the calculated energy barrier to pseudorotation of the cyclic oxyphosphorane, (23) p. 20, by the extended Huckel molecular orbital method[86]. A maximal barrier of about 6 kcal·mole^{-1} to pseudorotation is estimated from the NMR spectrum for methyltetrafluorophosphorane where pseudorotation may readily occur employing the methyl group as pivot. The Huckel calculation of energy barriers to pseudorotation does not optimize the geometry of the tetragonal pyramid transition state or intermediate in the pseudorotation process. It also markedly depends on the model for the trigonal bipyramidal intermediate, *i.e.* positions of the covalently bound protons. Estimates of the barrier to pseudorotation for the trigonal bipyramidal intermediate derived from nucleophilic attack of water on methyl ethylene phosphate range from 5–6 kcal·mole^{-1} to an upper limit of 12–15 kcal·mole^{-1}. Thus it appears the rate coefficient for pseudorotation of (30) to place the methoxyl group apical may be as low as 10^2–10^3 sec^{-1}.

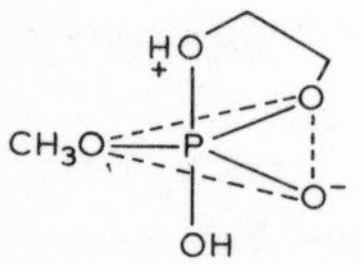

(30)

We will return to this point below. This same set of calculations finds that (*a*) the barrier for pseudorotation is lowered if the phosphoryl oxygen remains in an equatorial position, (*b*) protonation of the apical oxygens gives the more stable trigonal bipyramidal intermediate (the phosphoryl oxygen being equatorial and the ring-spanning apical and equatorial positions), and (*c*) the nucleophile approaches the phosphate ester on the back-side of one of the P–O ring bonds along an apical line[87]. These calculations corroborate the main features of the mechanism advanced to explain the reactivities of the cyclic compounds toward hydrolysis and oxygen exchange.

We have noted above that the intermediate formed in the spontaneous hydrolysis of methyl ethylene phosphate must undergo pseudorotation in order to yield products resulting from exocyclic cleavage. It is also reasonable to expect that the departing group will be methanol and not methoxide anion. If one depicts the initial intermediates as (23) or (30) then a proton transfer between an equatorial and apical position is required prior to hydrolysis. It is doubtful that proton transfer will accompany the initial attack by water, which results in formation of an oxonium ion at an equatorial position. Since the rate coefficient of pseudorotation may be as low as 10^2–10^3 sec^{-1}, it is conceivable that proton transfer may become rate-determining. Of course, other pathways to product are feasible in addition to the one illustrated; however, the present intent is merely to raise the possibility of a proton-transfer step becoming rate-determining and, perhaps, controlling the nature of the products.

The related hydrolysis of phosphonium salts has been considered recently in some detail by McEwen[3]. In general the alkaline hydrolysis of these salts gives mainly phosphine oxides and hydrocarbons, and is first-order in hydroxide ion and substrate, although a second-order dependence of rate on hydroxide ion concentration is sometimes observed. The stereochemical consequences depend on whether the phosphorus atom is incorporated in a four- or five-membered ring or in an acyclic system. In the latter case inversion of configuration at phosphorus is observed with the leaving group being the most stable hydrocarbon anion[88,89a,b]. These findings are usually accommodated in a mechanism which involves the rate-determining collapse of an intermediate trigonal bipyramidal phosphorane, *viz.*

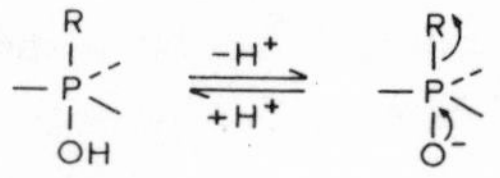

(31)

with the loss of the anion from the apical position. If, however, the phosphorus is present in a small ring system, then the differences in bond angles between apical and equatorial (90°) and diequatorial (120°) positions could lead to restraints on the conformation of the intermediate. It is anticipated that the ring system would span a less-strained apical–equatorial position and lead to ring opening or ring expansion depending upon the nature of the other substituents attached to phosphorus[90,91]. Alkaline hydrolysis of 9,9-dimethyl-9-phosphoniafluorene proceeds exclusively with opening of the heterocyclic ring to give the corresponding biphenyl-2-ylphosphine oxide[92] (32), *viz.*

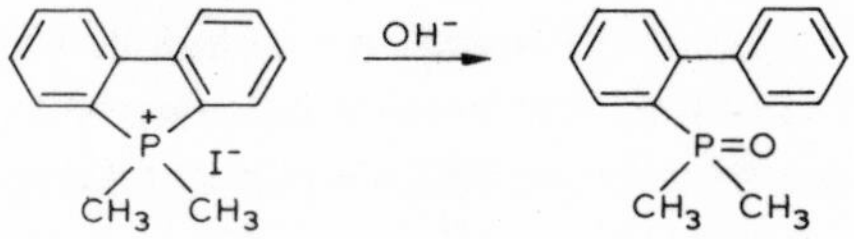

(32)

whereas alkaline hydrolysis of (33) gives 9-methyl-9-phosphafluorene 9-oxide with retention of the five-membered ring[93], *viz.*

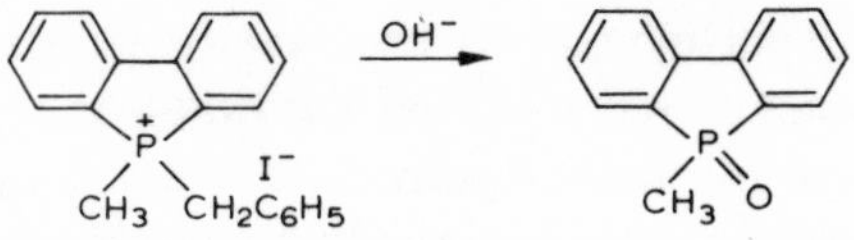

(33)

Moreover, replacement of the benzyl group by an iodomethyl substituent leads to the ring-expanded product upon alkaline hydrolysis (34), *viz.*

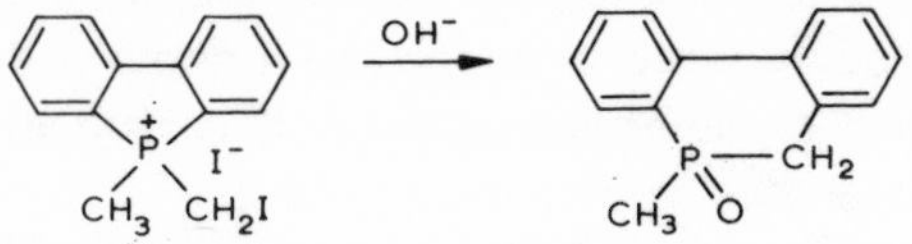

(34)

The course of these reactions can be interpreted in terms of an intermediate phosphorane (35)

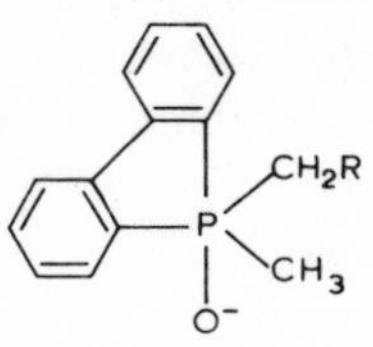

(35)

which may decompose with (*a*) ring cleavage directly, (*b*) pseudorotation and loss of RCH_2^-, or (*c*) intramolecular aryl migration. In principle, loss of RCH_2^- should occur with retention of configuration at phosphorus assuming an apical–equatorial location of the ring in the intermediate. This indeed, has been shown in the case of base-catalyzed hydrolysis of 1-benzyl-1-phenyl-2,2,3,4,4-pentamethylphosphetanium iodide (36).

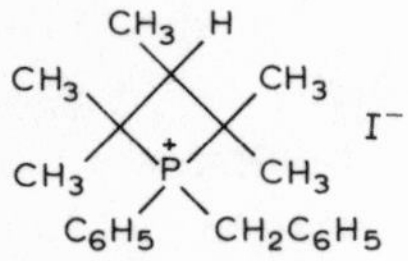

CH_3 H
CH_3 CH_3
I^- (36)
CH_3 $\overset{+}{P}$ CH_3
C_6H_5 $CH_2C_6H_5$

the *cis* and *trans* isomers of 1-ethoxy-1-phenyl-2,2,3,4,4-pentamethylphosphetanium hexachloroantimonate and benzyl phospholanium salts[94–96]. It should be noted that while a single pseudorotation step would not lead to loss of stereospecificity, a series of pseudorotations would eventually lead to racemization[97]. Consequently, products of retained configuration are under kinetic control. It also appears possible that the above reaction might involve apical attack and equatorial cleavage, although in most cases one can consider this a violation of microscopic reversibility. Note that, in all cases, the Muetterties rule is violated.

It is interesting that the alkaline rate of hydrolysis of the five-membered methyl phenyl phosphonium salt is some 1,300 times faster than the six-membered or acyclic counterparts[98]. This hydrolysis proceeds with loss of benzene and preservation of the five-membered ring. On the other hand, alkaline hydrolysis of ethylenephosphinate is comparable to its acyclic analogue. This disparity may be caused by steric effects in the phosphorane intermediate, a problem presently under active study[99]. The enhanced reactivity of the five-membered ring system relative to its acyclic reference apparently extends to the phosphoramidites (tervalent phosphorus) although the rate enhancements are only of the order of $1–3 \times 10^2$. Since the heats of hydrolysis for both systems are similar, the acceleration is caused by some factor other than ring strain; *i.e.* the 3–4 $kcal \cdot mole^{-1}$ discussed earlier[100].

At this point it is profitable to consider the hydrolysis or alcoholyses of the cyclic oxyphosphoranes themselves. The trimethoxyphospholene (37) undergoes a nearly instantaneous reaction with one molar equivalent of water in benzene solution at 0°C[101]. The products of this reaction are mainly dimethyl phosphoracetoin and the cyclic methyl phosphate of acetoinenediol, *viz.*

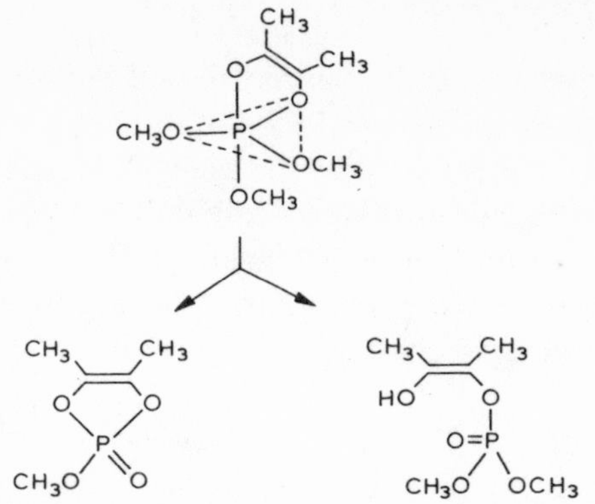

(37)

Hydrolysis at higher temperatures yields mainly trimethyl phosphate and acetoin. A tentative reationale for these results features a trigonal bipyramidal intermediate which may be represented with the hydroxyl group apical, *viz.*

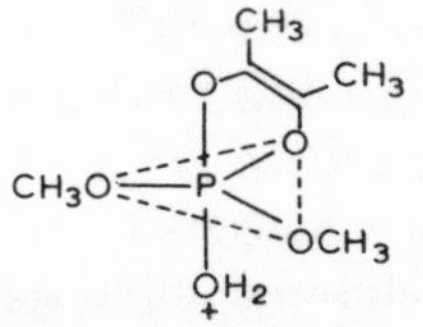

(38)

therefore requiring a single pseudorotation in order to expel methanol and form cyclic phosphate. Cleavage of the apical P–O bond in the present intermediate would lead to dimethyl phosphoacetoin product. The two products would equilibrate through these intermediates. At higher temperatures ring cleavage to form a zwitterionic species (39)

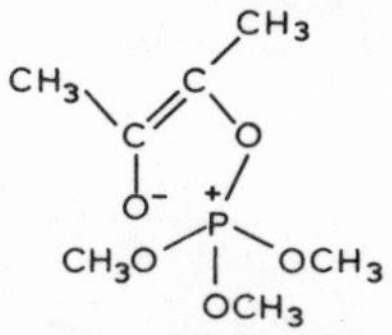

(39)

similar to those encountered in the NMR studies discussed earlier, followed by hydrolysis would lead to acetoin and trimethyl phosphate. It is unlikely that the trimethyl phosphate is derived from dimethylphosphoacetoin under these conditions[102]. An intriguing problem is the mechanism for formation of (38). Possible pathways include (*a*) a concerted displacement reaction at phosphorus *via* a transition state of possible octahedral symmetry; (*b*) a stepwise substitution involving an intermediate with octahedral configuration[103–105]; (*c*) preionization mechanisms involving rupture of a ring or exocyclic P–O bond. If however, an ionization mechanism is operational to explain the change in hydrolysis products at higher temperature, then (*a*) or (*b*) would represent the anticipated mechanisms for substitution at pentacovalent phosphorus[106].

The chemistry developed from the above studies of the cyclic phosphates

References pp. 51–56

has a large number of biological implications. In a direct sense nucleoside 2′,3′-phosphates are labile intermediates in the ribonuclease catalyzed hydrolysis of ribonucleic acid[107] whereas the six-membered cyclic phosphate, adenosine 3′,5′ cyclophosphate, exerts an important regulatory role through binding to induce conformational changes in critical proteins involved in glycogen degradation and synthesis, and lipolysis[108]. More important, however, will be the application of the discussed concepts to problems of enzymic catalysis of phosphate transfer.

2.4 POLYPHOSPHATES

Kinetic studies of the nucleotide analogs, γ-phenylpropyl di- and triphosphate, have been undertaken to define the role of the adenosine residue in the chemical and enzymic reactions of adenosine triphosphate. A catalytic function associated with binding of metal ions at the adenine nitrogens has been ascribed to the adenosine moiety in phosphate transfer reactions in which adenosine di- or triphosphates function as the phosphate source[109–112]. The pH-rate profile (Fig. 6) for the hydrolysis of γ-phenylpropyl diphosphate

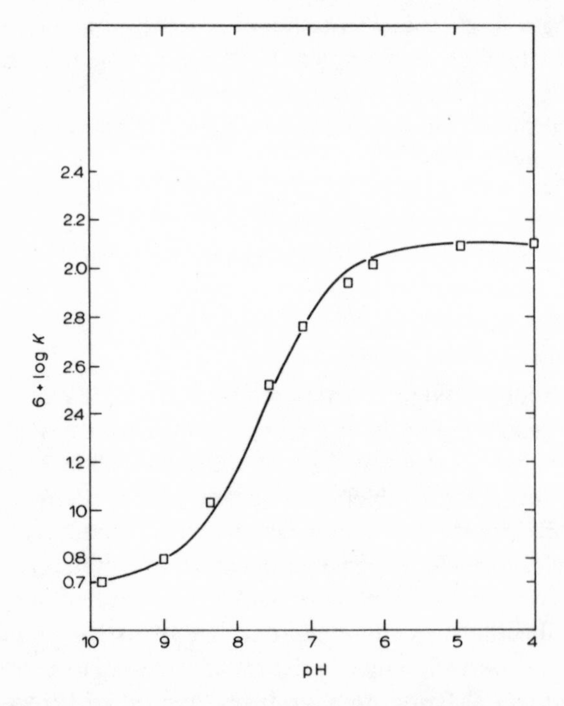

Fig. 6. Rates of hydrolysis of γ-phenylpropyl diphosphate at 95°C.

reveals that all species undergo hydrolysis at rates similar to that observed with adenosine diphosphate[113]. Furthermore, the rate of hydrolysis of γ-phenylpropyl triphosphate is similar to that of adenosine triphosphate over a pH range of 4–9. Thus the mechanisms for these hydrolyses are presumably the same with the adenosine residue having no role. Since hydrolysis of monoprotonated γ-phenylpropyl diphosphate is some two thousand times faster than hydrolysis of the symmetrically substituted P,P′-di-γ-phenylpropyl pyrophosphate, a monomeric metaphosphate mechanism appears operative for this species (40), *viz.*

$$C_6H_5CH_2CH_2CH_2O-\overset{\overset{\displaystyle O}{\|}}{\underset{\underset{\displaystyle O^-}{|}}{P}}-O-\overset{\overset{\displaystyle O}{\|}}{\underset{\underset{\displaystyle OH}{|}}{P}}-O^- \rightleftharpoons C_6H_5CH_2CH_2CH_2O-\overset{\overset{\displaystyle O}{\|}}{\underset{\underset{\displaystyle OH}{|}}{P}}-O-\overset{\overset{\displaystyle O}{\|}}{\underset{\underset{\displaystyle O^-}{|}}{P}}-O^- \quad (40)$$

$$\downarrow$$

$$C_6H_5CH_2CH_2CH_2O-\overset{\overset{\displaystyle O}{\|}}{\underset{\underset{\displaystyle OH}{|}}{P}}-O^- + [PO_3^-]$$

Verification for this hypothesis is derived from synthesis of the unsymmetrical pyrophosphate diester, P_1,P_1-diethyl pyrophosphate, which undergoes rapid hydrolysis as the dianion[114]. The close agreement between the observed rate coefficient in the latter case and the calculated rate coefficient for the hydrolysis of monoprotonated diphosphate monoester, after correction for tautomerization, strongly supports the above scheme. Yet, it is noteworthy that solvent competition studies with mixtures of methanol and water reveal that no methyl phosphate is formed when pyrophosphate monoanion is solvolyzed. Although this does not rule against the metaphosphate mechanism, it is necessary to assume that the product-forming transition state is selectively solvated by water[115]. This does not appear, however, to be caused by the adjacent phosphoryl moiety (see later discussion).

Since the requirement of a divalent metal ion is common to all reactions of nucleoside triphosphate, metal-ion-catalyzed phosphorylations with these analogs is of interest[116,117]. Calcium and cadmium ions catalyze the reaction of inorganic phosphate with γ-phenylpropyl triphosphate to generate inorganic pyrophosphate and the corresponding diphosphate at rates comparable to that for adenosine triphosphate. The mechanistic details are obscured by experimental difficulties. In contrast the cupric-ion-catalyzed hydrolysis of the model compound is considerably slower than for adenosine, indicating an interaction between the adenine moiety and the metal ion as previously suspected. Substitution of γ-phenyltriphosphate for adenosine triphosphate in hydrolytic reactions catalyzed by the enzymes potato apyrase, muscle myosin, and inorganic pyrophosphatase (Zn^{2+}) reveals the two to be interchangeable and to be hydrolyzed at comparable rates. Consequently the structural involvement of the adenine group in these enzymic reactions, also must be minimal[118]. On the basis of these facts a tentative mechanism may be written for cleavage of the terminal phosphoryl moiety in polyphosphates as catalyzed by metal ions. The metal ion may function to promote a metaphosphate elimination (depending on the degree of acceptor nucleophile involvement) *via* chelation to the α,β phosphoryl groups, *viz.*

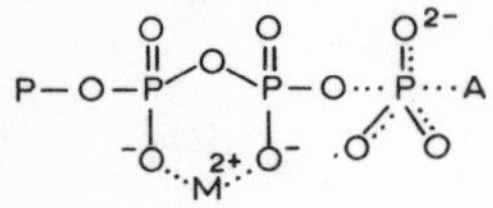

$$\mathrm{P{-}O{-}\overset{O}{\overset{\|}{P}}(\text{-}O^{-}){-}O{-}\overset{O}{\overset{\|}{P}}(\text{-}O^{-}){-}O\cdots \overset{O^{2-}}{\overset{\vdots}{P}}\cdots A} \quad (O^{-}\cdots M^{2+}\cdots O^{-};\ \cdot O \ \ O) \tag{41}$$

This attractive hypothesis may prove difficult to substantiate owing to the high probability of forming nonproductive complexes or dimeric chelates[119–121]. Supporting evidence for a similar pathway in phosphate monoesters will be discussed under Section 2.5.

2.5 CATALYSIS

Catalysis by neighboring groups in phosphate ester hydrolysis apparently falls into two general categories: (*a*) reactions which involve nucleophilic attack on phosphorus leading to the formation of cyclic five-membered rings and are general acid-base catalyzed, and (*b*) reactions which involve general-acid catalysis of phosphate group expulsion. Exceptions to the latter are found when phosphorus is di- or trisubstituted with reasonably good leaving groups in which case nucleophilic attack on phosphorus leads to the formation of six-membered rings. The reactions in category (*a*) generally exhibit very large rate accelerations when compared to their bimolecular counterparts, if indeed one can be found.

The presence of an α-carbonyl group greatly enhances the base-catalyzed rate of hydrolysis of a variety of phosphate and phosphonate esters[122,123]. The hydrolysis may take either or both of two competing pathways depending on the nature of the leaving group[124]. A third route involving nucleophilic displacement of the phosphate diester and formation of the epoxide is of major importance in methoxide–methanol systems[125]. With dimethyl phosphoacetoin 3% of the reaction proceeds by path (*a*), indicating the ratio of rates for routes (*b*)/(*a*) is *ca.* 30; with diphenyl phosphoacetoin the major pathway (98+%) is (*a*), *viz.*

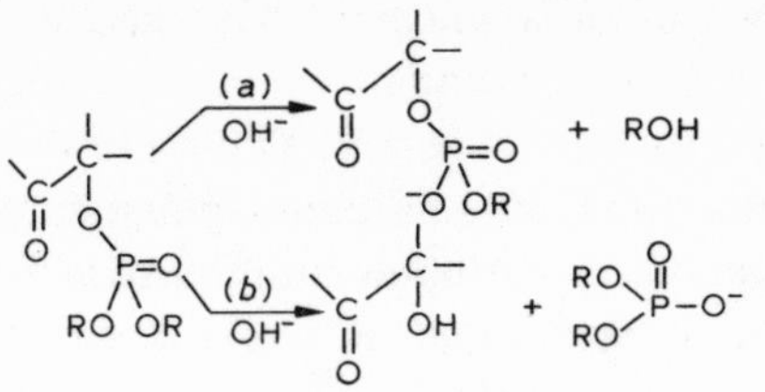

(42)

Substitution of a methyl group for a methoxyl at phosphorus causes a dramatic change in the product distribution with pathway (*a*) now accounting for 95% of reaction. This apparently is not caused by a large increase in the rate of pathway (*a*) but by a greater than 200-fold decrease in that of (*b*). Frank and Usher[126a] have suggested the following mechanism to rationalize these results, *viz.*

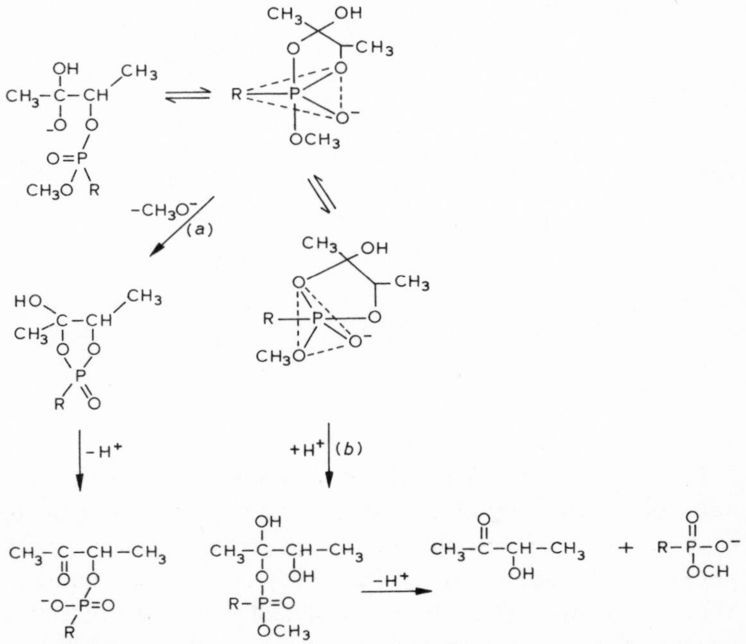

(43)

Intramolecular migration of the phosphoryl group in the phosphonate derivative requires that either the methyl group or the ionized phosphoryl oxygen take up an apical position in the pentacovalent species. Since either process is energetically unfavorable, the rate of pseudorotation apparently becomes competitive with the rate of product formation so that product arising from pathway (*b*) decreases relative to (*a*). An indication of the magnitude of the acceleration obtained with the neighboring keto group, which further favors the argument for a pentacovalent phosphorus intermediate, is the 10^5–10^6 increase in rate over that for trimethyl phosphate[124].

Recently Benkovic and Schray[126b] and Clark and Kirby[126c] have investigated the hydrolysis of dibenzylphosphoenolpyruvic acid and mono-benzylphosphoenolpyruvic acid which proceed *via* stepwise loss of benzyl alcohol (90%) and the concomitant formation of minor amounts (10%) of dibenzylphosphate and monobenzylphosphate, respectively. The pH–rate profiles for release of benzyl alcohol reveal that the hydrolytically reactive species must involve a protonated carboxyl group or its kinetic equivalent. In the presence of hydroxylamine the course of the reaction for the dibenzyl ester is diverted to the formation of dibenzyl phosphate (98%) and pyruvic acid oxime hydroxamate but remains unchanged for the monobenzyl ester except for production of pyruvic acid oxime hydroxamate. The latter presumably arises from phosphoenolpyruvate hydroxamate. These facts were rationalized according to scheme (44) for the dibenzyl ester, *viz.*

$$HOOC-C(=CH_2)-O-P(=O)(OBz)_2 \rightleftharpoons \text{cyclic oxyphosphorane} \xrightarrow{-BzOH} \text{cyclic acyl phosphate} \xrightarrow{H_2O} HOOC-C(=CH_2)-O-P(=O)(OBz)OH \quad (44)$$

$$\text{cyclic oxyphosphorane} \rightleftharpoons CH_2=C(OH)-C(=O)-O-P(=O)(OBz)_2 \xrightarrow{NH_2OH} CH_3C(=NOH)CONHOH + (BzO)_2PO(O^-)$$

$Bz = C_6H_5CH_2^-$

Hydroxamate formation cannot simply be trapping of a cyclic tetravalent phosphate as this requires prior release of benzyl alcohol which is not observed. Hydroxylamine and, to a smaller extent, water must trap an intermediate prior to the cyclic phosphate, most logically the cyclic pentacovalent intermediate or the acyl phosphate. In either case the intermediacy of the cyclic oxyphosphorane is required to rationalize the formation of acyl phosphate through pseudorotation leading to an apical enolic oxygen. The formation of this species is also consistent with the rapid catalyzed expulsion of benzyl alcohol, since on chemical grounds the acyl phosphate would not be anticipated to lose benzyl alcohol more rapidly than starting material. A possible competing pathway is a direct displacement of benzyl alcohol whose contribution to the hydrolysis scheme cannot be readily assessed.

In the case of the monobenzyl ester the experiments with hydroxylamine do not require intermediates analogous to the above. This is in accord with the decreased stability of a monoanionic pentacovalent intermediate whose zwitterion (45)

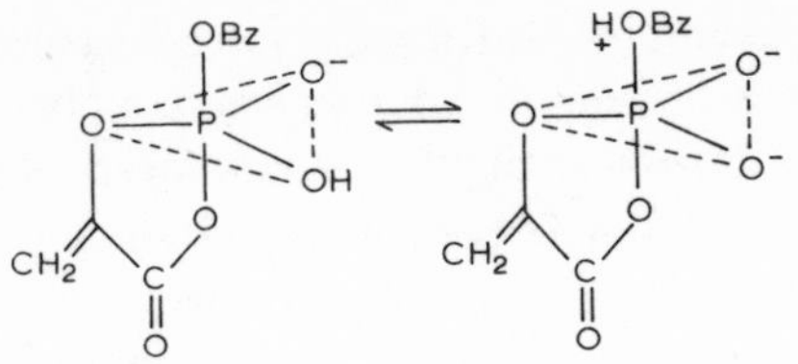

(45)

possesses a greater potential driving force for expulsion of benzyl alcohol than the neutral species. Production of the small amount of monobenzyl phosphate is probably through formation of the respective acyl phosphate, a process which, as in the case of the dibenzyl ester, requires pseudorotation in order to avoid apical–equatorial displacement. Furthermore, the lack of changes in product composition for the monobenzyl system in the presence of hydroxylamine implies that the rate of benzyl alcohol expulsion is nearly ten times greater than formation of acyl phosphate, whereas for the dibenzyl ester, enol expulsion is readily reversible. The fact that in aqueous solution only a small fraction of the total product is dibenzyl phosphate would be a consequence of the ratio of hydroxylaminolysis to hydrolysis for acyl esters with excellent leaving groups.

In order to define further the origin of the hydroxamate products, similar experiments were conducted with the phosphonate derivative (46).

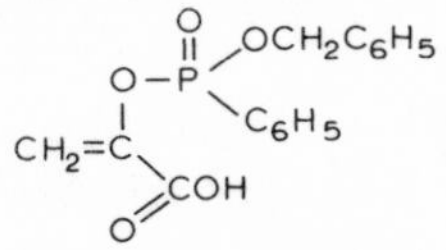

(46)

The hydrolytic products are benzyl alcohol and the corresponding enol phosphonate produced through carboxyl group participation. In the presence of hydroxylamine the products are diverted to benzyl phenylphosphonate and pyruvate oxime hydroxamate. These experimental results contrast markedly with those observed for the phosphoacetoin system. A simple explanation for the formation of benzyl phenylphosphonic acid is generation of the pentacovalent species (47)

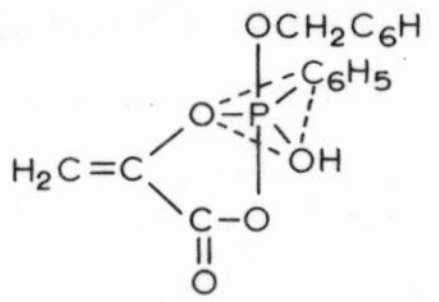

(47)

which, unlike (43), can readily pseudorotate, proton transfer and expel enol apically. The data thus indicates that the energy difference between the zwitterion and neutral species is not as large as that calculated. Apparently the same type of scheme as discussed above, (44) p. 30, is operative and no distinction can be made as to the site of trapping.

A possible means of resolving the dilemma is to study the rate of hydroxylaminolysis of a similar system which is incapable of forming acyclic acyl phosphate. The hydrolysis of diethyl 2-carboxyphenylphosphonate proceeds through carboxyl group cyclization with expulsion of ethanol[127], *viz.*

$$\text{2-}(HOOC)C_6H_4P(O)(OC_2H_5)_2 \longrightarrow C_6H_4\{P(O)(OC_2H_5)\text{-}O\text{-}C(=O)\} + C_2H_5OH \quad (48)$$

The cyclic phosphonate may be intercepted with hydroxylamine to yield the corresponding hydroxamate. Importantly, the rate of hydroxylaminolysis is approximately the same as ethanol loss, thus it appears that the cyclic phosphonate and not a precursor pentacovalent species is being trapped. Insofar as the analogy is appropriate, hydroxylamine is trapping an acyclic acyl phosphate in the above case, albeit at a rate faster than benzyl alcohol loss. It is also noteworthy that for both the phosphonate and phosphate esters there is no net transfer of oxygen from the carboxyl to phosphorus during the entire course of hydrolysis, *i.e.* loss of two moles of alcohol.

One may also obtain a partial insight into the configuration and the leaving tendencies of various groups attached to the pentacovalent intermediates by analysis of the above data. Recall that the alkaline hydrolysis of methyl ethylene phosphate gives entirely ring cleavage product, whereas the base-catalyzed hydrolysis of dimethylphosphoacetoin yields almost entirely the products of phosphoryl migration. Since the product of alkaline hydrolysis of phenyl ethylene phosphate is mainly phenol, the leaving group order is $C_6H_5O^- >$ ring $CH_2O^- > CH_3O^-$. With the mono-benzyl ester the dominance of benzyl alcohol expulsion might arise from a strong preference for the electronegative carboxyl moiety to remain apical. In none of these cases are the pseudorotations leading to either of the two products, ring or exocyclic cleavage, restricted.

The detection of minor amounts of monobenzyl phosphate, which presumably arise from hydrolysis of the monobenzyl acyl phosphate, is an example of the intramolecular migration of monoalkyl hydrogen phosphate. This process resembles the acid-catalyzed migration of a phosphoryl moiety adjacent to a vicinal hydroxyl function[128–130], *viz.*

$$R\text{-}CH(OH)\text{-}CH(R')\text{-}O\text{-}P(=O)(OR)(O(R)H) \quad (49)$$

Note that the available proton is an important feature since its presence allows the monobenzyl ester to simulate a diester. Furthermore, in these cases proton transfer may be concomitant with the pseudorotation process required for ring opening, in order to avoid energetically unfavorable intermediates.

Intramolecular five-membered ring formation occurs in the alkaline-catalyzed hydrolysis of a series of esters of 2-hydroxypropylphosphate, where a linear free-energy relationship is found for the rate of alkoxide expulsion as a function of the pK_a of the alcohol. No evidence for phosphoryl migration was obtained consistent with the fact that pseudorotation of the presumed intermediate leading to ring opening would be inhibited (50), *viz.*

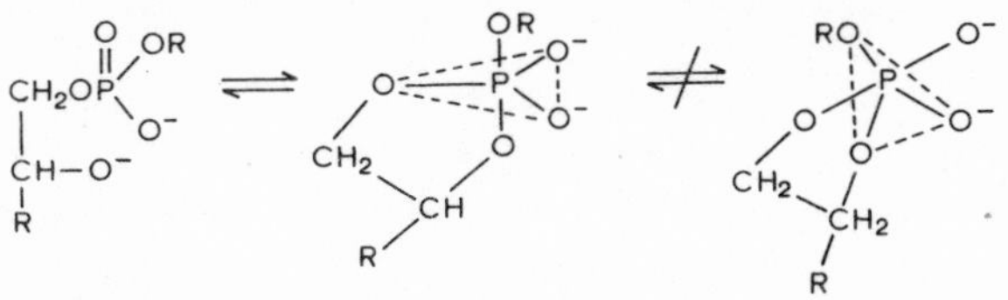

(50)

Esterification of one of the oxyanions in a similar system allows the efficient conversion of (51) to (52).

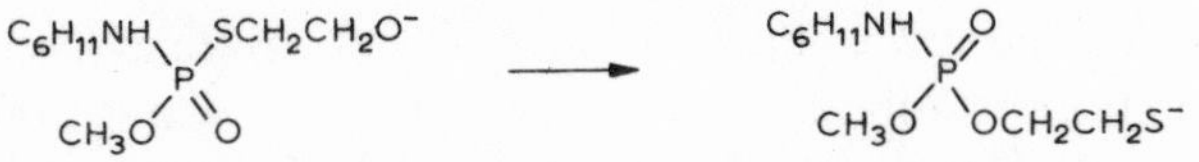

(51) (52)

This process appears to be an additional example of a phosphoryl migration mediated by a pentacovalent species[131].

Recently the related cyclization of the phenyl ester of *cis*-tetrahydrofuran-3,4-diol monophosphate to the corresponding five-membered phosphate with loss of phenol has been shown to be subject to general catalysis by imidazole[132]. This reaction serves as a model for the first step in the action of ribonuclease which leads to the formation of the nucleoside 2′,3′-cyclic phosphate. The actual details of the transition state leading to the cyclic phosphate as catalyzed by the enzyme are presently the subject of some debate. One possibility is the "in-line" mechanism (53)

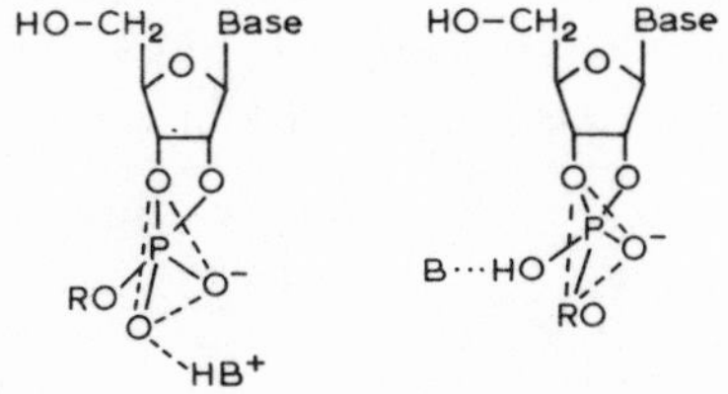

(53)

which envisages the pentacovalent intermediate forming with the departing group (nucleoside-5′-O^-) apical; the other possibility, the "adjacent" mech-

References pp. 51–56

anism (53) has the departing group in an equatorial position, hence pseudorotation is required for nucleoside expulsion. Based on X-ray diffraction and NMR investigation of the mode of binding of a product of the reaction, cytidine-3′-monophosphate, Roberts *et al.*[133] have concluded that only an "in-line" mechanism is possible if the substrate binds in an analogous way. On the other hand, Harris *et al.*[134] have shown that uridine-2′,3′-cyclic phosphonate hydrolyzes at least a thousand times more slowly than the corresponding phosphate. The slower rate of hydrolysis of the phosphonate is consistent with a mechanism requiring pseudorotation (54) since the methylene group initially occupies an apical position, *viz.*

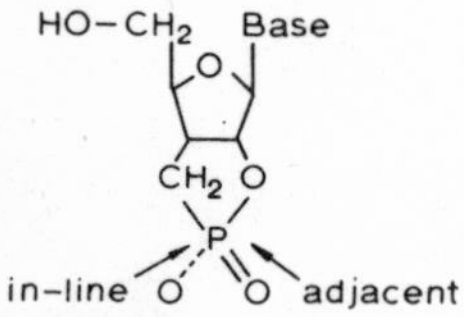

(54)

If ring opening is the microscopic reverse of cyclization, then additional studies obviously are needed to reconcile the evidence.

Locating a nucleophile and phosphoryl center so as to mandate six-membered ring formation for nucleophilic attack, generally results in the operation of an alternate mode of catalysis. For example, salicyl phosphate hydrolyzes with the carboxyl moiety functioning as a general-acid catalyst of monomeric metaphosphate expulsion. Conversion to the diester, however, leads to carboxylate participation with expulsion of the exocyclic leaving group and formation of the cyclic phosphate (55) followed by the latter's hydrolysis[135,136], *viz.*

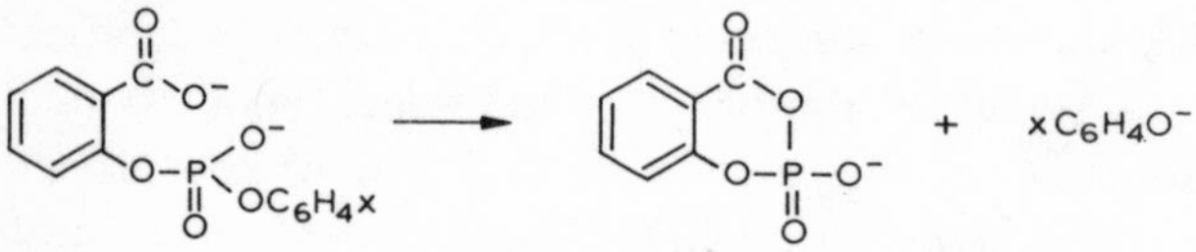

(55)

$x = H, 3\text{-}NO_2, 4\text{-}NO_2$

A rate enhancement of about 10^7 is observed over the slow spontaneous hydrolysis of the corresponding di-3-nitrophenyl phosphate. The specific displacement of the exocyclic phenolate may be rationalized in terms of a pentacovalent intermediate (56)

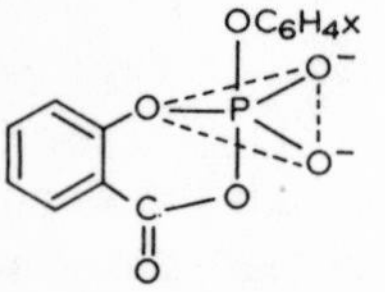

(56)

Decomposition of the species with loss of the apical phenolate presumably would be more rapid than pseudorotation since any such process brings a negatively charged oxygen atom into an apical position. The stereospecific nature of the hydrolysis may also be viewed as arising from the linear steric requirements of nucleophilic displacements on phosphorus and may not involve the actual intermediacy of a pentacovalent species. A large rate enhancement (10^7) also has been discovered in the oxime-catalyzed hydrolysis of structurally similar phosphonate diester (57)[137].

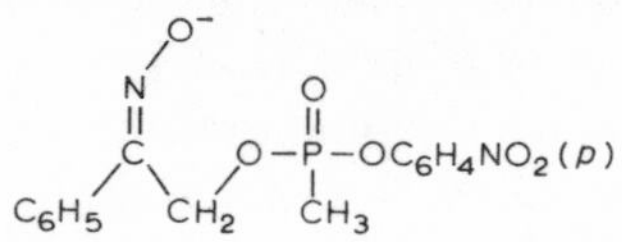

$C_6H_5C(=NO^-)CH_2O-P(=O)(CH_3)-OC_6H_4NO_2(p)$ (57)

The catalysis may not depend on the configuration of the oxime, since both *syn* and *anti* function with similar efficacy[138]. For this reason a mechanism involving the oxime anion acting as a general base catalyst for solvent attack has been favored. However, the observed large rate enhancement is more in accord with a cyclization mechanism, and, until an *anti–syn* interconversion prior to hydrolysis can be unequivocally eliminated, the direct attack of the oximate anion on phosphorus remains a strong possibility.

Catalysis may also be observed *via* transition states or intermediates which are more than six-membered. An example is the hydrolysis of glucose-6-phosphate dianion which surprisingly is more rapid (5 times) than the mono-anion. Presumably the relatively acidic 1-hydroxyl group of glucose (pK_a = 10.8, 100°C) can act as a general-acid catalyst of phosphate group expulsion (58)[139] even though the required chair conformation has all substituents axial, *viz.*

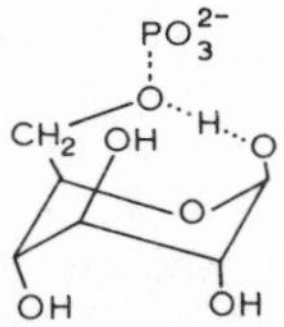

PO_3^{2-} … O(CH_2) … H … O; OH, OH, OH (58)

Other kinetically allowed mechanistic models, *i.e.* hydroxide ion attack on the monoanion, can be rejected on the grounds that the required rate coefficients far exceed that found for alkaline hydrolysis of phosphate triesters. At pH > 9 two new reactions appear, one yielding a 1,6-anhydro sugar by nucleophilic attack through a five-membered transition state of the 1-alkoxide ion upon C-6 with expulsion of phosphate trianion. The second is apparently general-base catalysis by 1-alkoxide of water attack on C-6 or phosphorus through greater than six-membered cyclic transition states.

The important search for viable models for the requirement of divalent metal ions in enzymic phosphate transfer reactions may also benefit from the application of the previously discussed concepts. The importance of Mn(II) in the mechanism of action of phosphoenol pyruvate carboxykinase, phosphoenol pyruvate carboxylase and pyruvate kinase which occurs in the formation of kinetically productive enzyme–metal–substrate bridge complexes has been demonstrated by a number of methods[140]. A ternary complex also has been implicated in the function of creatine kinase[141]. Although one may view the metal ion acting primarily as a template for the substrates and enzymic catalytic groups in the above reaction, a more direct role may also be ascribed to the metal ion. The most obvious proposal is that of charge neutralization, *i.e.* the reduction of the ionic charge of the phosphate anion by interaction with a metal ion of opposite charge *via* actual chelation or ion pairing[142]. Charge neutralization (actually a case of general acid catalysis) might render the phosphorus more susceptible to nucleophilic attack, *i.e.* in the limit the equivalent of esterification. A second related function, previously alluded to during the discussion on polyphosphate hydrolysis, is charge neutralization or chelation leading to promotion of metaphosphate expulsion. However, studies of metal-ion catalysis of hydrolysis of phosphate esters not possessing a potential chelating function in the alcohol moiety have not revealed large rate accelerations (usually less than ten fold)[143].

In contrast, the rate of Cu(II)-catalyzed hydrolysis of 3-pyridylmethylphosphate shows an enhancement of twenty-fold at a pH where the reactive species is probably the 1 : 1 complex represented in (59)[144,145].

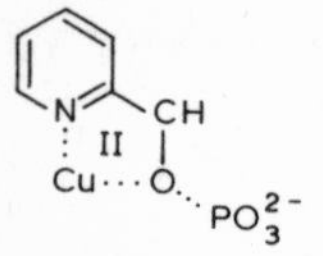

N CH
II
Cu···O···PO_3^{2-}

(59)

In the author's own laboratory the Cu(II)-catalyzed hydrolysis of the phosphate ester derived from 2-[4(5)-imidazolyl]phenol recently has been investigated[146]. The pertinent results are: (*a*) the pre-equilibrium formation of a hydrolytically labile Cu(II)–substrate complex (1 : 1), (*b*) the occurrence of catalysis with the free-base form of the imidazolyl and phosphate moieties and (*c*) the extraordinary rate acceleration at pH 6 (10^4) relative to the uncatalyzed hydrolysis[146]. The latter recalls the unusual rate enhancement encountered above with five-membered cyclic phosphates and suggests a mechanism in which the metal ion, at the center of a square planar complex or a distorted tetrahedral complex, might induce strain in the P–O ester bonds (60), *viz.*

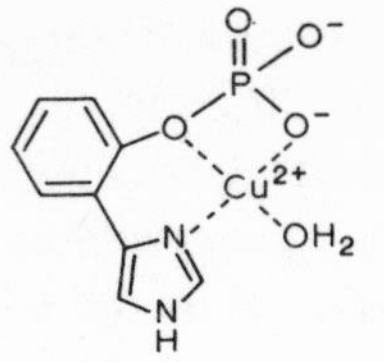

(60)

No evidence, other than the observed rate acceleration, however, can be cited at present to support this contention.

The bonding of methyl phosphate to the triethylenetetraminecobalt(III) ion similarly leads to an increased rate of hydrolysis (*ca.* 10^2) for the monoanionic species[147]. The possibility that catalysis in this case is of the bidentate type is supported by the inactivity of related Co(III) complexes where bidentate coordination is precluded owing to the slow exchange of amine ligands in Co(III) complexes. The proposed minimal mechanisms involve (61) and (62) with the former invoking the concept of induced strain and the latter the facility of five-membered ring cyclization, *viz.*

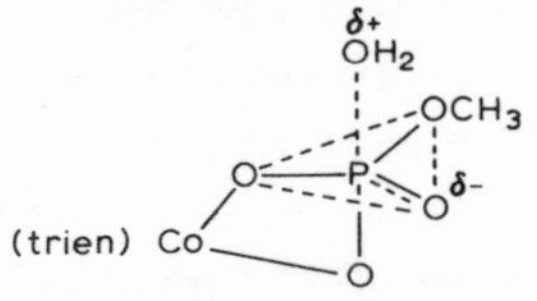

(61)

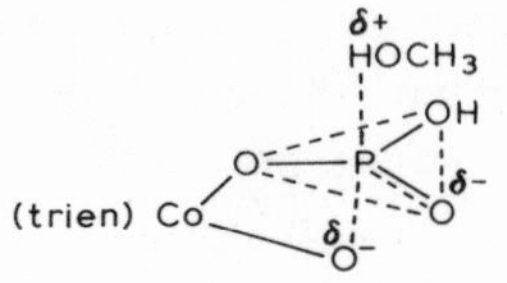

(62)

Although the transition states are depicted in accord with the preference rules, the presence of pentacovalent phosphorus is conjecture.

The preceding discussion has emphasized catalysis; nevertheless, metal ions may also significantly inhibit the rate of hydrolysis of phosphate esters through chelation at phosphorus. A pertinent example is the sixty-fold decrease in the rate of hydrolysis of 2-aminoethylphosphorothioate in the presence of excess Fe(III)[148,149]. Such a phenomenon underscores the exacting requirements for observation of metal-ion catalysis and implies that charge neutralization *per se* is not responsible. One should also note the ineffectiveness of Mg(II) or Zn(II) as catalysts in the above systems, the latter required for the activity of alkaline phosphatase (*E. coli.*)[150]. An attractive, but as yet experimentally untested hypothesis, is that such metal ions may catalyze pseudorotation processes which otherwise would violate the preference rules.

Another approach to catalysis of phosphate transfer is to vary the medium of the reaction. An interesting adaptation of this theme is the incorporation of

References pp. 51–56

the substrate into a micellar phase, a process which is analogous to enzymic catalysis. The spontaneous hydrolysis of 2,4- and 2,6-dinitrophenylphosphate dianions is increased by twenty-five-fold (at saturating detergent) upon the incorporation of these esters into the cationic micelle of cetyltrimethylammonium bromide[151]. Catalysis is neither observed in the hydrolysis of the corresponding monoanions where proton transfers are involved nor in the reaction between hydroxide ion and the aryl phosphate dianions. These results suggest that the fragmentation of the dianion into two monoanions is energetically favorable on the micelle surface and/or incorporation of the substrate into a micellar environment simulates an aqueous organic solvent which is known to accelerate dianion hydrolysis rapidly[8]. The lack of micellar assistance to hydroxyl ion attack can be explained in terms of the competitive inhibition of micellar catalysis observed in the presence of salts containing a negatively charged bulky organic residue *e.g.* sodium tosylate and *p-t*-butyl phenyl phosphate. Evidently the inhibiting anions compete with the substrate for identical sites and repel the approach of additional substrate molecules. The reaction of the triester, *p*-nitrophenyl diphenylphosphate, with fluoride or hydroxide ion is also catalyzed by incorporation into micelles of cetyltrimethylammonium bromide, a phenomenon which is again inhibited by added anions[152,153]. With decreasing pH the reaction of hydroxide ion with the substrate becomes unimportant and the intramicellar attack of these anions on the substrate can be detected. The detergent enhances the nucleophilicity of phenyl phosphate dianion *ca.* 1,000-fold and of inorganic phosphate dianion *ca.* 40-fold. These results show that the anions and the *p*-nitrophenyl diphenylphosphate must be able to orient themselves so that they form the activated complex in the micellar phase without losing the hydrophobic and electrostatic bonding between the reagents and the cationic micelle. The greater enhancement of the aryl phosphate dianion nucleophilicity—a reversal of the situation in the absence of detergent—emphasizes the importance of hydrophobic interactions. One may also imagine distortion of P–O bonds within these complexes to facilitate catalysis but this again is conjecture. Whatever the cause, considerable catalysis may be observed by altering the immediate environment around the phosphoryl moiety, a factor probably involved to some degree in the enzyme-catalyzed phosphate-transfer reactions.

The above discussion has relied to a considerable extent on the concepts developed in the chemistry of cyclic phosphates. Obviously pentacovalent species are not required on the basis of the available experimental evidence in many of the above reactions and alternate explanations are plausible. Nevertheless, many diverse observations are explained with equal facility by application of these concepts. Although this certainly does not constitute a proof, the approach offers a unifying framework for productive experimental design with both chemical and biological systems. One relevant example is the previously described investigation of the hydrolysis of dibenzylphosphoenolpyruvate which now has led to the observation that the parent phosphoenolpyruvic acid undergoes reversible cyclization in aqueous solution[154], *viz.*

$$CH_2{=}C(COOH){-}O{-}P(=O)(OH){-}O^- \underset{-H_2O}{\overset{+H_2O}{\rightleftharpoons}} \text{cyclic } CH_2{=}C{-}O{-}P(=O)(O^-){-}O{-}C(=O) \qquad (63)$$

Mechanisms may be readily envisaged where this reaction may lead to phosphoryl transfer through nucleophilic attack with ring opening on phosphorus followed by hydrolysis of the acyclic acyl phosphate. In biological systems this may function as an alternative to the metaphosphate reaction.

3. Inorganic esters of sulfur

3.1 THE SULFUR TRIOXIDE MECHANISM FOR ACYCLIC ESTERS

The hydrolysis of sulfate monoester monoanions appears to proceed *via* two mechanistic pathways; (*a*) at pH < 4, a rapid acid-catalyzed reaction is encountered and (*b*) a plateau region, pH 4–10, associated with a spontaneous rate of hydrolysis is observed with excellent leaving groups. The acid-catalyzed reaction for a series of aryl sulfates is characterized by (*i*) $\alpha\rho \approx +0.5$ suggesting that the leaving group is protonated, (*ii*) activation entropies, zero or slightly positive in accord with a unimolecular rate-determining step, and (iii) a ratio of $k_{H_2O}/k_{D_2O} < 0.5$ also consistent with an A-1 process[155–157], *viz.*

$$\begin{aligned} &ROSO_3^- + H_3O^+ \rightleftharpoons ROSO_3H + H_2O \\ &ROSO_3H \rightarrow ROH + SO_3 \end{aligned} \qquad (64)$$

In contrast, the acid-catalyzed hydrolysis of alkyl selenates is A-2[158]. The actual species which undergoes decomposition to alcohol and sulfur trioxide is probably the zwitterion as in the case of phosphate monoester monoanions. Evidence for sulfur trioxide as the reactive initial product of the A-1 solvolysis is obtained from the product compositions arising with mixed alcohol–water solvents. The product distribution is identical to that found for sulfur trioxide solvolysis, with the latter exhibiting a three-fold selectivity for methanol. Although the above entropies of activation and solvent deuterium isotope effects do not distinguish between the conventional A-1 mechanism and one involving rate-limiting proton transfer, a simple calculation, based on the pK_a of the sulfate moiety and the fact that its deprotonation is diffusion controlled,

References pp. 51–56

militates against rate-determining proton transfer. Owing to the limited data available for alkyl sulfate hydrolysis, one cannot calculate with reasonable accuracy a linear free-energy relationship of the form, log k (rate of acid-catalyzed hydrolysis) *versus* pK_a of the departing alcohol, except for the phenols (Fig. 7; slope −0.25). Crude computations, however, indicate that the

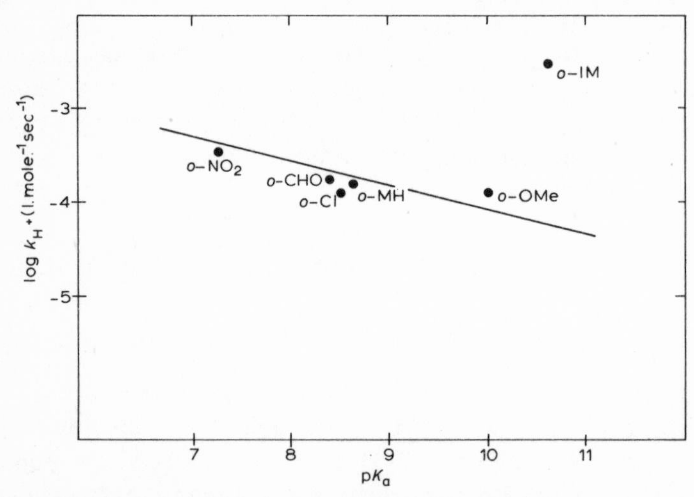

Fig. 7. Plot of log k_{H^+} for the hydrolysis (at 48.6°C) of a series of substituted phenyl sulfates vs. the pK_a of the corresponding phenols. The solid line represents *meta*- and *para*-substituted substrates; *ortho*-substituted compounds of interest are indicated individually.

acid-catalyzed hydrolysis of methyl sulfate is slower than predicted[156,159], suggesting more complete separation of sulfur trioxide and alcohol in the transition state than for the corresponding phosphate monoesters.

It is of considerable interest to note that the rate of acid-catalyzed solvolysis of aryl and alkyl sulfates is subject to striking accelerations in systems of low water content. Batts[160] has reported that the rate of acid-catalyzed hydrolysis of methyl sulfate is increased 10^7 times in changing from pure water to 1.9% H_2O–dioxan. The unusual dioxan–water solvent effect is in the opposite direction to those commonly observed for reactions in which a water molecule attacks a formally unsaturated electrophilic atom, as in the spontaneous hydrolysis of derivatives of organic and inorganic acids[161]. Probably the major contributing factor is the more effective interaction of the organic solvent with a sulfur trioxide transition state rather than the initial sulfate ground state, thus reducing the number of solvent–solute or solvent–solvent interactions.

The spontaneous reaction of aryl sulfates is particularly detectable when the aryl moiety bears strongly electron withdrawing substituents. A logarithmic plot of rate against pK_a of the corresponding phenol is linear with a slope of

-1.2[162]. Thus the effects of substituents are quantitatively similar to those found for the spontaneous hydrolysis of phosphate dianions which feature unimolecular P–O bond fission. The most likely mechanism thus involves rate-determining unimolecular sulfur–oxygen bond fission with the elimination of sulfur trioxide. Extensive studies measuring the effects of added electrolytes and solvents on the molar activity coefficients of 2,4-dinitrophenyl sulfate and its hydrolysis transition state eliminated the alternate mechanism involving bimolecular nucleophilic attack of water on sulfur[162]. Nevertheless, the extensive solvation of negatively charged esters of inorganic elements by water tends to yield conflicting experimental parameters typical of borderline behavior.

Comparison of the various mechanistic facets of the above sulfate reactions to those of the acid-catalyzed hydrolysis of the Bunte salts ($RSSO_3^-$) suggests that the mechanisms are identical, *i.e.* A-1[163]

$$RSSO_3^- + H_3O^+ \rightleftharpoons R\overset{+}{\underset{|\atop H}{S}}SO_3^- \rightarrow RSH + SO_3 \rightarrow \text{solvolysis products} \quad (65)$$

In (65) only electrophilic catalysis of the scission of the S–S bond is involved, a situation also encountered in the acid-catalyzed hydrolysis in weakly nucleophilic media (acetic acid–1% water) of certain aryl sulfinyl sulfones substituted with electron-donating substituents (66)[164], *viz.*

$$p\text{CH}_3\text{OC}_6\text{H}_4\underset{\text{O}}{\underset{\|}{\text{S}}}\text{–}\underset{\text{O}}{\underset{\|}{\overset{\text{O}}{\overset{\|}{\text{S}}}}}\text{C}_6\text{H}_4\text{OCH}_3p + \text{H}_3\text{O}^+ \rightleftharpoons p\text{CH}_3\text{OC}_6\text{H}_4\underset{\text{HO}}{\underset{|}{\overset{+}{\text{S}}}}\text{–}\underset{\text{O}}{\underset{\|}{\overset{\text{O}}{\overset{\|}{\text{S}}}}}\text{–C}_6\text{H}_4\text{OCH}_3p$$

$$p\text{CH}_3\text{OC}_6\text{H}_4\underset{\text{HO}}{\underset{|}{\overset{+}{\text{S}}}}\text{–}\underset{\text{O}}{\underset{\|}{\overset{\text{O}}{\overset{\|}{\text{S}}}}}\text{–C}_6\text{H}_4\text{OCH}_3p \longrightarrow p\text{CH}_3\text{OC}_6\text{H}_4\text{S}\overset{+}{\text{O}} + p\text{CH}_3\text{OC}_6\text{H}_4\text{SO}_2\text{H} \quad (66)$$

$$\downarrow$$

solvolysis products

Heterolytic scission of the –S–S– bond in which only electrophilic assistance is involved is the exception rather than the rule in reactions involving bond fission of this type. Kice *et al.*[165,166] have demonstrated that a variety of S–S bond cleavages involve concomitant electrophilic and nucleophilic catalysis including (*a*) the formation of aryl thiolsulfones from aryl thiolsulfinates and aryl sulfinic acids and (*b*) the hydrolysis (acetic acid – 1% water and 60%

dioxan) of aryl sulfinyl sulfones, with both (*a*) and (*b*) being catalyzed by hydronium ion and alkyl sulfides. A representative example is the reaction of phenylbenzene thiolsulfinate with an arylsulfinic acid in the presence of alkyl sulfide (solvent, acetic acid–H_2O). The reaction is (*a*) first-order in both thiolsulfinate and alkyl sulfide but the rate is independent of sulfinic acid concentration, (*b*) specific acid catalyzed and (*c*) dependent on the structure of the catalyzing sulfide ($\rho^* -2.0$). The suggested mechanism involves rate-determining nucleophilic attack of the sulfide on the sulfinyl-protonated thiolsulfinate, *viz*.

$$C_6H_5\underset{\underset{O}{\|}}{S}SC_6H_5 + H_3O^+ \rightleftharpoons C_6H_5\underset{\underset{OH}{|}}{\overset{+}{S}}SC_6H_5 + H_2O$$

$$R_2S + C_6H_5\underset{\underset{OH}{|}}{\overset{+}{S}}SC_6H_5 \xrightarrow{\text{slow}} R_2\overset{+}{S}SC_6H_5 + C_6H_5SOH \tag{67}$$

$$R_2\overset{+}{S}SC_6H_5 + ArSO_2H \longrightarrow Ar\underset{\underset{O}{\|}}{\overset{\overset{O}{\|}}{S}}SC_6H_5 + R_2S + H^+$$

The essential features of this mechanism apply to the sulfide-, halide- and thiocyanate-catalyzed racemization of phenylbenzene thiolsulfinate in acidic aqueous dioxan[6], and the hydrolysis in aqueous dioxan or acetic acid of aryl sulfinyl sulfones. In the latter, proton transfer to the departing $ArSO_2$ group is apparently synchronous with attack by sulfide, *i.e.* general-acid catalysis by hydronium ion[166,167]. In the absence of alkyl sulfides the reaction of thiolsulfinates with aryl sulfinic acids to form aromatic thiolsulfonates evidently proceeds with general-base catalysis involving (68) as anticipated from the aforementioned behavior of the $ArSO_2$ group[168].

$$\left[B\cdots H\cdots O-\underset{\underset{O}{\|}}{\overset{\overset{Ar}{|}}{S}}\cdots\underset{\underset{C_6H_5}{|}}{S}\cdots\underset{\underset{OH}{|}}{\overset{\delta+}{S}}-C_6H_5 \right] \tag{68}$$

The rates of spontaneous hydrolysis of aryl α-disulfones and aryl sulfinyl sulfones in dioxan–water are similar in their response to reaction variables: solvent composition, electronic effects ($\rho = +3.5$), activation parameters and solvent deuterium isotope effects $k_{H_2O}/k_{D_2O} = 2.3$[169]. The difference in rate (the aryl α-disulfone hydrolyzing some 10,000 times more slowly) is principally the result of a more unfavorable energy of activation and not a change in mechanism. The proposed mechanism, *e.g.* for α-disulfone hydrolysis

$$H_2O + ArS(=O)_2-S(=O)_2-Ar \xrightarrow[\text{determining}]{\text{rate-}} \left[Ar-S(=O)_2(-O-H)\cdots S(=O)(=O)-Ar \;\; (O\cdots H\cdots O) \right] \rightarrow ArSO_3H + ArSO_2H \quad (69)$$

features a rate-determining proton transfer to the departing $ArSO_2$ group and is consistent with mechanism of the sulfide-catalyzed reaction of aryl sulfinyl sulfones discussed above. It is also possible that transfer of the proton to the leaving group is concerted with attack by the incoming water, *i.e.* a one-encounter mechanism[12]. It is noteworthy that no acid-catalysis term is detected in the rate expression for the hydrolysis of α-disulfones in aqueous dioxan owing to the low basicity of the sulfonyl group, whereas hydrolysis of sulfinyl-containing substrates in a similar solvent media partially proceeds through the conjugate acid form[166,169]. Although no change in mechanism is encountered in the sulfinyl–sulfonyl replacement, alterations in the leaving group ($ArSO_2^-$ to $ArSO_3^-$) lower the requirement for catalysis. Consequently aryl sulfonic anhydrides (70) hydrolyze *via* a nucleophilic attack by water that does not appear to involve proton transfer[170].

$$Ar-S(=O)_2-O-S(=O)_2-Ar \quad (70)$$

The finding of simple nucleophilic catalysis in the hydrolysis of sulfinyl sulfones[171] and α-disulfones[172] in addition to the electrophilic–nucleophilic catalysis observed with the thiolsulfinates furnishes the type of data necessary to establish an accurate relative scale of nucleophilicity for attack leading to S–S bond cleavage at various sulfur centers. For sulfenyl sulfur, the order of reactivity is $I^- > SCN^- \gg Br^-$ with the more polarizable nucleophiles being highly reactive, a response paralleling nucleophilic attack at peroxide oxygen. Toward sulfinyl sulfur, the order of reactivity is $Br^- > Cl^- \cong CH_3COO^- > F^- \gg H_2O$, an order both qualitatively and quantitatively similar to that found for S_N2 substitutions at sp^3 carbon. In contrast toward sulfonyl sulfur the relative reactivity is in the order $F^- \gg CH_3COO^- \gg Cl^- > Br^- > H_2O$. This series extended to other nucleophiles reveals a pattern of relative reactivities roughly the same as toward carbonyl carbon[173,174]. In essence substitution of oxygen for unshared electron pairs on sulfur confers increasing hardness on the electrophilic sulfur owing to the progressive decrease in the potential polarization of sulfur and an increase in its positive charge[175]. The ultimate hardiness should be manifest in nucleophilic reactions on the sulfur

References pp. 51–56

of sulfate, and indeed in nucleophilic catalysis of *p*-nitrophenylsulfate hydrolysis there is apparently very little bond formation between the incoming nucleophile ($\beta = 0.13$ for a series of amines) and the sulfur center (Fig. 8)[176].

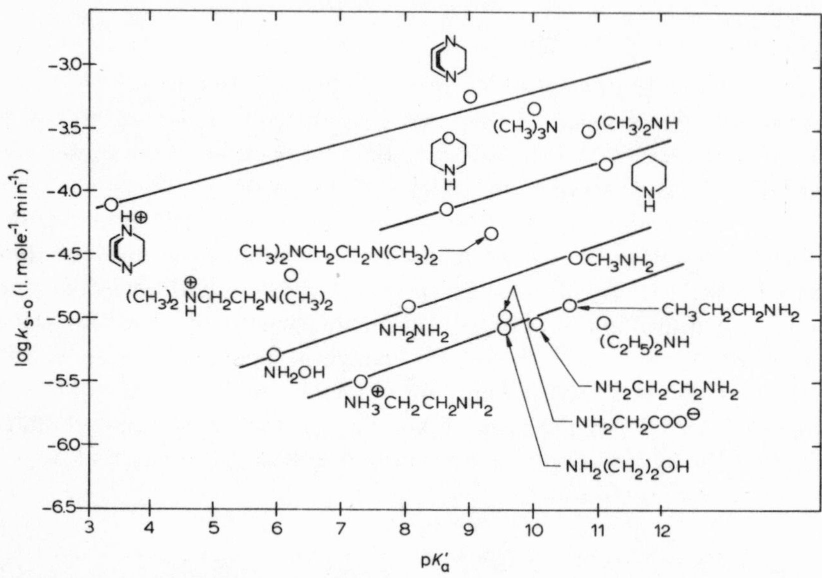

Fig. 8. Plot of the logarithm of the second-order rate coefficient (l·mole^{-1}·min^{-1}, 35°C, ionic strength = 1.0) for S–O bond cleavage *vs.* pK'_a. Rate k_{S-O} for tertiary amines assumed equal to calculated second-order rate coefficient. Rate coefficients for diamines statistically corrected.

Much less has been done on the hydrolytic cleavage of the S–O bond, the main reason being the difficulty in breaking this bond, especially if the sulfur atom is in a higher oxidation state[177], along with the tendency for groups, including $-SO_2Cl$ and SO_3H, to be expelled *via* C–O bond scission[178]. An exception is the well-studied acid-catalyzed hydrolysis of dialkyl and diaryl sulfites whose mechanism falls into the A-2 category[161]. Recently, preliminary results have been reported on the alkaline hydrolysis of esters of *p*-arylsulfinic acids[179]. The behavior is similar to that observed for the reaction of substituted benzyl bromides with nucleophiles and therefore, is in accord with the above discussion.

The above results have a number of implications for the functioning of biological systems. Concomitant electrophilic–nucleophilic catalysis has been suggested in the rupture of the sulfur–sulfur bond in thiosulfate, catalyzed by the enzyme rhodanese. The proposal envisages coordination of the substrate through its oxygens to an enzymic zinc ion. Attack by a nucleophile on the thiol sulfur (71) results in the formation of a sulfur-substituted rhodanese which is regenerated in a second displacement[180–182]. The low susceptibility

$$\sim N: \nearrow -S-\overset{O}{\underset{O}{\overset{\|}{\underset{\|}{S}}}}-O\text{------}Zn^{2+} \qquad (71)$$

of the sulfur of sulfate to nucleophilic attack further suggests that the hydrolytic enzymes, *e.g.* the aryl sulfatases, which cleave the S–O bond of phenolic sulfates may utilize a pathway that catalyzes sulfur trioxide expulsion[183]. A similar argument may apply to the reactions catalyzed by sulfotransferases where a sulfuryl moiety is transferred from 3′-phosphoadenylyl-5′-phosphosulfate to a phenol or alcohol (see Section 3.3)[184]. However, as with the phosphate esters, a mechanism involving an intermediate cyclic sulfate may also be operative owing to the latter's high susceptibility to hydroxide-ion attack. Finally, it is of interest that the increased susceptibility to nucleophilic attack of sulfur in a lower oxidation state can be utilized in the design of novel reagents for particular enzymic reactions. Methyl *p*-bromophenyl sulfite, for example, is an excellent substrate for pepsin. The enzymic reaction presumably involves formation of a mixed anhydride intermediate derived from phenol expulsion by an enzymic carboxyl group[185].

3.2 CYCLIC ESTERS CONTAINING SULFUR

In a manner characteristic of the five-membered cyclic phosphates, the five-membered aromatic *o*-phenylene sulfate, hydrolyses in base 2×10^7 times faster than does diphenyl sulfate[186]. Likewise the alkaline hydrolysis of *o*-hydroxy-α-toluene-sulfonic acid sultone shows a rate enhancement of 7×10^5 over its open-chain analogue[187]. Incorporation of sulfur into a larger six-membered ring decreases drastically the observed enhancement so that the alkaline hydrolysis of six-membered and acyclic sulfate esters are comparable within an order of magnitude[188]. All these reactions are examples of hydroxide-ion attack on sulfur leading to ring opening. Smaller kinetic accelerations (10^3) are also observed for both aliphatic and aromatic cyclic sulfites [73,74,189]. The difference between the rates of hydrolysis of the acyclic sulfate and the cyclic aliphatic ethylene sulfate is difficult to calculate owing to competing C–O bond cleavage with the cyclic ester and exclusive C–O bond scission with the acyclic ester[72].

Tillett *et al.*[190] have discussed the rapid hydrolysis of cyclic sulfur-containing esters in terms of the entropy strain principle. In the transition state for acyclic ester hydrolysis the molecule assumes a more ordered structure relative to the ground state, whereas with the five-membered cyclic esters the molecule is already held in a rigid structure with a corresponding reduction in entropy (*ca.* 10 eu). Unlike the cyclic five-membered phosphates, the corre-

References pp. 51–56

sponding sulfites are thermochemically ring-strain-free on the basis of such data[73,191]. The observation by Aksnes and Bergesen[81] that a major part of the rate acceleration observed for the phospholan ring is caused by a favorable entropy of activation, despite the presence of considerable ring strain, suggests that a maximal contribution of 10^3 to the observed rate enhancement might arise from this effect. The greater rate enhancements observed with the cyclic sulfates then originates from actual thermochemical ring strain as revealed by data showing an approximate 5–6 kcal·mole^{-1} difference in the heats of hydrolysis of dimethyl and ethylene sulfate[81]. The X-ray studies of vinylene sulfate and ethylene sulfate establish the O–S–O bond angle at 93.6° and 98.4°, respectively, relative to 101° in a six-membered sulfone, consistent with ring strain[192,193]. Whether one can actually dissect the rate factors in this fashion is problematic since one has the additional but now familiar problem of assessing the effect of cyclization on possible *p–d* π bonding[194].

It is of interest to consider whether the theory of pseudo-rotation is applicable to the hydrolysis of sulfur-containing esters. Although reactions involving attack on sulfur have been discussed in terms of a discrete trigonal bipyramidal intermediate with entering and leaving groups occupying apical or equatorial positions (alkoxy-sulfonium ions), there is no present evidence to support this contention[195]. In fact a recent experimental test for pseudorotation in the hydrolysis of a thietanium derivative (72) failed to disclose either pseudorotation or a pentacoordinate intermediate of a significant lifetime.

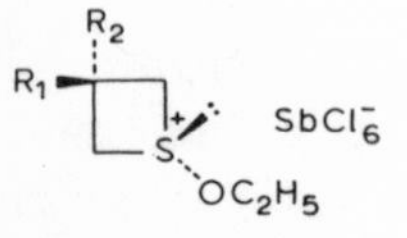

R_2 R_1 S^+ $SbCl_6^-$ OC_2H_5 (72)

Whereas the phosphetanium (Sect. 2.2) derivatives undergo base-catalyzed hydrolysis with essentially complete retention of configuration, the reaction with the related thietanium salts proceeds with complete inversion of configuration and can be described simply as a direct nucleophilic substitution[195]. As far as the author is aware, no exchange evidence of the type encountered in the acid-catalyzed hydrolysis of ethylene phosphate, has been found for the cyclic sulfate esters. Consequently despite the geometrical similarities between the cyclic sulfates and phosphates and the related hybridization of the central atoms, the situation in which a finite pentacovalent intermediate sulfur species exists has not been delineated (restricting the discussion to esters).

The above cyclic sulfates and 4-nitrocatechol cyclic sulfate are good substrates for α-chymotrypsin and carbonic anhydrase[196–198]. In the case of the chymotrypsin-catalyzed hydrolysis of (73) to the product acid, a rapid sulfonylation of the active site of the enzyme to give (74) occurs, *viz.*

O_2N–C$_6$H$_3$(CH$_2$SO$_2$O) (73) + E ⟶ O_2N–C$_6$H$_3$(CH$_2$–SO$_2$E)(OH) (74)

followed by slow decomposition of (74) to yield product and regenerate enzyme. Inhibition of the enzyme with α-toluene-sulfonyl fluoride also yields a sulfonyl enzyme which, however, does not desulfonylate[199]. This data appears to implicate the *o*-phenolic hydroxyl as a catalytic function in the desulfonylation of the enzyme. However, the 4-nitrocatechol cyclic sulfate also gives the sulfonyl enzyme intermediate but despite the presence of an *o*-hydroxyl function no desulfation of the enzyme is observed. The reasons for this change in behavior are unknown. The cyclic esters appear to react at the normal active site, serine hydroxyl of chymotrypsin.

3.3 CATALYSIS

The available studies imply that general catalysis will be operative in systems involving sulfate monoesters and potential six-membered ring transition states. Salicyl sulfate hydrolyzes at pH 4 *via* intramolecular carboxyl group participation involving pre-equilibrium proton transfer leading to sulfur trioxide expulsion (Fig. 9)[200], *viz.*

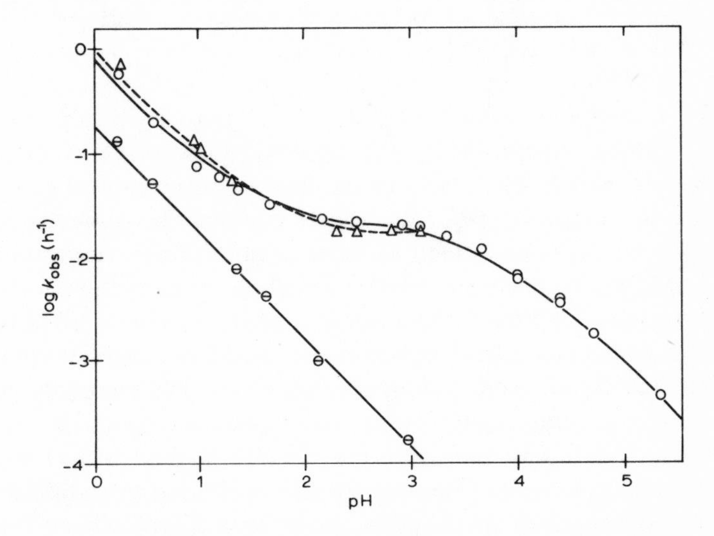

Fig. 9. The pH-rate profiles (35°C, ionic strength = 1.0) for the hydrolysis of salicyl sulfate (○, H_2O; △, D_2O) and *p*-carboxyphenyl sulfate (⊖, H_2O).

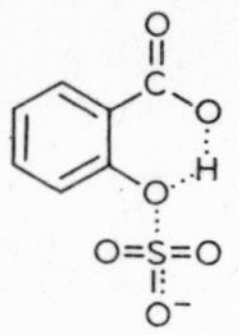

(75)

The data supporting this mechanism, in addition to the kinetic evidence, is as follows: (*a*) a solvent deuterium isotope effect, $k_{H_2O}/k_{D_2O} = 1.2$, accommodated by postulating nearly complete proton transfer prior to or in the transition state, (*b*) the inability to trap an intermediate acyl sulfate under conditions where acetyl sulfate is quantitatively converted to acetohydroxamic acid, and (*c*) a product distribution in methanol–water nearly identical to that obtained with sulfur trioxide solvolysis in the same mixed solvent, implying the transient formation of a product-generating species with sulfur trioxide character. The latter is not a property of acyl sulfates. The rate acceleration afforded by the neighboring carboxyl group is *ca.* 10^4 relative to a phenolic sulfate with comparable pK_a or 200 relative to the *p* isomer. Replacement of the *o*-carboxyl with an imidazolyl residue as in 2-[4(5)-imidazolyl] phenyl sulfate also leads to catalysis of sulfate ester hydrolysis with the pH–rate profile implicating imidazolium or its kinetic equivalent as the active species[201]. Again the possible mechanisms fall into two general categories with the imidazole group functioning either as a general acid or as a nucleophilic species after pre-equilibrium protonation. The rate of hydrolysis of the presumed intermediate, 4(5)-(2′-hydroxyphenyl)imidazole-N-sulfate, is some 70 times faster than hydrolysis of the O-sulfate in the pH region where catalysis is operative; hence the N-sulfate satisfies the kinetic requirement. Attempts, however, to trap the N-sulfate intermediate during the hydrolysis of the O-sulfate with sodium fluoride (specific for N-sulfate)[202] resulted in no formation of fluorosulfate. Consequently the general acid pathway appears to be the predominant one. This conclusion is further supported by the marked decrease in the catalytic efficiency with the rate of hydrolysis being only 49 times greater than for a phenolic sulfate of comparable pK_a. The effectiveness of general-acid catalysis can now readily be estimated since a linear relationship is found between $\Delta(\Delta G^{\ddagger})$ and ΔpK_a, where these parameters are obtained from the corresponding rates and pK_a values for the *o* and *p* isomeric sulfates. In order to insure the fact that ΔpK_a arises mainly from differences in hydrogen bonding, the comparison is limited to those *o*-substituted phenols whose rate of recombination with hydroxide ion is abnormally low[203]. The reasonable fit of the data for salicyl phosphate and sulfate and 2-[4(5)-imidazolyl] phenyl sulfate suggests that the rate enhancement with *o*-groups may be attributed to stabilization of an intermediate or transition state zwitterion by hydrogen bonding. Furthermore, the result implies a continuity of mechanism. The marked dependency of catalysis on the pK_a of the participating acid implies that groups more basic than imidazole would be largely ineffective as general-acid catalysts for this type of system. Thus, there is no evidence for participation by an *o*-hydroxyl group in the hydrolysis of a series of steroid sulfates[204]. Metal-ion catalysis (Cu(II)) is also efficient in promoting sulfate hydrolysis provided that chelation can increase the acidity of the leaving group. A productive complex is apparently obtained with 8-hydroxyquinoline sulfate although the rate enhancement obtained cannot be accurately assessed at present[205].

Several model systems have been designed to simulate sulfate transfer in biological reactions. Oxidative cleavage of quinol monophosphates with resultant transfer of phosphate suggests that similar reactions might be achieved with the corresponding sulfates. Periodate or bromine oxidation of hydroquinone monosulfate ester in methanol or dimethylformamide–methanol results in a 60–80% conversion of methanol to methyl sulfate presumably through the intermediacy of sulfur trioxide[206]. In aqueous solution the oxidation is apparently diverted to C–O bond cleavage. This is in accord with earlier results on oxidation of quinol monophosphates and the reaction of Hg(II) with phosphoenolpyruvic acid in which the intermediate cation is effectively quenched by water more rapidly than decomposition *via* P–O to generate a phosphorylating species[10]. The sulfur analog of 1-phosphoimidazole (see Section 2.1), imidazolium-N-sulfonic acid, is a more general sulfating agent transferring SO_3 to amines, phenols and alcohols[207]. Although none of the above intermediates have been identified in biological sulfate transfer reactions, it is now evident that they are potential candidates.

The properties of 3′-phosphoadenosine-5′-phosphosulfate (PAPS) the established sulfate donor in the formation of chrondrointin sulfate, cerebrosulfatides and phenolic sulfates[208–210] has been investigated through an appropriate model system[211]. The pH–rate profile for phenyl phosphosulfate given below (76), indicates that both the dianion and monoanion are definitely

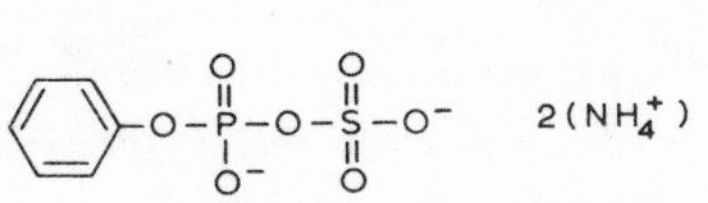

$$C_6H_5{-}O{-}P(=O)(O^-){-}O{-}S(=O)_2{-}O^- \quad 2(NH_4^+) \qquad (76)$$

hydrolytically labile (Fig. 10). The latter also undergoes acid-catalyzed hydrolysis. For all of these species S–O bond cleavage is the main reaction pathway (> 95%). The postulated mechanisms for dianion and monoanion hydrolysis are based on (*a*) deuterium solvent isotope effects and entropies of activation in accord with mechanisms involving a limited degree of water participation, (*b*) methyl sulfate production by both species in methanol–water solvents similar to that obtained with sulfur trioxide and (*c*) a correlation with the previously described linear free-energy relationship for the spontaneous hydrolysis of aryl sulfate monoanions. The accumulated evidence is in agreement with a unimolecular elimination of sulfur trioxide with increased

References pp. 51–56

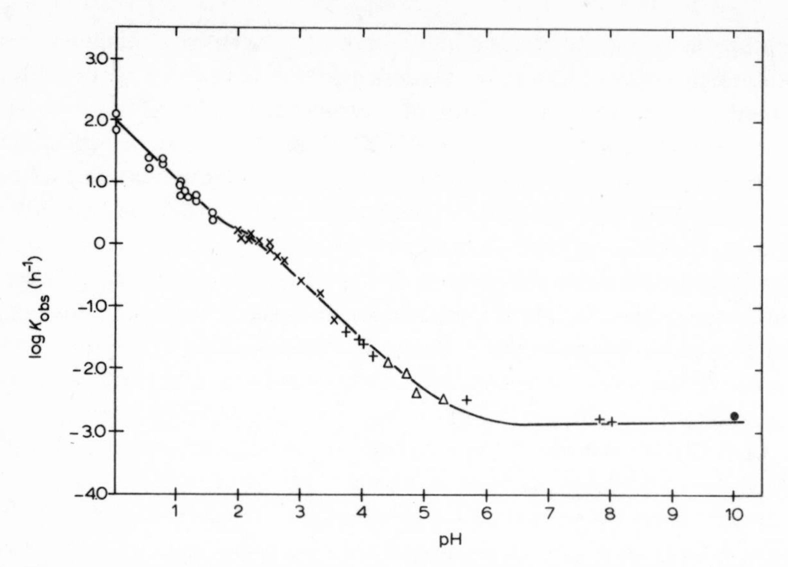

Fig. 10. The pH–rate profile for the hydrolysis of phenyl phosphosulfate (55°C, ionic strength = 1.0).

restriction of solvent translational modes in the transition state (cleavage of S–O bond precedes solvent-S bond formation) for both (77) and (78), *viz.*

$$C_6H_5-O-P(=O)(O^-)-O^{\delta-}\cdots S(=O)(=O)(O^{\delta-})\cdots OH_2 \quad (77)$$

$$C_6H_5-O-P(=O)(OH)-O^{\delta-}\cdots S(=O)(=O)(O^{\delta-})\cdots OH_2 \quad (78)$$

Hydrolysis of the monoanion is thought to proceed through the thermodynamically more stable (78) rather than a zwitterionic species based on the analogy to pyrophosphate hydrolysis (see Section 2.4). Comparison of the rates of hydrolysis in the present system with those of substituted pyrophosphates reveals that a very small margin of reactivity, no more than a maximal ten-fold, designates phenyl phosphosulfate as a sulfating rather than a phosphorylating reagent.

The results of the above study allow one to speculate on the significance of the structure chosen biologically to transport active sulfate. Although many "high energy" compounds found in biochemical systems are acid anhydrides, for example, ATP, PAPS, their chemical make-up constitutes a unique group. Whereas anhydrides like acetic anhydride and phosphoryl chloride are readily hydrolyzed by water at physiological pH, the biologically important inorganic anhydrides are relatively stable to hydrolysis until enzymically activated. The structure of the phosphate–sulfate anhydride PAPS allows such an enzymic "triggering device" to be operative. At physiological pH, PAPS would exist as a dianion, a form known to be stable. When activation is required, complexation with an enzyme and "activation" by either protonation of the phosphoryl oxygen or complexation with a metal ion should produce a reactive sulfate ester capable of carrying out the desired sulfation. A similar process can be envisioned for the polyphosphates. The choice of PAPS rather than an acyl sulfate is predicated by the necessity of more rapid sulfur trioxide elimination relative to nucleophilic reactions at the electrophilic atom linking the sulfate group. The choice of PAPS may serve a second function, *i.e.* activation of the sulfate group to reduction[212].

REFERENCES

1 F. H. Westheimer, *Accounts Chem. Res.*, **1** (1968) 70.
2 A. J. Kirby and S. G. Warren, *The Organic Chemistry of Phosphorus*, Elsevier, London, 1967.
3 W. E. McEwen, in *Topics in Phosphorus Chemistry*, M. Grayson and E. J. Griffith (Eds.), Vol. 2, Interscience, New York, 1965.
4 J. R. Cox, Jr. and B. Ramsay, *Chem. Rev.*, **64** (1964) 317.
5 T. C. Bruice and S. J. Benkovic, *Bioorganic Mechanisms*, Vol. II, Benjamin, New York, N.Y., 1966.
6 J. L. Kice, *Accounts Chem. Res.*, **1** (1968) 58.
7 L. L. Schaleger and F. A. Long, *Advan. Phys. Org. Chem.*, **1** (1963) 1.
8 A. J. Kirby and A. G. Varvoglis, *J. Am. Chem. Soc.*, **89** (1967) 415.
9 C. A. Bunton, E. J. Fendler, E. Humeres and Kui-Un Yang, *J. Org. Chem.*, **32** (1967) 2806.
10 S. J. Benkovic and K. J. Schray, *Biochemistry*, **7** (1968) 4097.
11 S. Milstien and T. H. Fife, *J. Am. Chem. Soc.*, **89** (1967) 5820.
12 W. P. Jencks, *Catalysis in Chemistry and Enzymology*, McGraw-Hill, New York, N.Y., 1969.
13 J. D. Chanley and E. Feageson, *J. Am. Chem. Soc.*, **80** (1958) 2686.
14 J. D. Chanley and E. Feageson, *J. Am. Chem. Soc.*, **85** (1963) 1181.
15 S. J. Benkovic and P. A. Benkovic, *J. Am. Chem. Soc.*, **89** (1967) 4714.
16 I. Oney and M. Caplow, *J. Am. Chem. Soc.*, **89** (1967) 6972.
17 W. P. Jencks and M. Gilchrist, *J. Am. Chem. Soc.*, **86** (1964) 1410.
18 R. L. Collin, *J. Am. Chem. Soc.*, **88** (1966) 3281.
19 S. J. Benkovic and E. J. Sampson, *J. Am. Chem. Soc.*, **93** (1971) 4009.
20 W. P. Jencks and M. Gilchrist, *J. Am. Chem. Soc.*, **87** (1965) 3199.
21 E. Jampel, M. Wakselman and M. Vilkas, *Tetrahedron Letters*, **31** (1968) 3533.
22 A. J. Kirby and A. G. Varvoglis, *J. Am. Chem. Soc.*, **88** (1966) 1823.
23 C. A. Bunton, E. J. Fendler and J. H. Fendler, *J. Am. Chem. Soc.*, **89** (1967) 1221.
24 S. J. Benkovic and P. A. Benkovic, *J. Am. Chem. Soc.*, **90** (1968) 2646.
25 A. F. Gerrard and N. K. Hamer, *J. Chem. Soc. B*, (1968) 539.
26 P. S. Traylor and F. H. Westheimer, *J. Am. Chem. Soc.*, **87** (1965) 553.
27 E. W. Crunden and R. F. Hudson, *J. Chem. Soc.*, (1962) 3591.
28 D. Levine, T. W. Reid and I. B. Wilson, *Biochemistry*, **8** (1969) 2374.
29 H. Neumann, L. Boross and E. Katchalski, *J. Biol. Chem.*, **242** (1967) 3142.

30 R. Breslow and I. Katz, *J. Am. Chem. Soc.*, **90** (1968) 7376.
31 T. Wieland and H. Aquila, *Chem. Ber.*, **101** (1968) 3031.
32 T. Higuchi and K. H. Gensch, *J. Am. Chem. Soc.*, **88** (1966) 5486.
33 D. O. Lambeth and H. A. Lardy, *Biochemistry*, **8** (1969) 3395.
34 W. S. Brinigar, D. B. Knaff and J. H. Wang, *Biochemistry*, **6** (1967) 36.
35 G. DiSabato and W. P. Jencks, *J. Am. Chem. Soc.*, **83** (1961) 4393.
36 A. J. Kirby and W. P. Jencks, *J. Am. Chem. Soc.*, **87** (1965) 3209.
37 A. J. Kirby and A. G. Varvoglis, *J. Chem. Soc. B*, (1968) 135.
38 C. A. Bunton, E. J. Fendler and J. H. Fendler, *J. Am. Chem. Soc.*, **89** (1967) 1221.
39 D. M. Brown and D. A. Usher, *J. Chem. Soc.*, (1965) 6558.
40 A. J. Kirby and M. Younas, personal communication.
41 C. A. Bunton and S. J. Farber, *J. Org. Chem.*, **34** (1969) 767.
42 A. J. Kirby and M. Younas, personal communication.
43 S. A. Khan and A. J. Kirby, personal communication.
44 C. A. Bunton and E. Humeres, *J. Org. Chem.*, **34** (1969) 572.
45 L. Kugel and M. Halmann, *J. Org. Chem.*, **32** (1967) 642.
46 A. Lapidot, D. Samuel and M. Weiss-Broday, *J. Chem. Soc.*, (1964) 637.
47 J. R. Cox, Jr. and M. G. Newton, *J. Org. Chem.*, **34** (1969) 2600.
48 J. Michalski and M. Mikolajczyk, *Chem. Ind.* (*London*), (1964) 661.
49 J. Michalski, M. Mikolajczyk, B. Mlotkowska and J. Omelanczuk, *Tetrahedron*, **25** (1969) 1743.
50 M. Mikolajczyk, *Tetrahedron*, **23** (1967) 1543.
51 M. Green and R. F. Hudson, *J. Chem. Soc.*, (1963) 3883.
52 M. Green and R. F. Hudson, *Angew. Chem.*, **75** (1963) 47.
53 N. K. Hamer, *J. Chem. Soc. C*, (1966) 404.
54 N. K. Hamer, *J. Chem. Soc.*, (1965) 46.
55 H. G. Khorana, G. M. Tener, R. S. Wright and J. G. Moffatt, *J. Am. Chem. Soc.*, **79** (1957) 430.
56 E. Cherbuliez, H. Probst and J. Rabinowitz, *Helv. Chim. Acta*, **42** (1959) 1377.
57 P. C. Haake and F. H. Westheimer, *J. Am. Chem. Soc.*, **83** (1961) 1102.
58 E. A. Dennis and F. H. Westheimer, *J. Am. Chem. Soc.*, **88** (1966) 3432.
59 R. S. Berry, *J. Chem. Phys.*, **32** (1960) 933.
60 L. O. Brockway and J. Y. Beach, *J. Am. Chem. Soc.*, **60** (1938) 1836.
61 H. S. Gutowsky and C. J. Hoffman, *J. Chem. Phys.*, **19** (1951) 1259.
62 M. Rouault, *Compt. Rend.*, **207** (1938) 620.
63 The designation, equatorial or basal, has been applied to the three shorted bonds; equatorial appears to be the more widely acid connotation. Similarly the terms apical and axial, are used interchangeably.
64 G. M. Whitesides and H. L. Mitchell, *J. Am. Chem. Soc.*, **91** (1969) 5384.
65 E. L. Muetterties and R. A. Schunn, *Quart. Rev. (London)*, **20** (1966) 245 and references therein.
66 P. C. Van Der Voorn and R. S. Drago, *J. Am. Chem. Soc.*, **88** (1966) 3255 and references therein.
67 H. Bent, *Chem. Rev.*, **61** (1961) 275.
68 F. Ramirez, *Accounts Chem. Res.*, **1** (1968) 168.
69 D. Gorenstein and F. H. Westheimer, *Proc. Natl. Acad. Sci. US*, **58** (1967) 1747.
70 This short discussion does not cover the available literature concerning NMR studies of oxyphosphorane derivatives but is meant merely to furnish experimental examples of the operation of the preference rules and pseudorotation.
71 J. R. Cox, R. E. Wall and F. H. Westheimer, *Chem. Ind. (London)*, (1959) 929.
72 E. T. Kaiser, M. Panar and F. H. Westheimer, *J. Am. Chem. Soc.*, **85** (1963) 602.
73 (a) P. B. D. de la Mare, J. G. Tillett and H. F. van Woerden, *Chem. Ind. (London)*, (1961) 1533.
(b) R. E. Davis, *J. Am. Chem. Soc.*, **84** (1962) 599.
74 C. A. Bunton, P. B. D. de la Mare, A. Lennard, D. R. Llewellyn, R. B. Pearson, J. G. Pritchard and J. G. Tillett, *J. Chem. Soc.*, (1958) 4761.

75 J. G. Tillett, *J. Chem. Soc.*, (1960) 5138.
76 C. A. Bunton and G. Schwerin, *J. Org. Chem.*, **31** (1966) 842.
77 D. A. Usher, E. A. Dennis and F. H. Westheimer, *J. Am. Chem. Soc.*, **87** (1965) 2320.
78 M. G. Newton, J. R. Cox, Jr. and J. A. Bertrand, *J. Am. Chem. Soc.*, **88** (1966) 1503.
79 G. M. Blackburn, J. S. Cohen and Lord Todd, *Tetrahedron Letters*, **39** (1964) 2873.
80 K. A. R. Mitchell, *Can. J. Chem.*, **46** (1968) 3499.
81 G. Aksnes and K. Bergesen, *Acta Chem. Scand.*, **20** (1966) 2508.
82 A. Eberhard and F. H. Westheimer, *J. Am. Chem. Soc.*, **87** (1965) 253.
83 E. A. Dennis and F. H. Westheimer, *J. Am. Chem. Soc.*, **88** (1966) 3431.
84 R. Kluger, F. Covitz, E. Dennis, L. D. Williams and F. H. Westheimer, *J. Am. Chem. Soc.*, **91** (1969) 6066.
85 R. Kluger and F. H. Westheimer, *J. Am. Chem. Soc.*, **91** (1969) 4143.
86 D. B. Boyd, *J. Am. Chem. Soc.*, **91** (1969) 1200.
87 This pathway is favorable because the lowest empty molecular orbital of the cyclic methyl ethylene phosphate ester consists principally of lobes on the back sides of the P–O ring bonds which can readily interact with the approaching electron pair from water.
88 K. E. DeBruin, K. Naumann, G. Zon and K. Mislow, *J. Am. Chem. Soc.*, **91** (1969) 7031.
89 (a) R. F. Hudson, *Structure and Mechanism in Organophosphorus Chemistry*, Academic Press, New York, 1965, p. 206.
(b) An exception to this rule has been found in the alkaline hydrolysis of acyclic *t*-butylphosphonium salts where steric crowding in the phosphorane intermediate leads to retention of configuration at phosphorus. N. J. De'ath and S. Trippett, *Chem. Commun.*, (1969) 172.
90 S. E. Fishwick, J. Flint, W. Hawes and S. Trippett, *Chem. Commun.*, (1967) 1113.
91 W. Hawes and S. Trippett, *Chem. Commun.*, (1968) 295.
92 D. W. Allen, F. G. Mann, and I. T. Millar, *J. Chem. Soc. C*, (1967) 1869.
93 D. W. Allen and I. T. Millar, *J. Chem. Soc. C*, (1969) 252.
94 J. R. Corfield, J. R. Shutt and S. Trippett, *Chem. Commun.*, (1969) 789.
95 K. E. DeBruin, K. Naumann, G. Zon and K. Mislow, *J. Am. Chem. Soc.*, **91** (1969) 7027.
96 K. L. Marsi, *J. Am. Chem. Soc.*, **91** (1969) 4724.
97 M. J. Gallagher and I. D. Jenkins, in *Topics in Stereochemistry*, E. L. Eliel and N. L. Allinger (Eds.), Vol. III, Interscience, 1968, Chap. 1.
98 G. Aksnes and K. Bergesen, *Acta Chem. Scand.*, **19** (1965) 931.
99 W. Hawes and S. Trippett, *Chem. Commun.*, (1968) 577.
100 R. Greenhalgh, J. E. Newbery, R. Woodcock and R. F. Hudson, *Chem. Commun.*, (1969) 22.
101 D. Swank, C. N. Caughlan, F. Ramirez, O. P. Madan and C. P. Smith, *J. Am. Chem. Soc.*, **89** (1967) 6503 and references therein.
102 F. Ramirez O. P. Madan and C. P. Smith, *J. Am. Chem. Soc.*, **87** (1965) 670.
103 G. Wittig, *Bull. Soc. Chim. France*, (1966) 1162.
104 D. Hellwinkel, *Chem. Ber.*, **98** (1965) 576.
105 H. R. Allcock, *J. Am. Chem. Soc.*, **86** (1964) 2591.
106 F. Ramirez, K. Tasaka, N. B. Desai and C. P. Smith, *J. Am. Chem. Soc.*, **90** (1968) 751.
107 J. P. Hummel and G. Kalnitsky, *Ann. Rev. Biochem.*, **33** (1964) 15.
108 Fr. Huijing and J. Larner, *Proc. Natl. Acad. Sci. US*, **56** (1966) 647.
109 M. Cohn and T. R. Hughes, Jr., *J. Biol. Chem.*, **237** (1962) 176.
110 H. Moll, P. W. Schneider and H. Brintzinger, *Helv. Chim. Acta*, **47** (1964) 1837.
111 H. Brintzinger, *Biochim. Biophys. Acta*, **77** (1963) 343.
112 M. Tetas and J. M. Lowenstein, *Biochemistry*, **2** (1963) 350.
113 D. L. Miller and F. H. Westheimer, *J. Am. Chem. Soc.*, **88** (1966) 1507.
114 D. L. Miller and T. Ukena, *J. Am. Chem. Soc.*, **91** (1969) 3050.
115 C. A. Bunton and H. Chaimovich, *Inorg. Chem.*, **4** (1965) 1763.
116 M. Cohn, *Biochemistry*, **2** (1963) 623.
117 D. L. Miller and F. H. Westheimer, *J. Am. Chem. Soc.*, **88** (1966) 1514.

118 D. L. Miller and F. H. Westheimer, *J. Am. Chem. Soc.*, **88** (1966) 1511.
119 T. G. Spiro, W. A. Kjellstrom, M. Zeydel and R. A. Butow, *Biochemistry*, **7** (1968) 859.
120 B. Cooperman, *Biochemistry*, **8** (1969) 5005.
121 J. van Steveninck, *Biochemistry*, **5** (1966) 1998.
122 F. Ramirez, B. Hansen and N. B. Desai, *J. Am. Chem. Soc.*, **84** (1962) 4588.
123 C. N. Lieske, E. G. Miller, Jr., J. J. Zeger and G. M. Steinberg, *J. Am. Chem. Soc.*, **88** (1966) 188.
124 D. M. Brown and M. J. Frearson, *Chem. Commun*, (1968) 1342.
125 H. Witzel, A. Botta and K. Dimroth, *Chem. Ber.*, **98** (1965) 1465.
126 (a) D. S. Frank and D. A. Usher, *J. Am. Chem. Soc.*, **89** (1967) 6360.
(b) S. J. Benkovic and K. J. Schray, *J. Am. Chem. Soc.*, **91** (1969) 5653.
(c) V. M. Clark and A. J. Kirby, *J. Am. Chem. Soc.*, **85** (1963) 3705.
127 G. M. Blackburn and M. J. Brown, *J. Am. Chem. Soc.*, **91** (1969) 525.
128 W. D. Fordham and J. H. Wang, *J. Am. Chem. Soc.*, **89** (1967) 4197. The reported data may be readily interpreted in terms of a pentacovalent intermediate which expels H_2O > ring opening.
129 L. Kugel and M. Halmann, *J. Am. Chem. Soc.*, **89** (1967) 4125.
130 L. Kugel and M. Halmann, *J. Am. Chem. Soc.*, **88** (1966) 3566.
131 See ref. 39.
132 D. G. Oakenfull, D. I. Richardson, Jr. and D. A. Usher, *J. Am. Chem. Soc.*, **89** (1967) 5491.
133 G. C. K. Roberts, E. A. Dennis, D. H. Meadows, J. S. Cohen and O. Jardetzky, *Proc. Natl. Acad. Sci. US*, **62** (1969) 1151.
134 M. R. Harris, D. A. Usher, H. P. Albrecht, G. H. Jones and J. G. Moffatt, *Proc. Natl. Acad. Sci. US*, **63** (1969) 246.
135 J. Arai, *J. Biochem. (Tokyo)*, **20** (1934) 474.
136 S. A. Khan, A. J. Kirby, M. Wakselman, D. P. Horning and J. M. Lawlor, personal communication.
137 C. N. Lieske, J. W. Hovanec, G. M. Steinberg and P. Blumbergs, *Chem. Commun.*, (1968) 13.
138 C. N. Lieske and J. W. Hovanec, *Chem. Commun.*, (1969) 976.
139 C. A. Bunton and H. Chaimovich, *J. Am. Chem. Soc.*, **88** (1966) 4082.
140 R. S. Miller, A. S. Mildvan, H. Chang, R. L. Easterday, H. Maruyama and M. D. Lane, *J. Biol. Chem.*, **243** (1968) 6030.
141 W. J. O'Sullivan and M. Cohn, *J. Biol. Chem.*, **241** (1966) 3104.
142 C. H. Oestreich and M. M. Jones, *Biochemistry*, **6** (1967) 1515 and references therein.
143 C. H. Oestreich and M. M. Jones, *Biochemistry*, **5** (1966) 2926.
144 Y. Murakami, J. Sunamoto and H. Sadamori, *Chem. Commun*, (1969) 983.
145 Y. Murakami and M. Takagi, *J. Am. Chem. Soc.*, **91** (1969) 5130.
146 S. J. Benkovic and L. Dunikoski, *J. Am. Chem. Soc.*, **93** (1971) 1526.
147 F. J. Farrell, W. A. Kjellstrom and T. G. Spiro, *Science*, **164** (1969) 320.
148 S. J. Benkovic and E. Miller, *Bioinorg. Chem.*, in press.
149 Y. Murakami and M. Tanaka, *Bull. Chem. Soc. Japan*, **39** (1966) 122.
150 G. H. Tait and B. L. Vallee, *Proc. Natl. Acad. Sci. US*, **56** (1966) 1247.
151 C. A. Bunton, E. J. Fendler, L. Sepulveda and K. Yang, *J. Am. Chem. Soc.*, **90** (1968) 5512.
152 C. A. Bunton and L. Robinson, *J. Org. Chem.*, **34** (1969) 773.
153 C. A. Bunton, L. Robinson and L. Sepulveda, *J. Am. Chem. Soc.*, **91** (1969) 4813.
154 K. J. Schray and S. J. Benkovic, *J. Am. Chem. Soc.*, **93** (1971) 2522.
155 S. J. Benkovic, *J. Am. Chem. Soc.*, **88** (1966) 5511 and references therein.
156 J. L. Kice and J. M. Anderson, *J. Am. Chem. Soc.*, **88** (1966) 5242.
157 G. N. Burkhardt, C. Horrex, and D. I. Jenkins, *J. Chem. Soc.*, (1936) 1654.
158 C. A. Bunton and B. N. Hendy, *J. Chem. Soc.*, (1963) 3130.

159 B. D. Batts, *J. Chem. Soc. B*, (1966) 551.
160 B. D. Batts, *J. Chem. Soc. B*, (1966) 547.
161 C. A. Bunton and G. Schwerin, *J. Org. Chem.*, **31** (1966) 842.
162 E. J. Fendler and J. H. Fendler, *J. Org. Chem.*, **33** (1968) 3852.
163 J. L. Kice, J. M. Anderson and N. E. Pawlowski, *J. Am. Chem. Soc.*, **88** (1966) 5245.
164 J. L. Kice and G. Guaraldi, *J. Org. Chem.*, **31** (1966) 3568.
165 J. L. Kice, C. G. Venier and L. Heasley, *J. Am. Chem. Soc.*, **89** (1967) 3557.
166 J. L. Kice and G. Guaraldi, *J. Am. Chem. Soc.*, **89** (1967) 4113.
167 J. L. Kice and G. Guaraldi, *J. Am. Chem. Soc.*, **88** (1966) 5236.
168 J. L. Kice and G. B. Large, *J. Org. Chem.*, **33** (1968) 1940.
169 J. L. Kice and G. J. Kasperek, *J. Am. Chem. Soc.*, **91** (1969) 5510.
170 N. H. Christensen, *Acta Chem. Scand.*, **20** (1966) 1955; **21** (1967) 899.
171 J. L. Kice and G. Guaraldi, *J. Am. Chem. Soc.*, **90** (1968) 4076.
172 J. L. Kice, G. J. Kasperek and D. Patterson, *J. Am. Chem. Soc.*, **91** (1969) 5516.
173 J. L. Kice and G. B. Large, *J. Am. Chem. Soc.*, **90** (1968) 4069.
174 W. P. Jencks and M. Gilchrist, *J. Am. Chem. Soc.*, **90** (1968) 2622.
175 J. O. Edwards and R. G. Pearson, *J. Am. Chem. Soc.*, **84** (1962) 16.
176 S. J. Benkovic and P. A. Benkovic, *J. Am. Chem. Soc.*, **88** (1966) 5504.
177 E. Buncel and J. P. Millington, *Can. J. Chem.*, **47** (1969) 2145.
178 P. B. Brindley, *J. Chem. Soc. C*, (1966) 163.
179 C. Brown and D. R. Hogg, *Chem. Commun.*, (1967) 38.
180 K. Leininger and J. Westley, *J. Biol. Chem.*, **243** (1968) 1896.
181 R. Mintel and J. Westley, *J. Biol. Chem.*, **241** (1966) 3381.
182 J. Westley and T. Nakamoto, *J. Biol. Chem.*, **237** (1962) 547.
183 A. B. Roy, *Advan. Enzymol.* **22** (1960) 204.
184 R. K. Banerjee and A. B. Roy, *Biochim. Biophys. Acta*, **151** (1968) 573.
185 T. P. Stein and D. Fahrney, *Chem. Commun.*, (1968) 555.
186 E. T. Kaiser, I. R. Katz and T. F. Wulfers, *J. Am. Chem. Soc.*, **87** (1965) 3781.
187 O. R. Zaborsky and E. T. Kaiser, *J. Am. Chem. Soc.*, **88** (1966) 3084.
188 E. T. Kaiser, K. Kudo and O. R. Zaborsky, *J. Am. Chem. Soc.*, **89** (1967) 1393.
189 J. G. Tillett, *J. Chem. Soc.*, (1960) 37.
190 P. A. Bristow, J. G. Tillett and D. E. Wiggins, *J. Chem. Soc. B*, (1968) 1360.
191 N. Pagdin, A. K. Pine, J. G. Tillett and H. F. Van Woerden, *J. Chem. Soc.*, (1962) 3835.
192 F. P. Boer, J. J. Flynn, E. T. Kaiser, O. R. Zaborsky, D. A. Tomalia, A. E. Young and Y. C. Tong, *J. Am. Chem. Soc.*, **90** (1968) 2970.
193 E. B. Fleischer, E. T. Kaiser, P. Langford, S. Hawkinson, A. Stone and R. Dewar, *Chem. Commun.*, (1967) 197.
194 D. P. Craig and T. Thirunamachandran, *J. Chem. Phys.*, **43** (1965) 4183.
195 C. R. Johnson and D. McCants, Jr., *J. Am. Chem. Soc.*, **87** (1965) 5404.
196 J. H. Heidema and E. T. Kaiser, *J. Am. Chem. Soc.*, **90** (1968) 1860.
197 K. W. Lo and E. T. Kaiser, *Chem. Commun*, (1966) 834.
198 G. Tomalin, M. Trifunac and E. T. Kaiser, *J. Am. Chem. Soc.*, **91** (1969) 722.
199 A. M. Gold and D. E. Fahrney, *Biochemistry*, **3** (1964) 783.
200 S. J. Benkovic, *J. Am. Chem. Soc.*, **88** (1966) 5511.
201 S. J. Benkovic and L. K. Dunikoski, Jr., *Biochemistry*, **9** (1970) 1390.
202 B. E. Fleischfresser and I. Lauder, *Australian J. Chem.*, **15** (1962) 251.
203 M. Eigen, *Angew. Chem. Intern. Ed. Engl.*, **3** (1964) 1.
204 M. Miyazaki and J. Fishman, *Steroids*, **12** (1968) 465.
205 R. W. Hay and J. A. G. Edmonds, *Chem. Commun.*, (1967) 969.
206 S. W. Weidman, D. F. Mayers, O. R. Zaborsky and E. T. Kaiser, *J. Am. Chem. Soc.*, **89** (1967) 4555.
207 D. F. Mayers and E. T. Kaiser, *J. Am. Chem. Soc.*, **90** (1968) 6192.
208 J. D. Gregory and F. Lipmann, *J. Biol. Chem.*, **229** (1957) 1081.

209 E. Meezan and E. A. Davidson, *J. Biol. Chem.*, **242** (1967) 1685.
210 A. S. Balasubramanian and B. K. Bachhawat, *Indian J. Exptl. Biol.*, **1** (1963) 179.
211 S. J. Benkovic and R. Hevey, *J. Am. Chem. Soc.*, **92** (1970) 4971.
212 K. Torii and R. S. Bandurski, *Biochim. Biophys. Acta*, **136** (1967) 286.

Chapter 2

Hydrolysis and Formation of Esters of Organic Acids

A. J. KIRBY

1. Introduction

The hydrolysis and formation of esters have been studied more often than any other chemical reaction. The history of this work is a history of our growing understanding of reactions in solution, for the reaction has been a traditional proving ground for theories of chemical action[1-3]. As a result there is available a vast wealth of data, widely scattered over time and journal. It is the aim of this chapter to bring together in one place a representative selection of these data, to set them in the perspective of an overall view of the subject, and to offer the interpretation of each type of reaction that seems most acceptable in the light of our present knowledge. I have tried to be as generous with the original data as with its interpretation, so that this part of the chapter may be of lasting value. As to the interpretation, I have not hesitated to draw my own conclusions, or to extend those of other authors. What emerges is a close-to self-consistent picture of the reaction, from hydrolysis in alkaline solution to the goings-on in very strong acid, at least at the level of over-simplification that the writer finds most congenial.

2. Ester hydrolysis and formation in strongly acidic media

Simple carboxylic acids and their esters, in common with other neutral organic compounds of oxygen, are weak bases. As such they are completely protonated only in very strongly acidic media, like 100% sulphuric acid. The concentrations of the protonated forms may be kinetically significant at pHs as high as 3–4, but the hydrolysis reactions are only fast enough for convenient measurement in the presence of hydronium ion concentrations of the order of 0.1 *M* or above. A wealth of kinetic data is available for both hydrolysis and ester formation (which is generally considerably faster) in this acidity region, and these results are summarized in the discussion of structure and reactivity, below. But the results which appear to allow the closest insight into the details of the mechanism of the reaction come from recent investigations of ester hydrolysis in concentrated solutions of the strong mineral acids.

We do not fully understand the behaviour in strong acids even of compounds

References pp. 202–207

which do not react in them. So it is not surprising that the kinetics of reactions in such media, with the added uncertainties associated with the differences between molecules and transition states, should present some difficulties of interpretation.

The source of the problem is well recognised. In the study of reactions in dilute aqueous solution it is possible to make the variations in such factors as reagent concentration which are necessary for systematic kinetic investigations, while maintaining a relatively constant environment. As long as there is enough solvent to go round, to solvate fully every particle of solute, the activity coefficient of each species in solution is essentially independent of the nature and concentration of every other. But in sufficiently concentrated solutions of electrolytes, like the strong acids, there is not enough solvent to solvate fully every particle of solute. In these circumstances there is competition for the available molecules of solvent. Since different species have different affinities for solvent, which do not necessarily bear any predictable relation to their function in their primary role in the reaction, changes in the concentration of any one solute may lead to changes in the extent of solvation, and thus in the activity coefficients, of other species present. Thus the usual technique of isolating a single variable under otherwise constant conditions fails with strongly acidic media: the acid-catalyzed reactions of weak bases, such as carboxylic acids and esters, must perforce be studied in an environment which is continually changing.

Although the quantitative theory of reactions in moderately concentrated solutions of strong acids is unsatisfactory, we do have a good qualitative idea of the processes involved in the acid-catalyzed hydrolysis and formation of esters. Under conditions where the degree of protonation of the substrate is small it is not possible to separate with confidence the factors which affect the solvolytic process and those which affect the preliminary protonation equilibrium. But there have been a number of recent studies of the behaviour of carboxylic acids and esters in very strongly acidic media, in which they are essentially completely protonated. Under these conditions it is possible to observe the behaviour of the protonated species directly. It is appropriate to summarize the results of this research before discussing the reactions under more normal solvolytic conditions.

2.1 CARBOXYLIC ACIDS AND ESTERS AS BASES

Evidence that carboxylic acids are protonated in concentrated solutions of strong acids has existed in the literature for many years. This consisted originally of the results of cryoscopic measurements. For example Hantzsch[4] and Treffers and Hammett[5] showed that several substituted benzoic acids, in common with many other simple carbonyl compounds, exhibit *i*-factors of 2 in concentrated sulphuric acid, and therefore act as monoacidic bases.

$$PhCO_2H + H_2SO_4 \rightleftharpoons PhCO_2H_2^+ + HSO_4^-$$

Esters show similar behaviour. Hantzsch found the *i*-factor for ethyl acetate to be close to 2, and accurate determinations by Leisten[6] also gave values close to 2 for ethyl and methyl benzoate and *p*-nitrobenzoate. In a solvent containing sufficient water, hydrolysis of the ester occurs. The reaction of methyl benzoate has a time of half-change of a few hours at 25°C in sulphuric acid containing about 0.07 *M* water, and Leisten[6] actually used the increase in the *i*-factor, from 2 to 3, to follow the hydrolysis reaction, *viz.*

$$PhCO_2Me + H_2SO_4 \rightleftharpoons PhCO_2HMe^+ + HSO_4^-$$
$$PhCO_2HMe^+ + H_2SO_4 \rightleftharpoons PhCO_2H_2^+ + MeHSO_4$$

Thus it is possible to study the hydrolysis reactions of esters under conditions where the substrate is completely protonated. The properties of the protonated ester, however, are more conveniently examined using more strongly acidic media, in the absence of water, where bimolecular reactions are reduced to insignificance. At sufficiently low temperatures under these conditions the rates of exchange of the added protons are slow, and the detailed structures of protonated carboxylic acids and esters can be investigated, particularly by proton NMR techniques.

2.1.1 *The structures of protonated carboxylic acids and esters*

In principle, a carboxylic acid or ester could be protonated on either of its two oxygen atoms.

$$R'\!-\!C(\!=\!\overset{+}{O}H)\!-\!OR \;\underset{}{\overset{H^+}{\rightleftharpoons}}\; R'\!-\!C(\!=\!O)\!-\!OR \;\overset{H^+}{\rightleftharpoons}\; R'\!-\!C(\!=\!O)\!-\!\overset{+}{O}(H)\!-\!R$$

(1) (2)

There is general agreement that in fact protonation occurs predominantly at the carbonyl oxygen as in (1). This is expected to give a more stable product than (2) because of the possibility of delocalisation, *viz.*

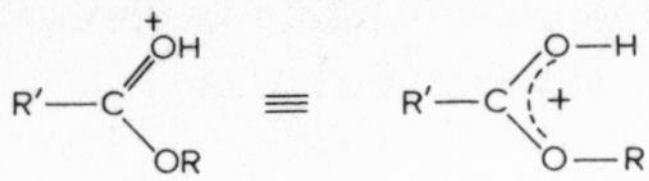

$$R'-C(=\overset{+}{O}H)-OR \equiv R'-C(\cdots O-H)(\cdots O-R)^{+}$$

Experimentally, the close correlation between the basicities of substituted benzoic acids and the corresponding acetophenones[7] is most simply explained if protonation occurs on the carbonyl oxygen in each case: as, of course, it must for the ketones.

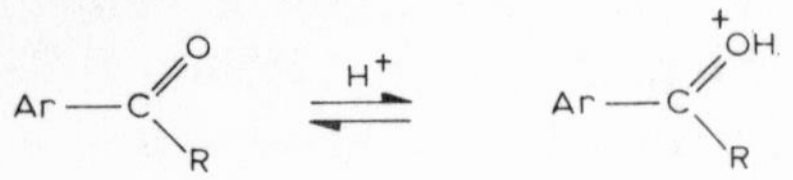

$$Ar-C(=O)-R \underset{}{\overset{H^+}{\rightleftharpoons}} Ar-C(=\overset{+}{O}H)-R$$

Further, the chemical shifts for the protons concerned, which correlate well with the acidity of the conjugate acids, fall in overlapping regions for protonated ketones and acids, at considerably lower field than for the corresponding protons of the conjugate acids of ethers and alcohols[8].

In the very low field region, about 12–13 ppm downfield from tetramethylsilane, protonated carboxylic acids generally show two peaks of equal intensity due to the OH protons. These peaks are clearly resolved at −60°C[8,9], but collapse to a single resonance line at higher temperatures (about −30°C[8]). This observation rules out a completely symmetrical structure for the protonated acid, and again suggests that the protons are on different oxygen atoms, as in (1); since there would be expected to be free rotation about the C–O single bond of (2)[10]. Birchall and Gillespie[9] felt that intramolecular hydrogen-bonding, as in (2a) below, might account for the results, and preferred not to rule out (2) on these grounds, although they concluded that protonation on the carbonyl oxygen atom is more likely, by analogy with the protonated esters (see below).

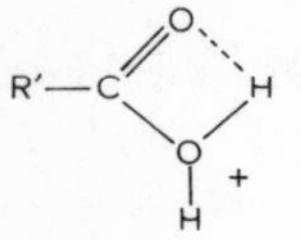

(2a) 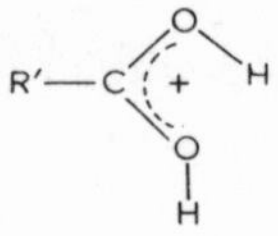 (1a)

Olah and White[8] have published the NMR spectra of a number of protonated carboxylic acids, taken at −60°C in FSO_3H–SbF_5 diluted with SO_2. They interpret the spectra in terms of structure (1a), and were able to uncover more structural details by investigating the protonation of formic acid, (1a, R′ = H) where spin–spin coupling occurs between the C–H and O–H protons. Two forms of protonated formic acid were observed. In one case (3a) the spin–spin splitting was in the region expected for a transoid arrangement of CH and OH groups. In the second, the predominant form (by a factor of 2 : 1), two different values of the coupling constant were observed, one that expected for (3a), the second that expected for a *cisoid* arrangement. It was concluded, therefore, that protonated formic acid consists of a 1:2 mixture of forms (3a) and (3b).

(3a) (3b)

Hogeveen and Bickel[11] have confirmed this result, finding that protonated formic acid in $HF–BF_3$ at $-75°C$ consists of about 23% (3a) and 77% (3b). On extending this work to other aliphatic acids, Hogeveen[12] found, in agreement with Olah and White[8], that most protonated aliphatic acids exist as a single form, corresponding to (3b), with two types of proton, but that trimethylacetic acid gives only one peak. Hogeveen suggests that the *cisoid* forms become progressively less stable as the size of the acyl substituent R′ increases (a small proportion, about 3%, of the all *cisoid* form, corresponding to (3a), of protonated acetic acid is revealed by careful experiments[8,13]) and that the single peak observed for the *t*-butyl compound may indicate that it exists as the all *transoid* isomer. Since the usual two peaks are observed for this acid in $FSO_3H–SbF_5–SO_2$[8], however, it seems more likely that the steric effect manifests itself in an increased rate of proton exchange, rather than a different species in solution. A similar explanation would account for Birchall and Gillespie's observation[9] of only a single OH peak for protonated benzoic acid in $FSO_3H–SbF_5$ at $-84.5°C$.

The protonation behaviour of esters is closely similar to that of the parent acids. Birchall and Gillespie[9] reported the observation of protonated ethyl acetate in $HSO_3F–SbF_5$ solution at $-75°C$, and concluded that protonation occurs on the carbonyl oxygen atom as in (4b), since they did not observe the spin–spin coupling between the added proton and the methylene group of the ether oxygen which would be expected if both were attached to the same atom, as in (4a).

(4a) (4b)

References pp. 202–207

The predominant isomer of protonated formate esters has been assigned structure (5) by the methods used to distinguish the isomers of the protonated acid, described above: the coupling constants are almost identical in the two cases.

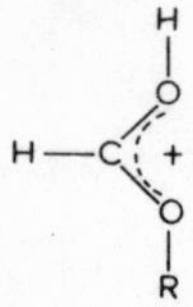

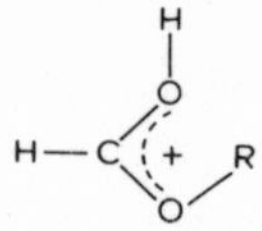

(5a) or (5b)

It is not possible to assign the structures definitely as (5a) or (5b). Olah *et al.*[14] write (5a) without any strong preference. Hogeveen[12] prefers (5b) since non-bonded repulsions will be at a minimum in the structure with H and R′ *trans* to each other: and the CH_3 and H groups of protonated acetic acid show a very definite preference to be *transoid*, compared with formic acid (see above). Neutral esters exist predominantly in the *transoid* conformation, and the small spin–spin coupling constant between the HC and OCH_3 groups of protonated methyl formate is almost the same as that measured for the neutral ester[10]. Since the reasons for preferring the conformation of type (5b) become stronger as the size of the groups concerned increases, and since all-*trans* conformations (6) have never been observed, and are presumably sterically unfavourable[14], it seems reasonable to suppose that the favoured conformation of a protonated ester is that represented by (7)*.

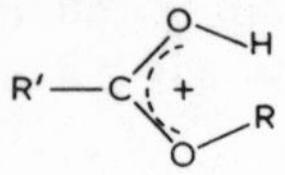

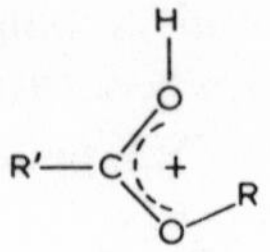

(6) (7)

Finally it is relevant to mention at this point that esters are weakly nucleo-

* On the basis of more detailed results White and Olah[308], as well as Borsch[309], agree that this conformation is also that of the predominant isomer of protonated methyl formate (5b). The minor component will be the conformer with the ester methyl group and added proton reversed (5c), *viz.*

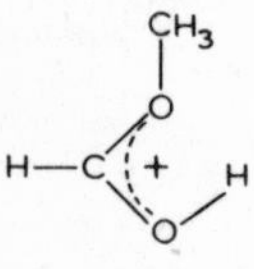

(5c)

philic as well as weakly basic, and can be alkylated by sufficiently strongly electrophilic reagents. In particular the reactions with carbonium ions generate cations closely related to the protonated forms of esters and acids. The action of silver tetrafluoroborate on *trans*-2-acetoxycyclohexyl bromide (8) gives the *cis*-cyclohexene acetoxonium ion (9)[15]. More general routes for the preparation of dialkoxycarbonium ions, which together leave no doubt of the structure of the product, involve hydride abstraction from acetals, *e.g.* (10), and the action of Lewis acids on orthoesters, *e.g.* (11)[15].

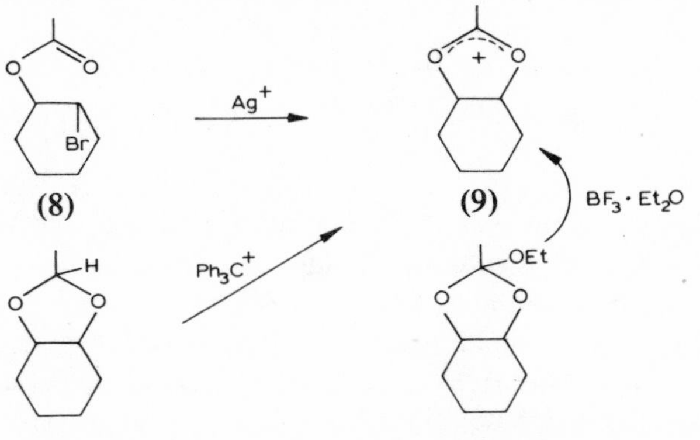

(8) $\xrightarrow{Ag^+}$ (9)

(10) $\xrightarrow{Ph_3C^+}$ (9)

(11) $\xrightarrow{BF_3 \cdot Et_2O}$ (9)

This last method has been used by Ramsey and Taft[16] to prepare a series of simple alkoxymethyl cations. The NMR chemical shifts, δ, of the protons of methylated methyl acetate and formate, given by Ramsey and Taft (measured in 30% SO_3 in H_2SO_4) are compared below with the figures for the corresponding protonated esters, measured in $FSO_3H–SbF_5–SO_2$[14].

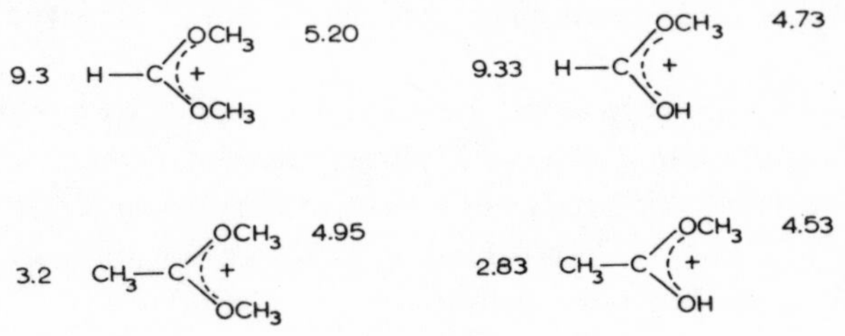

9.3 $H–C^+(OCH_3)(OCH_3)$ 5.20

9.33 $H–C^+(OCH_3)(OH)$ 4.73

3.2 $CH_3–C^+(OCH_3)(OCH_3)$ 4.95

2.83 $CH_3–C^+(OCH_3)(OH)$ 4.53

Clearly the protons are in closely similar magnetic environments in the two classes of compound, and there can be no doubt that the protonated forms of carboxylic acids and esters are now well-characterized organic compounds.

2.1.2 Reactions of protonated carboxylic acids and esters

The important reactions of protonated acids and esters were deduced many years ago, largely on the basis of the anomalous cryoscopic behaviour of a few compounds. Treffers and Hammett[5], while extending the work of Hantzsch[4] on cryoscopic measuremnets in concentrated sulphuric acid, found that 2,4,6-trimethylbenzoic acid gives an *i*-value of 4, twice the usual figure. The only reasonable interpretation of this result was in terms of an ionization, as proposed by Hantzsch to account for the similar behaviour of triphenylcarbinol, *viz.*

$$\text{—COOH} + 2\,H_2SO_4 \rightleftharpoons \text{—C}{\equiv}\overset{+}{O} + H_3O^+ + 2\,HSO_4^-$$

Consistent with this interpretation was the observation that the acid could readily be esterified by pouring the solution in sulphuric acid into methanol, although it is highly unreactive under normal conditions[17].

Under the same conditions, the methyl ester is also converted to the trimethylbenzoyl cation. The *i*-value in this case is 5, since the molecule of methanol lost is converted to methyl sulphate[18], *viz.*

$$ArCOOCH_3 + 3\,H_2SO_4 \rightleftharpoons ArC{\equiv}O^+ + CH_3SO_4H + H_3O^+ + 2\,HSO_4^-$$

For steric reasons the 2,6-disubstituted benzoic acids and esters are particularly susceptible to this type of cleavage reaction, and also particularly unreactive in the usual bimolecular solvolytic processes, and they have proved very convenient substrates for the study of the $A_{AC}1$ mechanism. The kinetic work is discussed in a later section: we are concerned at this point only with the qualitative behaviour of protonated esters and acids, and of the structures of the cationic species.

A second type of anomalous cryoscopic behaviour is exhibited by *t*-butyl esters[19]. The benzoate and acetate both give varying *i*-values between 3 and 4, and by extrapolating the value for *t*-butyl benzoate to zero time Kuhn[19] obtained a value close to 3.0. This he interpreted as evidence for alkyl–oxygen cleavage, and the formation of the *t*-butyl cation, *viz.*

$$C_6H_5COOC(CH_3)_3 + H_2SO_4 \rightleftharpoons C_6H_5C(OH)OC(CH_3)_3^+ + HSO_4^-$$
$$C_6H_5C(OH)OC(CH_3)_3^+ \rightleftharpoons C_6H_5COOH + (CH_3)_3C^+$$
$$C_6H_5COOH + (CH_3)_3C^+ + H_2SO_4 \rightleftharpoons C_6H_5COOH_2^+ + (CH_3)_3CSO_4H$$

With the availability of methods which allow the direct observation of protonated carboxylic acids and esters, it has become possible to follow these reactions directly, and to identify the species formed.

(a) Acyl–oxygen cleavage

The earliest direct study of the formation of acyl cations from protonated carboxylic acids was that of Deno *et al.*[20]. These authors used the changes in the chemical shift of the protons on the α-carbon atom of a series of aliphatic carboxylic acids to follow the equilibria

$$\text{R–COOH} \overset{\text{H}^+}{\rightleftharpoons} \text{RC(OH)}_2^+ \rightleftharpoons \text{R–C≡O}^+ + \text{H}_2\text{O}$$

in solutions of varying acidity, from 0 to 100% H_2SO_4 and from 0 to 80% SO_3 in H_2SO_4, at 35°C. For example, the chemical shift, δ, of the methyl protons of acetic acid is 2.11 in 50% H_2SO_4, 2.67 in 96% H_2SO_4, and 3.93 in 60% SO_3 in H_2SO_4, and these were taken to represent the chemical shifts of acetic acid, its conjugate acid, and the acetyl cation, respectively. These assignments are clearly correct: the figure for the conjugate acid is similar to those established since in more detailed NMR investigations (see above), and the figure for the acetyl cation is close to that reported previously for CH_3CO^+ in SbF_5[21].

Acyl cations are now well-established chemical species. They can be prepared in quantity in solution, and several have been isolated as their crystalline salts. Recent papers have described their formation from carboxylic acids and esters under strongly acidic conditions[8,13,14], but they are most conveniently available from the reactions of acyl halides with Lewis acids[21]: well-defined Lewis acid–base complexes are formed, which decompose in a second stage to the acyl cations[21–23], *viz.*

$$\text{R–C(=O)F} + \text{BF}_3 \rightleftharpoons \text{R–CO–F---BF}_3 \rightleftharpoons \text{R–C≡O}^+ + \text{BF}_4^-$$

Acetylium tetrafluoroborate was prepared in this way as early as 1943[24], and the structures of at least two acyl cations have been investigated by X-ray diffraction methods[25–27]. As expected, CH_3–C≡O^+ is linear about the central carbon, with a CO bond length of 1.12 A, just shorter than in carbon monoxide, and a very short C–C bond length of 1.38 A[25,26]. Structurally, therefore, the acyl cations closely resemble the isoelectronic nitriles.

Protonated carboxylic acids and esters are stable at −60°C in solution in FSO_3H–SbF_5–SO_2, but on warming above about −30°C the protonated acids and methyl esters are converted at measurable rates into the corresponding acyl cations[8,14,28,29]. In this way it is possible to investigate the crucial single step of the series of processes constituting the $A_{AC}1$ cleavage of esters. Al-

References pp. 202–207

though these reactions have to be followed under conditions very different from those usually employed for ester hydrolysis and formation, the few results available are of considerable interest and importance for the study of the $A_{AC}1$ mechanism. The results are summarised in Tables 1 and 2.

TABLE 1

KINETIC DATA FOR THE FORMATION OF ACYL CATIONS FROM PROTONATED CARBOXYLIC ACIDS

Acid	*$10^4 k_{obs}$ at 26°C[a] (sec^{-1})*	*Rate relative to CH_3COOH at −26°C*	*Rate relative to CH_3COOH at 30°C[b]*	*$\Delta H^‡$ (kcal·mole^{-1})*	*$\Delta S^‡$ (eu)*
CH_3COOH	1.5	1.00	1.00	16.3 ± 1.5[b]	−13.1 ± 5[b]
				17.8[a]	−4.8[a]
CH_3CH_2COOH	2.8	1.87	1.46	14.8 ± 0.5[b]	−18.6 ± 3[b]
$(CH_3)_2CHCOOH$	3.7	2.47	1.23	14.7 ± 0.3[b]	−19.0 ± 3[b]
$(CH_3)_3CCOOH$	3.1	2.04	0.56	15.9 ± 0.5[b]	−15.5 ± 3[b]

[a] In $HF–BF_3$. Data from ref. 12.
[b] In $FSO_3H–SbF_5–SO_2$. Data from ref. 8.

TABLE 2

KINETIC DATA FOR THE FORMATION OF ACYL CATIONS FROM PROTONATED METHYL ESTERS

Ester	*$10^4 k_{obs}$ at 30°C (sec^{-1})*	*Rate relative to CH_3COOCH_3*	*Rate relative to $RCOOH$[a]*	*$\Delta H^‡$ (kcal·mole^{-1})*	*$\Delta S^‡$ (eu)*
CH_3COOCH_3	8.31	1.00	44.5	22.6	+1.4
$CH_3CH_2COOCH_3$	12.6	1.52	20.3	20.5	−3.9
$(CH_3)_2CHCOOCH_3$	12.2	1.47	17.3	19.8	−6.2
$(CH_3)_3CCOOCH_3$	3.27	0.39	30.7	18.7	−12.6

[a] Data from refs. 8 and 14, compared at −13°C by extrapolation of the Arrhenius plot.

The cleavage of protonated esters and acids in $FSO_3H–SbF_5–SO_2$ is a simple first-order reaction. In Hogeveen's experiments[12] using solvent $HF–BF_3$ a measurable equilibrium is set up, and the reverse reaction has to be taken into account.

$$RC(OH)_2^+ \underset{k_2}{\overset{k_1}{\rightleftharpoons}} RCO^+ + H_2O$$

(An estimate for the rate coefficient, k_2, for the addition of water to the acetylium cation, gave a value of 2×10^3 l·mole·sec^{-1} at 0°C.)

The trends of reactivity apparent from the data of Tables 1 and 2 are most interesting. The enthalpies of activation are generally some 6 kcal·mole^{-1} higher for the cleavage of the methyl esters, qualitatively as expected since the inductive electron-donating effect of the methyl group stabilizes the protonated ester slightly relative to the protonated acid[14]. However, this considerable difference is very largely compensated for by the higher entropies of activation, which are 13–14 eu more favourable for the reactions of the methyl esters, so that the esters are in fact cleaved only some 20–40 times more slowly than the acids. The entropies of activation for the ester reactions are positive or near zero, in the region expected for unimolecular reactions[30]: the more negative figures for the carboxylic acids are more likely to represent differences in solvation than a change in mechanism[14].

Changing the alkyl group of the carboxylic acid from methyl to ethyl to isopropyl to *t*-butyl, has only a very small effect on the rate of the cleavage reaction, which passes through a maximum, for the propionate or isobutyrate, depending on the medium. The separate effects on the enthalpy and entropy of activation are consistent: increased branching in the alkyl group results in a small decrease in the enthalpy of activation, together with an almost exactly compensating decrease in the entropy of activation. It should be noted, however, that there is a sharp difference in rate between these substituted acetic acid derivatives and formic acid esters.

These effects of different alkyl groups on the rates of cleavage are reminiscent of the Baker–Nathan order[31], and indeed Ehrenson[32] has suggested that hyperconjugation might be an important factor in carbonium ion reactions. Furthermore the considerable shortening of the C–C bond in the acetyl cation (1.38 A[25,26]) compared with C–C single bonds in general, and with the C–CO$^+$ single bond of the isobutyryl cation (1.44 A[27]) in particular, might be taken as an indication of hyperconjugative stabilization. However, arguments about hyperconjugation are never conclusive, and the fact that the propionic and isobutyric acids are slightly more reactive than acetic could be taken as evidence that the effect responsible for the Baker–Nathan order is weaker than usual in this weakly solvating medium, and thus that it is a result of differential solvation[31].

Deno *et al.*[20], on the basis of their equilibrium studies, have concluded that substituents which stabilize carbonium ions also stabilize $RCO_2H_2^+$ relative to RCO_2H, and RCO^+ relative to $RCO_2H_2^+$. In fact the differences are small for simple aliphatic acids. There is, however, a marked effect in the expected direction for chloroacetic acid, which is half converted to the protonated form only in 100% H_2SO_4, and remains predominantly in this form up to 65% SO_3 in H_2SO_4[20].

Two other possible reactions of acyl cations, which do not seem to be of general importance under the conditions of ester hydrolysis and formation, are ketene formation

$$RCH_2{-}C{\equiv}O^+ \overset{H^+}{\rightleftharpoons} RCH{=}C{=}O$$

and decarbonylation, which is often observed for tertiary acyl cations under the conditions of the Friedel–Crafts reaction, *e.g.*

$$(CH_3)_3C{-}C{\equiv}O^+ \rightarrow (CH_3)_3C^+ + CO$$

The detailed mechanism of the formation of acyl cations is not yet clear. It seems certain that the leaving group is a neutral water or alcohol molecule, and thus that a proton transfer is involved. It is not possible at this time, however, to distinguish between a pre-equilibrium formation of the tautomeric form protonated on the alkyl-oxygen, which would then cleave in the rate-determining step (route A)[14,28] and the similar mechanism in which the proton transfer and cleavage reactions are concerted (route B)[14].

path A

path B

$$R'{-}C{\equiv}O^+ + ROH$$

(12)

There is no direct evidence for the formation of a tautomer of type (12), but this does not of course rule out path A, since (12) need not be present in high concentration if it is sufficiently reactive. A mechanism involving a second proton in the transition state, suggested as a possibility by Hogeveen[28] seems less likely, since there is no evidence that the protonated species is significantly basic and since the rate of cleavage in $FSO_3H{-}SbF_5$ is independent of acid concentration[14]. The fact that acyl-cleavage is favoured in very strongly acidic solutions is probably explained by the low activity of the water or alcohol resulting in a reduced rate of the reverse reaction.

(b) Alkyl–oxygen cleavage of esters

The first-order rate coefficient for the cleavage of protonated ethyl acetate in 1:1 $FSO_3H{-}SbF_5$, as determined by the disappearance of the acyl methyl proton singlet, was found to be 2.7×10^{-4} sec^{-1} at 0°C, ten times greater than that for the cleavage of protonated methyl acetate under the same conditions[14]. There is little doubt that the acyl–oxygen cleavage reactions discussed so far ought to be slower for ethyl esters than for the methyl derivatives, and this observation was interpreted by Olah *et al.*[14] as evidence for a change of mechanism, to alkyl–oxygen fission in the case of the ethyl ester, *viz.*

$$CH_3-C(OH)(OCH_2CH_3)^+ \xrightarrow{H^+} CH_3-C(OH)_2^+ + [CH_3CH_2^+]$$

In agreement with this, it was shown that protonated ethanol is not a product of the cleavage reactions: the carbon atoms of the ethyl group end up, eventually, as part of a *t*-butyl cation. This is a rare example of alkyl–oxygen cleavage of a primary aliphatic ester (for two others see Kershaw and Leisten[33]) and it is not a great deal more favourable than acyl–oxygen fission, which can in fact be observed with ethyl acetate in FSO_3H–SbF_5 mixtures containing more than 50% FSO_3H[14]. The esters of secondary alcohols, on the other hand, cleave exclusively by alkyl–oxygen fission between −20 and −40°C in FSO_3H–SbF_5, to give protonated acids and stable tertiary carbonium ions[14,28], *e.g.*

$$CH_3-C(OH)(O-CH(CH_3)CH_2CH_3)^+ \xrightarrow[H^+]{-30°C} CH_3-C(OH)_2^+ + [CH_3\overset{+}{C}HCH_2CH_3] \rightarrow (CH_3)_3C^+$$

Esters of tertiary alcohols are cleaved so rapidly that the protonated esters cannot be observed, their solutions showing only the signals for the protonated acids and tertiary carbonium ions, even at −80°C.

2.1.3 *The basicity of carboxylic acids and esters*

The single most crucial piece of information necessary for the interpretation of the behaviour of a specific acid-catalyzed reaction is the concentration of the protonated, reactive species; this depends on a knowledge of the basicity of the substrate. Until very recently this information has not been available for carboxylic esters, and consequently most work has necessarily been done using relatively weakly acidic media, where it can be assumed that the concentration of the protonated species is small.

The pK_a's of protonated benzoic acid and its derivatives are readily measured by conventional spectrophotometric methods. Data for over forty compounds have been obtained by Stewart and Granger[34] and Stewart and Yates[35] and the results have been summarized by Arnett[36]. Substituted benzoic acids are well-behaved Hammett bases, with slopes of the logarithm of the ionization ratio plots close to 1.0. But, as explained by Arnett[36], this is only true, for benzoic itself at any rate, in its region of protonation. In more dilute acid a sudden sharp change in activity coefficient occurs, which means that the observed

pK_a's are not necessarily directly comparable with those of compounds of much greater basicity. Nevertheless, the relative pK_a's of the conjugate acids of substituted benzoic acids are meaningful. They fall in the region −7.0 to −8.0, and are correlated by the Hammett equation, (using σ^+), with a ρ value of 1·09. Thus the values parallel those for the benzoic acids themselves for those substituents with σ^+ not significantly different from σ.

The protonation of aliphatic carboxylic acids can be followed by NMR techniques (as explained above) or spectrophotometrically, using a band near 190 nm in the ultraviolet. Some values are given in Table 3. The aliphatic

TABLE 3

BASICITIES OF CARBOXYLIC ACIDS AND ESTERS IN SULPHURIC ACID

Acid	*% H_2SO_4 for 50% protonation*	pK_a	*Method*	*Ref.*
CH_3COOH (35°C)	77	−6.86	NMR	20
CH_3COOH (20°C)	74	−6.41	UV	39
CH_3CH_2COOH (35°C)	80	−7.34	NMR	20
CH_3CH_2COOH (25°C)	79	−7.18	UV	40
$CH_3CH{=}CHCOOH$ (35°C)	76	−6.71	NMR	20
C_6H_5COOH	~ 83	~ −7.7	UV	a
2,4,6-trimethylbenzoic				
(room temp.)	~ 82	~ −7.7	UV	41
(60°C)	82	−7.8	UV	41
$ClCH_2COOH$	100	< −11	NMR	20
Ester				
CH_3COOCH_3		−7.25	UV	37
$CH_3COOC_2H_5$	77	−6.93	NMR, UV	38
$CH_3COOC_3H_7$		−7.18	UV	37
$CH_3COOCH(CH_3)_2$		~ −7.2[b]	UV	37
$C_6H_5COOCH_3$		~ −8.3	UV	c
$C_6H_5COOC_2H_5$	~ 82	~ −7.7	UV	42

[a] Data from refs. 34 and 35, corrected using H_0 values of Jorgenson and Hartter[43].

[b] Ionization measured over part of range only, since hydrolysis becomes very fast in strong acid.

[c] Figure given by Arnett[36], corrected using H_0 values of Jorgenson and Hartter[43].

acids are stronger bases than the aromatic compounds by about one pK unit, and their basicity is increased by substituents which stabilize carbonium ions[20]. Strongly electron-withdrawing substituents have a marked effect. Chloroacetic acid is several powers of ten less basic than acetic acid[20] and dichloroacetic acid is only partially protonated in 100% sulphuric acid according to cryoscopic determinations; trichloro- and trifluro-acetic acids, and oxalic acid, do not behave as bases at all[4].

There have been very few studies of the basicity of carboxylic esters, mainly because they are generally rapidly hydrolyzed in the region of acidity in which ionization occurs. Several values for aliphatic esters have recently been measured, however, using similar techniques to those described above for the aliphatic acids. In the case of the esters measurements have to be extrapolated to zero time to take account of hydrolysis. It is clear that the effect of the alkyl group on the basicity of a series of acetate esters is small. The methyl, *n*-propyl and isopropyl esters, investigated by Yates and McClelland[37], differ very little in basicity (Table 3), while Lane *et al.*[38] found that ethyl acetate is half-protonated in 77% H_2SO_4 at 25°C, as is acetic acid at 35°[20]. Benzoate esters appear, from the two figures quoted, to have basicities similar to those of the benzoic acids, and are thus also about one pK unit less basic than the aliphatic esters.

Aliphatic esters do not behave even approximately as expected for Hammett bases. Plots of the logarithms of the ionization ratios against $-H_0$ are linear, but have a slope of 0.62 for simple acetates[37]. Lane *et al.*[38] have found a similar slope, 0.65, for ethyl acetate[38].

2.2 KINETICS OF REACTIONS IN STRONGLY ACIDIC MEDIA

The hydrolysis and formation of esters in solutions of strong acids show very varied behaviour, and depend on a rather large number of different factors. Our understanding of the quantitative behaviour of these reactions has become much clearer in recent years, and since an attempt to follow the historical development of the subject might obscure parts of the pattern that has emerged, it seems best to base a discussion on recent results.

A large amount of data is available for the hydrolysis of simple esters in sulphuric acid. In Table 4 are listed observed first-order rate coefficients for the hydrolysis of a series of acetate esters in sulphuric acid solutions of

TABLE 4

RATE COEFFICIENTS FOR ACID-CATALYZED ESTER HYDROLYSIS AT 25°C[a]

H_2SO_4 (%)	10^4k_1 (sec^{-1})	H_2SO_4 (%)	10^4k_1 (sec^{-1})
Methyl acetate		*Ethyl acetate*	
14.1	2.50	11.9	1.88[b]
20.7	4.35	16.8	3.09[b]
28.3	7.04	24.2	5.61[b]
34.8	10.1	30.0	7.97[b]
40.4	13.6	33.4	10.4[b]
45.4	17.3	36.7	12.8[b]
50.2	19.0	47.9	19.2[b]

TABLE 4 (CONTINUED)

H_2SO_4 (%)	$10\ k_1$ (sec^{-1})	H_2SO_4 (%)	$10^4 k_1$ (sec^{-1})
55.2	22.2	61.0	16.7[c]
60.4	23.0	65.0	14.9[c]
65.2	19.8	71.4	6.29[c]
70.4	12.1	80.2	0.65[c]
74.1	6.39	84.4	0.31[c]
80.0	1.55	90.0	0.51[c]
		98.4	1.09[c]
		100.0	1.24[c]
n-Propyl acetate		*iso-Propyl acetate*	
14.1	2.45	14.1	1.48
20.7	4.20	25.3	3.32
30.2	8.71	34.8	5.55
34.8	11.3	40.4	7.03
40.4	14.1	45.4	8.28
45.4	16.3	50.2	8.96
50.2	19.0	55.2	9.14
55.2	19.2	60.4	7.55
60.4	17.3	65.2	6.00
65.2	13.7	70.4	3.00
70.4	7.46	74.1	1.90
74.1	3.57	80.0	2.46
89.7	0.34	81.7	3.84
95.0	0.54	84.7	7.80
98.6	0.75	88.2	18.7
		89.7	22.5
		90.8	28.4
t-Butyl acetate[d]		*Phenyl acetate*	
11.6	4.63	15.1	1.73
12.9	6.25	20.1	2.70
13.8	7.01	25.3	3.97
17.3	13.1	30.2	5.50
19.2	20.9	34.8	7.23
21.4	32.9	40.4	10.3
22.7	41.6	45.4	13.9
28.3	132	50.2	18.2
Benzyl acetate		55.2	24.2
10.1	1.23	60.4	32.0
25.3	4.20	65.2	37.7
30.2	6.73	70.4	45.3
34.8	7.70	74.1	48.9
40.4	10.1		
45.4	11.9	*p-Nitrophenyl acetate*	
50.2	16.4	15.1	1.25
52.8	15.5	20.1	1.87
55.2	16.5	25.3	2.53
60.4	14.2	30.2	3.93
62.5	15.4	34.8	5.15
65.2	16.8	40.4	7.12

TABLE 4 (CONTINUED)

H_2SO_4 (%)	10^4k_1 (sec^{-1})	H_2SO_4 (%)	10^4k_1 (sec^{-1})
66.8	22.8	45.4	9.52
67.3	26.2	50.2	13.1
69.0	37.2	55.2	17.3
		60.4	25.2
p-Chlorophenyl acetate		65.2	35.5
15.1	1.54	70.4	60.1
34.8	7.09	74.1	117
45.4	13.2	75.3	155
55.2	22.4	78.5	254
65.2	37.8	80.0	370
70.4	45.5		
74.1	63.6		

[a] Data from ref. 37, unless otherwise stated.
[b] Data from ref. 44.
[c] Data at 25.6°C, from ref. 45.
[d] Data from ref. 46.

various concentrations, measured at 25°C. Data for five representative esters are plotted in Fig. 1.

The rate of hydrolysis of a carboxylic ester in strong sulphuric acid generally shows one of the three types of dependence on acid concentration illustrated in Fig. 1. The simplest behaviour, a continuous increase in hydrolysis rate with increasing acid concentration, is shown by esters of tertiary alcohols, which are hydrolyzed very rapidly even in moderately concentrated acid, and by phenol esters, which are somewhat less reactive, but are hydrolyzed much faster than esters of simple primary and secondary alcohols with above about 60% H_2SO_4. Substituted phenyl acetates behave very much like the parent compound, the *p*-chlorophenyl ester being hydrolyzed at almost the same rate as the unsubstituted compound, while *p*-nitrophenyl acetate is somewhat less reactive at low acid concentrations, but more reactive in above 70% sulphuric acid.

Esters of primary aliphatic alcohols show an initial increase in rate with increasing acid concentration, but the curve, illustrated in Fig. 1 for ethyl acetate, passes through a maximum at an intermediate acid concentration (50–60% H_2SO_4), and the rate coefficient subsequently falls almost to zero in the region of 80% acid. Finally, the rate increases slightly between 80% and 100% H_2SO_4. Methyl acetate is rather more reactive than the ethyl ester in above about 50% H_2SO_4, but *n*-propyl acetate shows closely similar reactivity, so that the curve for ethyl acetate can be considered to represent the typical behaviour of a primary alkyl ester.

The third type of behaviour combines the very sharp increase in rate at moderate acid concentrations, typical of the first type of ester, with the rate

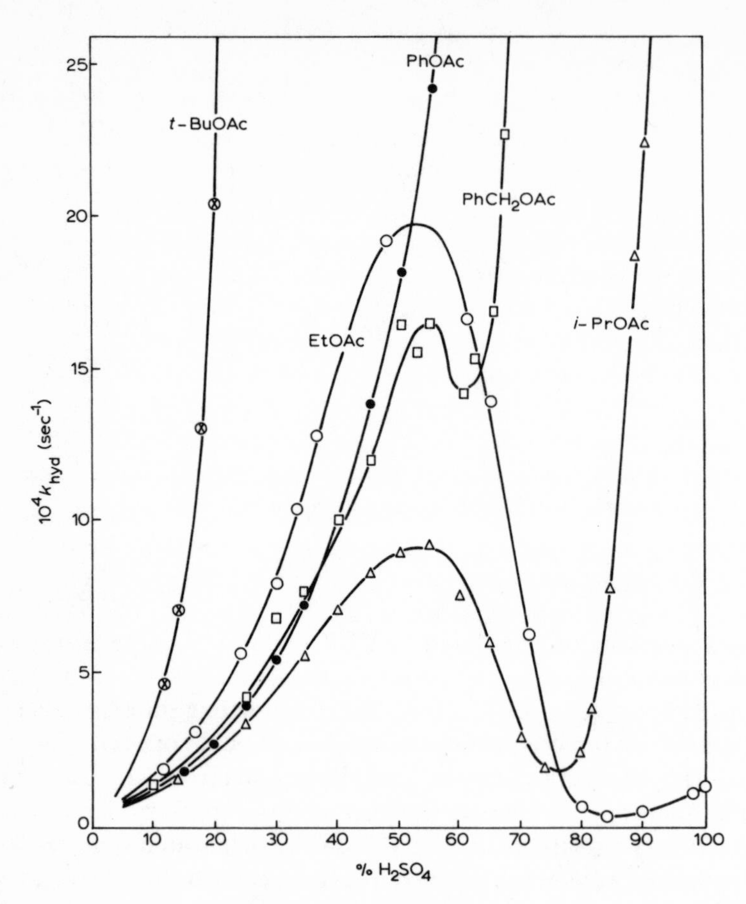

Fig. 1. Hydrolysis behaviour of acetate esters in sulphuric acid solutions at 25°C. Data from Table 4 and ref. 37; k in sec^{-1}.

maximum at intermediate acidities typical of the second. Esters showing this pattern are those of secondary aliphatic alcohols (the curves for isopropyl acetate, Fig. 1, and *sec*-butyl acetate are closely similar), and benzyl acetate.

The reactions of protonated carboxylic esters in the absence of significant amounts of water have been discussed above. Two types of cleavage are possible. Alkyl–oxygen cleavage is a very fast reaction for the conjugate acids of tertiary alcohols

$$R'-C\begin{smallmatrix}O-H\\ +\\ O-C(CH_3)_3\end{smallmatrix} \longrightarrow R'-C\begin{smallmatrix}OH\\ \\ O\end{smallmatrix} + \overset{+}{C}(CH_3)_3$$

It also occurs at a measurable rate at low temperatures for the protonated esters of secondary alcohols, but is not important for primary alkyl esters. Acyl–oxygen cleavage, on the other hand, becomes significant only in very concentrated sulphuric acid[20], and thus is important only with esters of primary alcohols.

Either of these cleavage mechanisms can be involved in reactions which are very fast even at moderate acid concentrations. For example, the alkyl–oxygen cleavage of the conjugate acid of *t*-butyl acetate appears to be so fast that the formation of only a small proportion of the protonated form leads to a rapid hydrolysis. The similar cleavage of protonated isopropyl acetate presumably accounts for the rapid increase in the rate of hydrolysis above about 80% H_2SO_4, but in this case the reaction does not become too fast for convenient measurement until the ester is nearly completely protonated – it is possible, for example, to measure the lower part of the ionization curve[28] – and thus the curve for isopropyl acetate might be expected to level out as protonation is completed, if it could be measured in the 100% H_2SO_4 region.

The final increase in the rate of hydrolysis of the esters of primary aliphatic alcohols, in 85–100% H_2SO_4, is presumably due to acyl–oxygen cleavage, since methyl acetate reacts faster than the ethyl ester, which would not be the case for alkyl–oxygen cleavage. So also may be the rapid hydrolysis of aryl acetates at moderate acidity, since aryl–oxygen cleavage is not expected to be so rapid a reaction.

Thus it is a comparatively simple matter to explain qualitatively why the hydrolysis of some esters becomes faster as the acidity of the medium is increased. It is not possible, however, to rationalize in terms of the two cleavage processes the behaviour of those esters whose hydrolysis rates decrease with increasing acid concentration in certain regions.

These can be explained if they involve a third reaction of the protonated ester, *i.e.* with a molecule or molecules of water, a reaction not, of course, observed in media such as SbF_5–FSO_3H. The activity of water falls rather sharply as the sulphuric acid concentration increases from 60–100%, and we know that most esters become essentially completely protonated in this region. Thus the situation can arise where the increase in the concentration of protonated ester produced by a given increase in acid concentration is proportionately smaller than the concomitant decrease in the activity of water, so that bimolecular (or higher molecularity) hydrolysis goes more slowly as the acidity is increased. Similar behaviour is observed when amides are hydrolyzed in strong acid solutions, but the rate maximum occurs at lower acid concentration, since amides are more basic than esters, and protonation is complete in solutions of lower acidity.

With solutions of low acidity the increase in the concentration of the conjugate acid of the ester outweighs the decrease in activity of water, and the rate of hydrolysis by the A2 mechanism increases with increasing acid concentration, as expected. Most esters are, in fact, hydrolyzed by the $A_{AC}2$ mechanism in less concentrated solutions of strong acids: all the esters in Fig. 1, for example, with the exception of *t*-butyl acetate, are thought to be hydrolyzed by the $A_{AC}2$ mechanism below 50% H_2SO_4. But although this is the commonest mechanism for acid-catalyzed hydrolysis, it is also the most complex, and difficult to study in detail, since it involves a reaction with the solvent. The unimolecular mechanisms can be isolated by an appropriate choice of conditions and substrate, and are relatively well understood. So at this point it is appropriate to discuss in turn the evidence for, and characteristics of, each of the important classes of ester hydrolysis, as defined by Ingold[1].

2.2.1 *The $A_{AC}1$ mechanism*

The rate coefficients for the hydrolysis and formation of esters in acid are generally many times smaller than those for their protonation (estimated at about 10^3 sec^{-1} on p. 124), and the usual assumption that the proton transfers are not involved in determining the rate of these reactions is valid. The detailed mechanism of the $A_{AC}1$ reaction is thus not in doubt[1].

$$R'COOR \overset{H^+}{\rightleftharpoons} R'\text{—}C\begin{matrix}\diagup O\text{—}H \\ + \\ \diagdown O\text{—}R\end{matrix}$$

$$R'\text{—}C\begin{matrix}\diagup O\text{—}H \\ + \\ \diagdown O\text{—}R\end{matrix} \underset{\text{fast}}{\overset{\text{slow}}{\rightleftharpoons}} R'\text{—}C\equiv O^+ + ROH$$

$$R'\text{—}C\equiv O^+ + H_2O \underset{\text{slow}}{\overset{\text{fast}}{\rightleftharpoons}} R'\text{—}C\begin{matrix}\diagup O\text{—}H \\ + \\ \diagdown O\text{—}H\end{matrix}$$

$$R'\text{—}C\begin{matrix}\diagup O\text{—}H \\ + \\ \diagdown O\text{—}H\end{matrix} \rightleftharpoons R'COOH + H^+$$

Acyl cations are today well-established chemical species. The evidence for their formation and identification in strongly acid media has already been discussed, and it is known that protonated carboxylic acids and esters are generally cleaved to acyl cations in sufficiently strong acid, providing no more favourable process supervenes. The reaction was first inferred from the cryoscopic behaviour of mesitoic acid and its methyl ester, and this acid and its derivatives have been frequently used since in the study of the $A_{AC}1$ mechanism. This is not only because alkyl–oxygen cleavage can be ruled out, but also because bimolecular attack on the carbonyl group is unimportant, since it is subject to severe steric hindrance. Furthermore the dissociation of mesitoic acid to the mesitoyl cation is 50% complete in 97% H_2SO_4[20,41], a lower acidity than for any other acid measured, so that not only is the alternative reaction particularly unfavourable, but the cleavage actually appears to be particularly favourable for derivatives of mesitoic acid, presumably because of steric assistance[47].

A number of authors have studied the hydrolysis of methyl mesitoate in concentrated sulphuric acid solutions, and the results from the various experimental approaches together give a clear and consistent picture of the mechanism. The cryoscopic behaviour of the ester has already been mentioned, and further evidence that acyl–oxygen fission is involved is available from experiments using the carbonyl-^{18}O-labelled methyl mesitoate[47]. These show that the carbonyl oxygen atom does not exchange with the oxygen atoms of the solvent at three different sulphuric acid concentrations (3.09, 5.93 and 11.7 *M*), *viz.*

$$\text{Mes–C}(=^{18}\text{O})\text{OCH}_3 + H_2O \xrightarrow{H_2SO_4} \text{Mes–C}(=^{18}\text{O})\text{OH} + CH_3OH$$

The result for the lowest sulphuric acid concentration is particularly significant, since exchange by a bimolecular mechanism would be expected to be rapid under these conditions for a normal ester.

The kinetic parameters obtained for the hydrolysis of methyl mesitoate in strong sulphuric acid conform with all the usual empirical criteria for a unimolecular rate-determining step. Chmiel and Long[48] found that the logarithms of the rate coefficients for hydrolysis at 90°C in aqueous solutions of perchloric and sulphuric acids (1–6 *M* acid) are linear functions of $-H_0$ (at 25°C), with slopes of about 1.2, but do not vary linearly with $[H_3O^+]$. This is the result expected for a unimolecular reaction from the Zucker–Hammett Hypothesis, and is one clear-cut result, supported in this case by much other evidence, which was regularly used as evidence for the correctness of the hypothesis. The plot of log k_{hyd} against $-H_0$ for the benzoate ester of glycerol, for example, thought to be an $A_{AC}2$ reaction, shows very different behaviour (Fig. 2).

Bunton *et al.*[49] found that the perchloric acid-catalyzed exchange of ^{18}O from solvent water into mesitoic acid is a very similar reaction. Here, too, the rate is proportional to h_0 (in dioxan–water, 60:40 v/v, at 100°C). In this case the plot of log k against $-H_0$ has a slope of 0.97.

Activation parameters have been measured for these reactions, and are also in agreement with the unimolecular mechanism, since the entropies of activation are all positive. For ^{18}O- exchange from solvent water into mesitoic acid, under the conditions described above, Bunton *et al.*[49] found a $\Delta S^{\ddagger}$ of

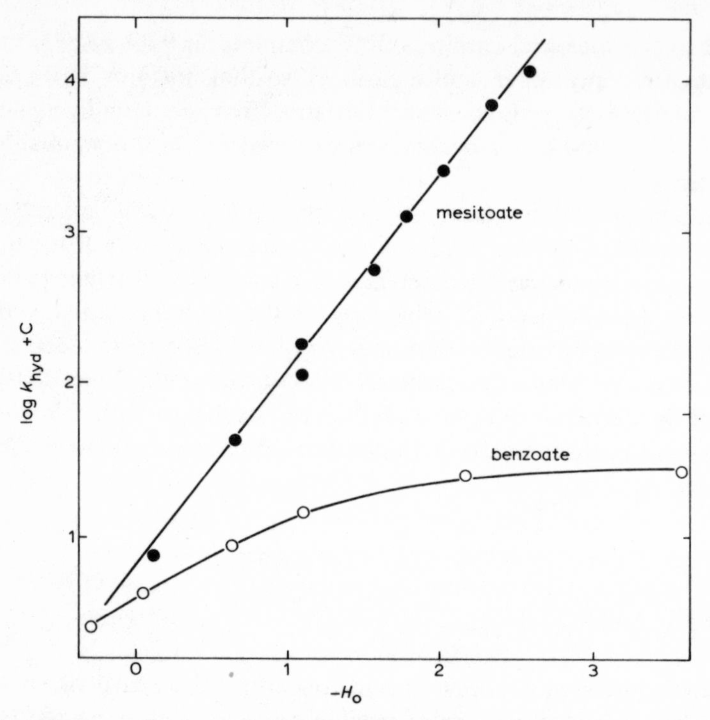

Fig. 2. Zucker–Hammett plot for the hydrolyses of methyl mesitoate and α-glyceryl benzoate, at 90°C. Data from ref. 48.

+9 eu, and $\Delta H^{\ddagger} = 33$ kcal·mole^{-1}. For the hydrolysis of methyl mesitoate these parameters were measured at two different concentrations of sulphuric acid[47], giving values of $\Delta S^{\ddagger} = 17 \pm 5$ and 6.0 ± 5 eu, $\Delta H^{\ddagger} = 28.4 \pm 1$ and 21.5 ± 1 kcal·mole^{-1}, with 9.8 and 11.5 *M* sulphuric acid, respectively.

Since the hydrolysis of methyl mesitoate conforms with the Zucker–Hammett hypothesis, it is not unexpected that the parameters calculated from the same data should meet Bunnett's more recent mechanistic criteria. The value of Bunnett's w for the reaction in sulphuric acid is calculated as -1.1, and in perchloric acid as -2.5[50], both values falling in the region (w between -2.5 and 0) characteristic of reactions not involving a molecule of solvent in the transition state. Bunnett's ϕ is also negative, as expected, values of -0.25 and -0.42[51] being found for the same two sets of data at 90°C[48].

Another class of compounds constrained by special structural features to react in strong acid by the $A_{AC}1$ mechanism are β-lactones. Cyclic esters of this type containing the highly strained four-membered ring, are very reactive towards hydrolysis by all mechanisms. Addition to the carbonyl group reduces the strain as the sp^2-hybridized carbonyl carbon is converted to the sp^3-hybrid, with bond angles that are naturally more acute, while mechanisms involving cleavage of either the acyl–oxygen or the alkyl–oxygen bonds in the rate-determining step are assisted directly by the release of strain consequent upon the partial opening of the ring in the transition state.

Olson and Miller[52], using optically active β-butyrolactone[13], showed that the hydroxybutyric acid produced on acid hydrolysis has the opposite configuration to that produced by the reaction in neutral water, and by this, and subsequent work using ^{18}O-labelled water[53], showed that the hydrolysis in concentrated solutions of strong acids proceeds with acyl–oxygen fission. A detailed kinetic investigation by Long and Purchase[54] of the similar reaction of β-propiolactone showed that the rate of hydrolysis at 25°C in up to 6-7 M perchloric, sulphuric and nitric acids, is not directly correlated with acid concentration, as would be expected for a normal ester hydrolysis reaction. The reaction is also characterized by a very large positive salt effect, as indicated by a single experiment with added sodium perchlorate. Long and Purchase show that their data are, in fact, correlated by h_0. A plot of log k_{obs} *versus* $-H_0$ is a good straight line, of slope close to 1.0. It seems clear that this is a case where the Zucker–Hammett hypothesis (see below, p. 108) leads to a correct conclusion, and that the hydrolysis of β-lactones involves a unimolecular reaction of the conjugate acid. Since acyl–oxygen fission is proved by the results described above, primary and secondary β-lactones are hydrolyzed by the $A_{AC}1$ mechanism. (The tertiary compounds are discussed on p. 93.)

Substituent effects on the $A_{AC}1$ reaction have been studied by Bender and Chen[55]. These authors measured the rates of hydrolysis of a series of 4-substituted 2,6-dimethylbenzoates in 9.70 M sulphuric acid at 25°C, and found that the values for the first-order coefficients with 4-methoxy, 4-methyl, 4-unsubstituted and 4-bromo-compounds (5.0, 0.37, 0.033 and 0.01×10^{-4} sec^{-1}, respectively) are satisfactorily correlated by the Hammett equation, following σ^+ with a slope $\rho = -3.22$. Since the esters are not fully protonated in 9.70 M H_2SO_4, part of this factor is due to the effects of the 4-substituent on the protonation equilibrium. ρ for the protonation of substituted benzoic acids is about -1[35], but is likely to be considerably smaller for di-*ortho*-substituted compounds, since the conjugative interaction of the *p*-substituents with the protonated carboxyl group requires coplanarity with the ring.

In fact, if steric inhibition of resonance is complete, the effect of a 4-substituent on the protonation equilibrium could be very small, and the observed ρ of -3.22 could be quite close to ρ for the cleavage reaction. Some indication that this might be so can be gleaned from the results of Kershaw and Leisten[33], who measured the rates of cleavage of six substituted methyl benzoates in 99.9% sulphuric acid. Under these conditions the esters are fully protonated—the rates were measured by following the change in the *i*-factor from 2 to 3—and so the effects of substituents on the relative rates of hydrolysis of the six compounds are specifically effects on the cleavage process. (Some of the esters concerned were di-substituted compounds, and the Hammett plot was made using $\Sigma\sigma$ rather than σ: but σ and σ^+ do not differ significantly for the substituents involved.) Kershaw and Leisten's data[33] for substituted methyl benzoates gave a ρ-value of about -3.25, almost the same as that found by Bender and Chen[55] for 4-substituted-2,6-dimethylbenzoates in 9.20 M H_2SO_4. Inasmuch as the two *ortho* methyl groups may affect the position of the transition state on the reaction co-ordinate, these figures should not be compared too closely. Nevertheless, it is clear that a substantial, negative value of ρ is associated with the acyl–oxygen cleavage reaction. Since the results also follow σ^+ rather than σ[55], the transition state must have a high degree of cationic character. If, in fact, steric hindrance of resonance is important in the protonated ground state for the *ortho*-dimethyl-substituted esters it must be presumed that bond-breaking is well-advanced in the transition state for these $A_{AC}1$ reactions, since overlap with the developing cationic centre requires the π-orbital concerned to be almost parallel with those of the π-system of the ring, *viz.*

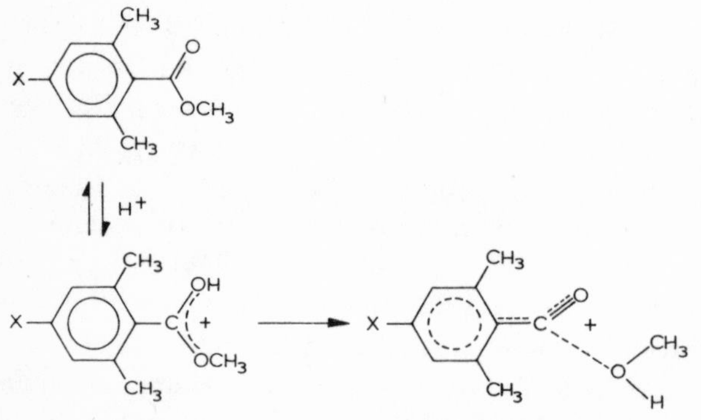

The most recent work on the $A_{AC}1$ reaction[56] includes a comparison of the catalytic efficiencies of the three strong acids commonly used in acid hydrolysis, *viz.* sulphuric, perchloric and hydrochloric. Perchlorates have a large positive salt effect on S_N1 hydrolysis reactions, even in hydroxylic solvents, and it is likely that the transition state in an $A_{AC}1$ reaction also has considerable carbonium ion character. So it is not unexpected that perchloric acid is catalytically more effective than the other strong acids in the $A_{AC}1$ hydrolysis of methyl mesitoate[56]. The order is $HClO_4 > HCl \simeq H_2SO_4$ (Table 5) and this result requires that the effect of varying the anion is an effect on the transition state, since the activity coefficient of methyl mesitoate in the different solvents decreases in the order $H_2SO_4 > HCl > HClO_4$ (Fig. 3). If only ground-state effects were important this sequence should apply also to reaction rates, as indeed it does for the $A_{AC}2$ hydrolysis of methyl benzoate, which shows qualitatively similar activity coefficient behaviour[56].

Bunton *et al.*[56] offer the generalization that anions of low charge density

TABLE 5

KINETIC DATA FOR THE ACID HYDROLYSIS OF METHYL MESITOATE AT 90°C

$HClO_4$ (M)	10^5k_{hyd} (sec^{-1})	HCl (M)	10^5k_{hyd} (sec^{-1})	H_2SO_4 (M)	10^5k_{hyd} (sec^{-1})
1.00	0.073[b]	3.56	0.55[a]	3.12	1.0[b]
1.93	0.42[b]	4.42	1.41[a]	4.16	3.1[b]
2.56	0.94[a]	5.42	4.31[a]	5.00	15.5[b]
2.75	0.92[a]	6.05	9.16[a]	6.14	42[b]
2.88	0.97[a]				
2.94	1.16[b]				
2.96	1.80[a]				
3.79	4.30[a]				
3.90	5.60[a]				
4.39	12.7[b]				
4.80	26.0[b]				
5.36	67.0[b]				
5.76	114				

[a] Ref. 56.
[b] Ref. 48.

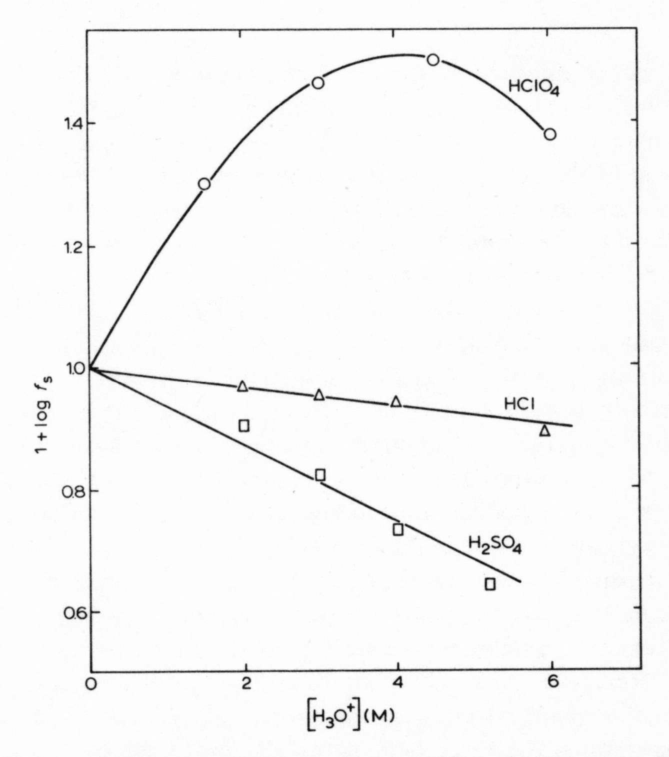

Fig. 3. The activity coefficient of methyl mesitoate as a function of acid concentration, in aqueous solutions of three strong acids. Data from ref. 56.

stabilize selectively transition states with carbonium ion character, and show that perchlorates assist the ionization of trianisylmethanol, and the $A_{AC}1$ hydrolysis of *t*-butyl acetate. One consequence of this selectivity is that it depends, apparently fortuitously, on the catalyzing acid, whether the results do or do not fit the predictions of the Zucker–Hammett hypothesis[56].

Direct evidence that hydrolysis reactions going by the $A_{AC}1$ mechanism are kinetically first-order can be obtained, at least in principle, for reactions in strongly acidic solution, because the activity of water varies significantly with the acid concentration. Graham and Hughes[57] showed that the hydrolysis of methyl benzoate in sulphuric acid at 20°C is first-order with respect to ester concentration, but zeroth-order with respect to water in concentrations up to 1 *M*. Leisten[6] showed further that the first-order rate coefficient for this reaction is almost independent of the initial concentration of the ester, and thus ruled out the possibility that a bimolecular attack by bisulphate ion is involved, since the ester is completely protonated in 100% sulphuric acid and the concentration of bisulphate ion depends on the concentration of the ester, *viz.*

$$PhC(=O)OCH_3 + H_2SO_4 \rightleftharpoons Ph{-}\overset{+}{C}(OH)(OCH_3) + HSO_4^-$$

A more recent approach to the problem of determining the kinetic order of ester hydrolysis reactions in strong acid is to plot the logarithm of the observed first-order rate coefficient for the hydrolysis of the protonated ester, against the logarithm of the activity of water in the medium. This method avoids the problems associated with the effects of variations in the medium on the protonation equilibrium, but depends, of course, on an accurate knowledge of the protonation behaviour of the substrate in the acidity range concerned. It was applied originally to the study of the $A_{AC}2$ reaction by Lane[38] (see below) but has given valuable information on the A-1 reactions also. Jaques[45] discovered that a plot of log $k_{hyd}/[BH^+]$ against log a_{H_2O} for the hydrolysis of ethyl acetate shows a discontinuity. Up to about 82% sulphuric acid it is linear, with slope near 2, as found by Lane[38], and interpreted by him in terms of a transition state containing two molecules of water. At higher acid concentrations, however, the first-order rate coefficient becomes almost completely independent of water activity, and the slope of the graph falls to -0.16, suggesting that water is no longer involved in the transition state. Yates and McClelland[37], using a similar method, derived from Bunnett's hydration parameter treatment[58], have found that this sort of behaviour is general. The protonated forms of most esters are hydrolyzed in weak to moderately concentrated sulphuric acid in reactions that depend strongly on the water activity, but at sufficiently high acid concentrations the rates become almost independent of the activity of water[37]. This is almost certainly a result of a change from $A_{AC}2$ to A-1 mechanisms. The acetate esters of secondary aliphatic alcohols, and benzyl alcohol, are hydrolyzed by the $A_{AL}1$ mechanism at high acid concentration, while primary alkyl acetates and the phenol esters are thought to change to the $A_{AC}1$ mechanism. Yates and McClelland use the full expression for the fraction of ester in the protonated form, and plot $\log k_1 - \log (h_0{}^m/h_0{}^m - K^m_{SH^+})$ * against $\log a_{H_2O}$. Their results for a typical primary alkyl acetate, *n*-propyl acetate, and for *p*-nitrophenyl acetate, are plotted in Fig. 4. Above about 80% H_2SO_4 the slope becomes small and negative, about −0.15 for the line in Fig. 4, similar to the figure of −0.16 obtained by Jaques for the comparable plot of his data for ethyl acetate. The plot for *p*-nitrophenyl acetate also levels out to a small negative slope above about 70% H_2SO_4. In each case it is suggested that this very weak dependence on water activity is evidence that water is not involved as a reactant in the transition state for hydrolysis, which is a unimolecular process involving acyl–oxygen fission. (Jaques suggests

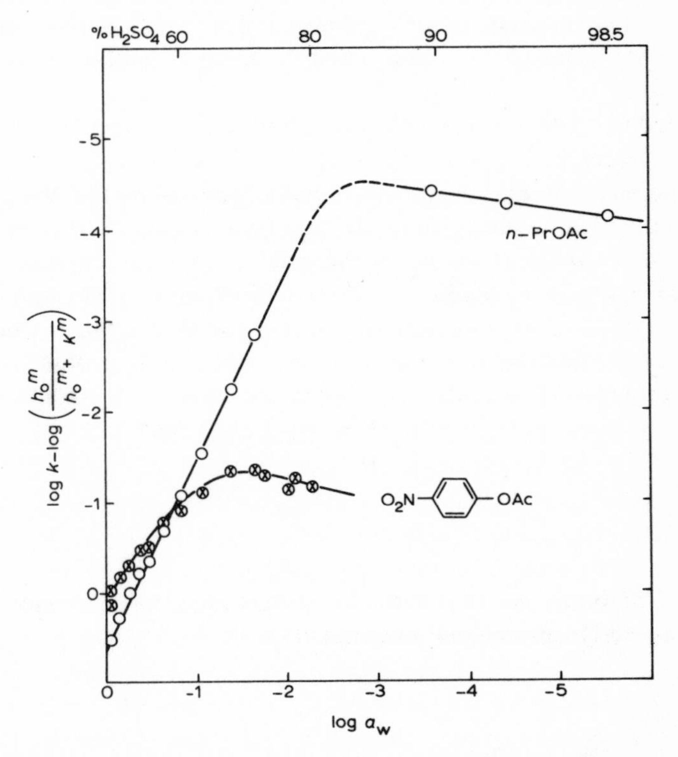

Fig. 4. Typical Yates–McClelland plots of the rate coefficients for hydrolysis of protonated aryl and primary alkyl acetates, in sulphuric acid solutions, against the activity of water. Data from ref. 37.

* The ionisation ratios of the ester substrates, S, are correlated by the expression $\log [SH^+]/[S] = -mH_0 + \text{constant}$. m is a constant, about 0.62 for these esters.

that ethyl acetate is hydrolyzed by the $A_{AL}1$ mechanism in this region, but this would be unusual for an ethyl ester, as discussed below, and the final negative slope of plots of the type shown in Fig. 4 is usually considerably more negative for alkyl–oxygen cleavage, see below.) Jaques[45] measured the activation parameters for the hydrolysis of ethyl acetate, and found $\Delta S^{\ddagger} =$ 11.5 eu, and $\Delta H^{\ddagger} = 23.6$ kcal·mole^{-1} with 98.4% H_2SO_4. $\Delta S^{\ddagger}$ is still positive (4.9 eu) for the 75.4% acid in the region of mechanism change, although the effect of changing temperature on the protonation equilibrium is a complication in the medium of lower acidity.

It is reasonable that the aryl acetates should be hydrolyzed faster than the primary alkyl esters by the $A_{AC}1$ mechanism, because phenols are considerably less basic, and thus better leaving groups, than are alcohols[37]. This effect may be weakened by the less favourable equilibrium constant for the protonation of the better leaving group, but the effect of substitution on the cleavage reaction is decisive, just as is substitution in the acyl group. Thus the rate coefficients for the hydrolysis of aryl acetates at high acid concentrations fall in the order *p*-nitrophenyl > *p*-chlorophenyl > phenyl (although the reverse order is found in the $A_{AC}2$ region).

The method described above makes it a relatively simple matter to determine the point of mechanism change, from $A_{AC}2$ to $A_{AC}1$ or $A_{AL}1$, with increasing acidity of the medium. Earlier work by Leisten[6] had shown that the mechanism actually involved in hydrolysis in the strongly acidic region depends in a predictable way on the structure of the ester concerned and he was able to demonstrate how the mechanism of ester hydrolysis in 100% sulphuric acid changes from $A_{AC}1$ to $A_{AL}1$ as either the acyl or the alkyl group is changed.

Under these conditions the ester is almost completely protonated, and the effects of proton substituents are simply predicted. Electron-withdrawing substituents on benzoate esters assist the $A_{AL}1$ cleavage[59], *viz.*

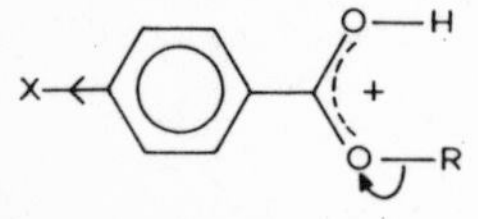

while they destabilize the acyl cation, and thus hinder $A_{AC}1$ cleavage, which is favoured by electron-donating substituents, *viz.*

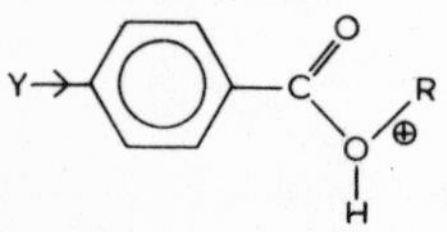

Leisten[6] found that the introduction of a *p*-nitro group into ethyl benzoate causes a 60-fold decrease in the rate of hydrolysis in 100% sulphuric acid at 65°C, while in the case of isopropyl benzoate it causes a 200-fold rate *increase*.

This suggests that the hydrolysis of the ethyl ester involves the $A_{AC}1$ mechanism, but that protonated isopropyl benzoate cleaves by alkyl–oxygen cleavage. A change in mechanism at this point would explain the order of reactivity towards hydrolysis of the various alkyl benzoates, observed by Kuhn[19].

Subsequently Kershaw and Leisten[33] were able to demonstrate that the same change in mechanism, from $A_{AC}1$ to $A_{AL}1$, can be brought about by appropriate substitution in ethyl benzoate. Figure 5 shows two Hammett

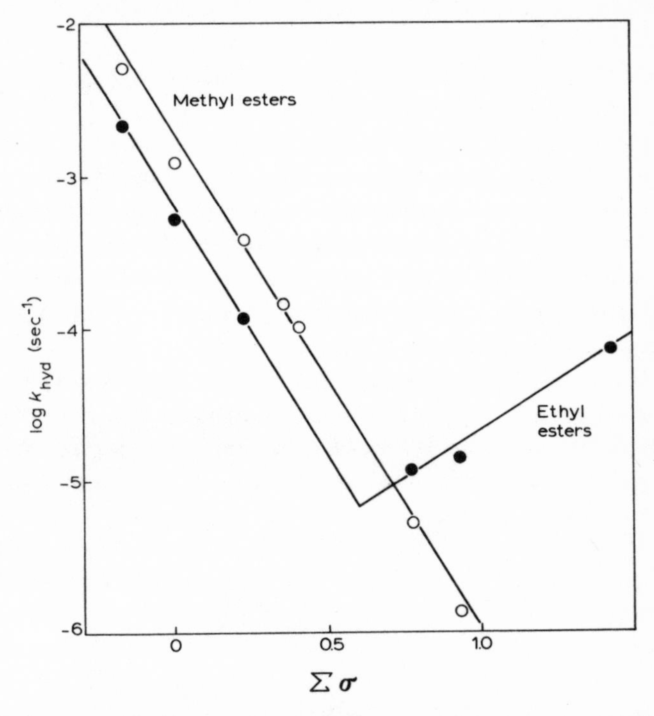

Fig. 5. Hammett plot for the hydrolysis of substituted methyl (○) and ethyl (●) benzoates at 45°C in 99.9% sulphuric acid. Data from ref. 33.

plots, correlating data for the hydrolysis at 45°C in 99.9% sulphuric acid of the esters of a series of six substituted benzoic acids (using $\Sigma\sigma$, since two di-substituted benzoates were used). The circles represent data for the methyl esters. These are correlated by a good straight line, of slope $\rho = -3.25$, of the sign and magnitude expected for a series of reactions involving the $A_{AC}1$ mechanism (see above). The full circles represent the data for the corresponding ethyl esters. While the reactivity of compounds with electron-donating or only weakly electron-withdrawing substituents parallels that of the methyl esters, there are successive *increases* in rate coefficient with ethyl 4-nitro, 4-chloro-3-nitro, and 3,5-dinitro-benzoates[33], correlated by a new value of ρ, in the region of $+2$. There seems little doubt that these three esters are hydrolyzed by the $A_{AL}1$ mechanism, and that the generalization made above, that the esters of primary aliphatic alcohols hydrolyze by the $A_{AC}1$ mechanism in very strongly acidic media requires a qualification. Sufficiently strongly electron-withdrawing substituents on the acyl group can make the formation of the acyl cation even less favourable than that of a primary carbonium ion, and under these circumstances even a primary alkyl ester may be hydrolyzed by the $A_{AL}1$ mechanism.

2.2.2 *The $A_{AL}1$ mechanism*

Ester hydrolysis involving alkyl–oxygen cleavage has been the subject of a number of investigations, usually using much less concentrated solutions of the strong acids than those required for significant rates of reaction by the $A_{AC}1$ mechanism. From Olah's work with FSO_3H–SbF_5 solutions (see above, p. 68), it is known that the alkyl–oxygen cleavage of the protonated forms of tertiary alkyl esters is a very fast reaction. This is clearly the case in aqueous acid also, since the hydrolysis of *t*-butyl acetate becomes too fast for convenient measurement in 20–30% H_2SO_4[37,56], where only a very small proportion of the protonated species is present. The reaction is not so fast, however, that the proton transfer step is rate-determining, and the mechanism of the $A_{AL}1$ hydrolysis and formation of esters must be that written in Scheme 2.

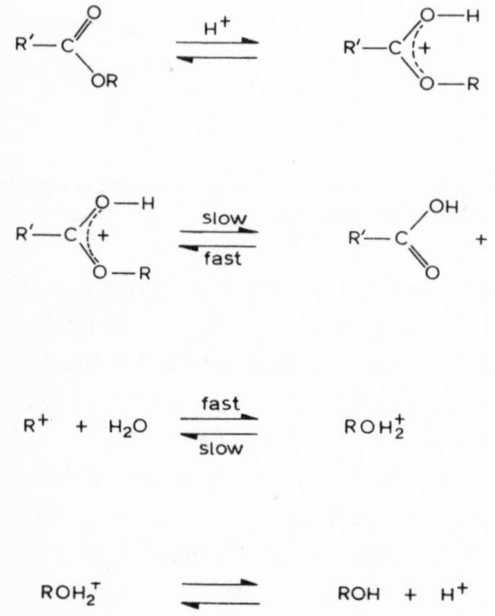

$$R'\!-\!C(=O)OR \overset{H^+}{\rightleftharpoons} R'\!-\!C^{+}(O\!-\!H)(O\!-\!R)$$

$$R'\!-\!C^{+}(O\!-\!H)(O\!-\!R) \underset{\text{fast}}{\overset{\text{slow}}{\rightleftharpoons}} R'\!-\!C(=O)OH + R^+$$

$$R^+ + H_2O \underset{\text{slow}}{\overset{\text{fast}}{\rightleftharpoons}} ROH_2^+$$

$$ROH_2^+ \rightleftharpoons ROH + H^+$$

Scheme 2.

Since the great majority of studies of ester hydrolysis involving the $A_{AL}1$ mechanism have been with tertiary alkyl esters, strongly acidic media have not been necessary; indeed some esters undergo alkyl–oxygen cleavage even in the absence of strong acid (see p. 146).

(*a*) *Evidence for alkyl–oxygen cleavage*

A large number of simple diagnostic tests is available to distinguish alkyl-oxygen from acyl–oxygen cleavage. Since most of these have been recognized for many years, there seems no need to review them here in detail. Most of the methods are applicable equally to reactions in acidic, neutral or basic media, and some are described in Section 3.1.1, since they have been most commonly used in the study of the $B_{AL}1$ reaction. The various types of evidence for alkyl-oxygen cleavage of esters have been reviewed by Davies and Kenyon[60].

Much of the evidence for alkyl–oxygen cleavage in acid-catalyzed ester hydrolysis comes from the systematic investigations of Bunton and his group, using ^{18}O as a tracer. Earlier product studies by Cohen and Schneider[61] established that *t*-butyl benzoate and mesitoate are solvolyzed with alkyl–oxygen cleavage under neutral and acidic conditions. Thus when a solution of *t*-butyl benzoate in anhydrous methanol is boiled for four days, *t*-butyl methyl ether is produced in over 60% yield, although the ether is not a product when sodium methoxide is added to the reaction mixture[61]. Together those two pieces of evidence make it possible to rule out a significant amount of neutral solvolysis, since this should not be affected by the addition of base, and establish that the main reaction in initially neutral solution involves a catalysis by the benzoic acid produced, *viz*.

$$PhCOOC(CH_3)_3 + CH_3OH \xrightarrow{H^+} PhCOOH + (CH_3)_3C{\cdot}O{\cdot}CH_3$$

The initial, slow, reaction probably involves acyl–oxygen fission, since *t*-butyl mesitoate reacts similarly, but much more slowly (about 12.5% of ether is produced after refluxing for 7 days[61]). On the other hand, the solvolysis of *t*-butyl mesitoate is complete after 30 min at 0°C in methanol containing 39.5% of HCl. Since methyl mesitoate is unchanged under these conditions it seems clear that solvolysis of the protonated ester must be involved, and that the mechanism must be $A_{AL}1$, since *t*-butyl derivatives are much less reactive than the corresponding methyl compounds in bimolecular processes. Bunton *et al.*[62] obtained more direct evidence for the involvement of a carbonium ion intermediate in the acid-catalyzed hydrolysis of an ester of a tertiary aliphatic alcohol by using optically active 1-ethyl-1,5-dimethylhexyl acetate (the optically active alcohol is available from the catalytic hydrogenation of linaloöl). In alkaline solution the ester is hydrolyzed by the $B_{AC}2$ mechanism, and the alcohol is regenerated optically pure, with complete retention of configuration. However, on hydrolysis with 0.01–0.06 *M* HCl in 70% aqueous acetone at 76.1°C, largely racemic alcohol is produced, and the sign of the residual optical activity indicates that a small amount of reaction results in inversion. The following rotations were found ($\alpha_D^{17°C}$, $l = 10$):

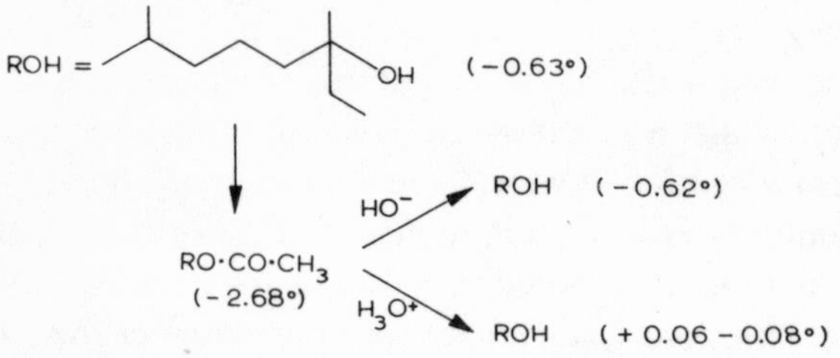

This is the behaviour expected if hydrolysis involves the $A_{AC}1$ mechanism.

(b) Kinetic behaviour

The acid-catalyzed hydrolysis of *t*-butyl acetate has been studied by Bunton and Wood[63] using ^{18}O-enriched water. Hydrolysis in 0.24 and 0.175 *M* HCl gives quantitative alkyl–oxygen fission, as measured by the incorporation of ^{18}O from the enriched solvent into *t*-BuOH. In experiments using the same acid concentrations in 70% aqueous dioxan, on the other hand, the proportion of alkyl–oxygen fission falls to about 36%[63]. The kinetic results make it clear that this is because the rate of the alkyl–oxygen cleavage reaction is very sensitive to changes in the solvent, although the reaction involving acyl–oxygen cleavage is not. Bunton and Wood give the following figures for the two reactions (Table 6).

TABLE 6

EFFECT OF SOLVENT ON THE ACID-CATALYZED HYDROLYSIS OF *t*-BUTYL ACETATE[63]

	$10^6 k_{H^+}$ ($l \cdot mole^{-1} \cdot sec^{-1}$) at 25°C		
Solvent	*Overall*	*Alkyl–oxygen cleavage*	*Acyl–oxygen cleavage*[a]
Water	128	*ca.* 124	*ca.* 4
70% aqueous dioxan	6.76	2.4	4.3

[a] Assumes 97% alkyl–oxygen cleavage in water.

A more detailed investigation has been reported by Stimson *et al.*[64]. It was anticipated that the hydrolysis of *t*-butyl acetate in 60% aqueous dioxan at 100°C would involve both $A_{AC}2$ and $A_{AL}1$ mechanisms, since the rates of hydrolysis of typical *t*-butyl esters by the latter mechanism and simple acetates by the former are very similar under these conditions. In fact the hydrolysis of *t*-butyl acetate is several times faster than that of *t*-butyl mesitoate or of ethyl acetate under these conditions, and does, indeed, involve a mixture of mechanisms. This was established in two ways.

The incorporation of ^{18}O, from solvent containing $H_2{}^{18}O$, into the alcohol produced on ester hydrolysis, is evidence for alkyl–oxygen cleavage, and the amount of incorporation, corrected for incorporation into *t*-butanol after hydrolysis, gives a measure of the amount of alkyl–oxygen fission. In water containing about 0.01 *M* HCl the percentage of alkyl–oxygen fission increases with the temperature (Table 7). From the kinetic results and these data it is

TABLE 7

PERCENTAGE OF ALKYL–OXYGEN CLEAVAGE OF *t*-BUTYL ACETATE IN AQUEOUS HCl, AS MEASURED BY ^{18}O-INCORPORATION[64]

Temperature (°C)	25	35	50	60	70
%	84.8	90.8	96.1	97.0	95.5

possible to calculate the activation parameters for both $A_{AC}1$ and $A_{AC}2$ reactions of *t*-butyl acetate[64]. These are

$$A_{AL}1: \Delta H^{\ddagger} = 26.8 \text{ kcal}\cdot\text{mole}^{-1}, \Delta S^{\ddagger} = +13.1 \text{ eu}$$
$$A_{AC}2: \Delta H^{\ddagger} = 16.6 \text{ kcal}\cdot\text{mole}^{-1}, \Delta S^{\ddagger} = -24.4 \text{ eu}$$

(The results for the $A_{AC}2$ reaction are considerably less accurate, since only a small proportion of the reaction involves this mechanism.) The higher proportion of reaction by alkyl–oxygen cleavage at higher temperatures is clearly a consequence of the much greater enthalpy of activation for the $A_{AL}1$ reaction. A similar change from mixed mechanism to near complete $A_{AL}1$ reaction occurs in mixed acetone–water solvents, as shown by the progressive increase of the enthalpy of activation with increasing temperature (Table 8). Note also that the enthalpy of activation at the highest temperatures is higher in the more polar solvent mixtures, the difference between that in water and that in 80% acetone–water being of the order of 4 kcal·mole^{-1}.

Similar results have been obtained by Yrjänä[65] for *t*-butyl acetate, and for a methoxy-substituted derivative, $CH_3OCH_2(CH_3)_2C\cdot OCOCH_3$. The effect of the methoxy group is to reduce the stability of the tertiary carbonium ion slightly, so that the $A_{AL}1$ cleavage is relatively less favourable, compared with $A_{AC}2$, than for *t*-butyl acetate. Again curved Arrhenius plots are obtained, but in this case it is $A_{AC}2$ which is the predominant mechanism, and alkyl–oxygen

TABLE 8

FAILURE OF THE ARRHENIUS EQUATION FOR THE HYDROLYSIS OF *t*-BUTYL ACETATE IN AQUEOUS ACETONE (0.5–2.0×10^{-2} *M* IN HCl)[64]

Acetone (%)	*Temperature range* (°C)	$\Delta H^{\ddagger}$ (*kcal·mole*$^{-1}$)
80	85.3–97.1	30.1
80	72.4–85.3	28.9
60	85.3–97.1	28.6
60	72.4–85.3	27.2
60	48.7–72.4	25.4
40	84.4–97.1	26.0
40	72.4–84.4	25.6
40	48.5–72.4	25.3

cleavage appears only at higher temperatures even in water[65]. Interestingly, the rate of hydrolysis of the dimethyl methoxymethylcarbinyl ester is not very sensitive to changing polarity of the solvent, the rates of its $A_{AC}2$ and $A_{AC}1$ reactions being about equal in water at 70° and in 50% aqueous dioxan at 75°C. It is possible that this atypical behaviour is a result of methoxyl-3 participation[66], *viz.*

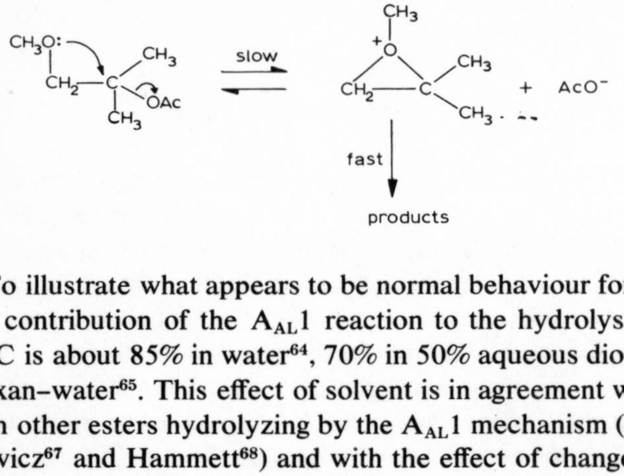

CH_3O:
CH_2—C(CH_3)(CH_3)—OAc
slow
CH_3
O^+
CH_2—C(CH_3)(CH_3) + AcO^-
fast
products

To illustrate what appears to be normal behaviour for the $A_{AL}1$ mechanism, the contribution of the $A_{AL}1$ reaction to the hydrolysis of *t*-butyl acetate at 25°C is about 85% in water[64], 70% in 50% aqueous dioxan[65], and 39% in 75% dioxan–water[65]. This effect of solvent is in agreement with observations made with other esters hydrolyzing by the $A_{AL}1$ mechanism (*e.g.* Bunton and Konasiewicz[67] and Hammett[68]) and with the effect of changes in the ionic strength (see above, p. 89). Cleavage is faster at higher ionic strengths and in more polar solvents. Although this is the behaviour expected for most S_N1 type reactions, it is not predicted for reactions involving the cleavage of positively charged species, in which the charge is dispersed in the transition state (ref. 1, p. 347), *viz.*

$$R\text{—}X^+ \longrightarrow \overset{\delta+}{R}\text{-----}\overset{\delta+}{X} \longrightarrow R^+ + X$$

In fact there is no anomaly here. Charge is indeed dispersed on going from the protonated substrate to the transition state, but in dilute acid the proportion of substrate protonated is very small. The proper comparison is thus between the transition state and the neutral ester. In other words the salt effect probably operates on the protonation equilibrium.

Activation parameters that have been measured for $A_{AL}1$ reactions are generally consistent with a unimolecular rate-determining step. The volume of activation for the hydrolysis of *t*-butyl acetate in 0.01 *M* HCl at 60°C is zero, within experimental error[70]. No significant change in rate is observed from atmospheric pressure up to 2 kbar, although this increase in pressure almost doubles the rate of hydrolysis of ethyl acetate in 0.1 *M* HCl at 35°C.

Bunnett's *w*-parameter is −1.2 for the hydrolysis of *t*-butyl acetate catalyzed by HCl[50], and his more recent parameter, ϕ, is −0.21, from the same data at 25°C[51]. In both cases the value falls in the region expected for a reaction not involving a molecule of water in the transition state.

The entropy of activation is generally positive or near zero for $A_{AL}1$ reactions. Values for several different esters hydrolyzed by the $A_{AL}1$ mechanism are summarized in Table 9. It must be remembered that the figures quoted in the first part of this table include the parameters for the protonation of the ester. Since the observed values of ΔS^{+} are strongly positive it is unlikely

TABLE 9

ACTIVATION PARAMETERS FOR THE $A_{AL}1$ HYDROLYSIS REACTIONS OF ESTERS

Ester	*Temperature range (°C)*	*Conditions*	ΔH^{+} *(kcal·mole⁻¹)*	ΔS^{+} *(eu)*	*Ref.*
t-butyl acetate	15–35	60% aq. HCl	26	+13.1	64
t-butyl benzoate	71–97	HCl, aq. acetone	28.5	+9.5	71
t-butyl mesitoate	72–97	HCl, 60% aq. acetone	29.8	+12	72
t-butyl mesitoate	72–97	HCl, 80% aq. acetone	30.3	+13	72
t-butyl mesitoate	72–97	HCl, 60% aq. ethanol	28.7	+9.9	73
t-butyl 2,4,6-triphenyl benzoate	65–80	HCl, 95% aq. methanol	28.7	+12.1	74
t-butyl trifluoroacetate[a]	25–45	70% aq. acetone	24.4	−4.25	75
t-butyl formate[b]	72–97	HCl, 60% aq. acetone	15.6	−19.3	76
t-butyl formate[b]	15–45	aq. HCl	14.2	−21.2	78
Ethyl formate[b]	15–45	aq. HCl	14.7[c]	−20.5	78

[a] Included for comparison: probably $B_{AL}1$ (see text).
[b] Included for comparison: probably $A_{AC}2$ (see text).
[c] 14.9 in 60% aqueous acetone[77].

that ΔS° for the protonation equilibrium is large and negative. In fact these values of ΔS^{+} are 10–20 eu more positive than those measured by Moffatt and Hunt[79] for the neutral hydrolysis of tertiary alkyl trifluoroacetates in 70% aqueous acetone by the $B_{AL}1$ mechanism, suggesting the possibility that the entropy change for the protonation is actually positive.

The figures for *t*-butyl formate are noteworthy. The entropy of activation is particularly favourable for reactions of formate esters involving nucleophilic attack at the carbonyl group, but ΔS^{+} for the acid-catalyzed hydrolysis of *t*-butyl formate is very much lower than that for the hydrolysis of *t*-butyl acetate. In fact, as shown by Stimson[76], and by Salomaa[78], the activation parameters for the hydrolysis of *t*-butyl formate are closely similar to those for the hydrolysis of ethyl formate. The inference is clear that the *t*-butyl ester is hydrolyzed in dilute HCl by the $A_{AC}2$ mechanism. It is probably the only tertiary aliphatic ester to be hydrolyzed by this mechanism in aqueous acid. Evidently the $A_{AL}1$ reaction is not so very much more favourable than $A_{AC}2$ for simple esters, and the exceptional reactivity of the formyl group is sufficient to make bimolecular attack at the carbonyl group the faster reaction in the case of *t*-butyl formate, just as the inhibitory effect of organic solvents on the $A_{AL}1$ process results in predominant acyl–oxygen cleavage even of *t*-butyl acetate in 70% aqueous dioxan (p. 90). Because the $A_{AL}1$ reactions have much higher enthalpies of activation, which largely compensate for their more favourable activation entropies, they are much more sensitive to changes in temperature, and Salomaa's data[78] show that the $A_{AL}1$ mechanism becomes important at higher temperatures even for *t*-butyl formate. The Arrhenius plot exhibits curvature at temperatures above 45°C, with the observed rate coefficients for hydrolysis becoming progressively larger than predicted. This can be explained by assuming that a second hydrolysis mechanism becomes increasingly important at the higher temperatures. The amount of reaction due to this additional reaction, and thus rough rate coefficients for hydrolysis due to it, can be calculated by subtracting from the observed rate coefficients those calculated from the Arrhenius plot, extrapolated to this region from measurements made at lower temperatures, where linearity is maintained. These derived rate coefficients are themselves correlated by the Arrhenius equation, and give rise to activation parameters ($\Delta H^{+} \simeq 28$ kcal·mole^{-1} and $\Delta S^{+} \simeq +17$ eu) typical of reactions known to involve the $A_{AL}1$ mechanism[78]. Stimson[76] found no evidence for the incursion of the $A_{AL}1$ mechanism in this temperature region, presumably because his solvent was 60% aqueous acetone, and alkyl–oxygen cleavage is relatively less favoured in the less polar solvent.

Two further types of esters which are clearly hydrolyzed by the $A_{AL}1$ mechanism in strongly acidic media are best treated as special cases. Salomaa[80] has shown that alkoxymethyl acetates are hydrolyzed in acid by the $A_{AL}1$ mechanism, and the corresponding formates by a mixture of $A_{AL}1$ and $A_{AC}2$

mechanisms. Since the acetal group is probably more basic than the ester group these substrates may be regarded as reacting as expected for acetals. As esters, alkyl–oxygen cleavage is not unexpected, since the methoxyl group can stabilize the carbonium ion very effectively, *viz.*

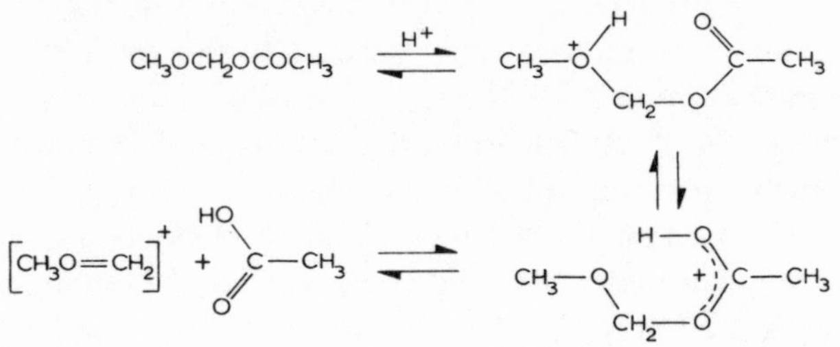

In fact this pathway cannot be a great deal more favourable than the normal ester hydrolysis ($A_{AC}2$) mechanism; since this appears in the case of methoxymethyl formate[80], is dominant in the hydrolysis of β-chloroethoxymethyl acetate[81], $ClCH_2CH_2OCH_2OCOCH_3$, and is apparently the sole important mechanism for the hydrolysis of methylene diacetate, $CH_3COO{\cdot}CH_2{\cdot}OCOCH_3$[82].

The second special case is that of β-isovalerolactone[83]. Primary and secondary β-lactones are hydrolyzed in strongly acidic media by the $A_{AC}1$ mechanism (see p. 79). The rate of hydrolysis of the tertiary lactone also follows H_0 rather than $[H_3O^+]$, and the plot of log k_{obs} *versus* $-H_0$ has the unit slope often characteristic of a unimolecular mechanism. However, although the addition of a single methyl group to β-propiolactone reduces the rate of hydrolysis slightly, as expected on the basis of the $A_{AC}1$ mechanism, β-isovalerolactone is hydrolyzed about a thousand times faster than β-butyrolactone. This is the classical type of observation indicating the appearance of a carbonium ion mechanism, and is convincingly assigned by Liang and Bartlett[83] to a change of mechanism to $A_{AL}1$ for β-isovalerolactone hydrolysis, *viz.*

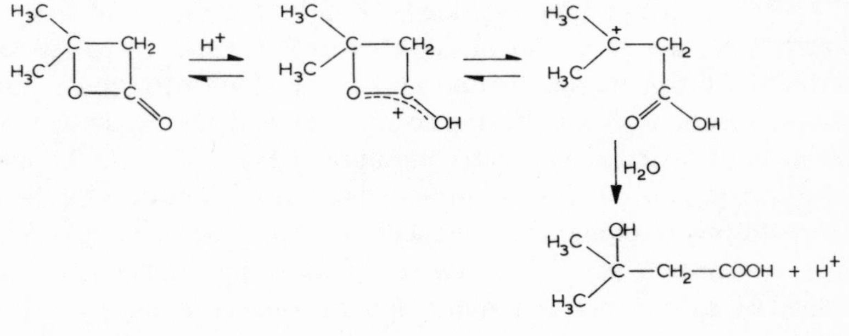

The acid-catalyzed hydrolysis of arylmethyl esters closely resembles that of the esters of aliphatic alcohols. Although the benzene ring is presumably more effective than an alkyl group in stabilizing the developing carbonium ion, as it is in simple S_N1 reactions (ref. 1, p. 325), this advantage is not apparent in practice since the aromatic compounds are generally less soluble in water. Consequently their reactions are studied in solvents containing a large proportion of organic diluent, under conditions unfavourable to the $A_{AL}1$ reaction. Under comparable conditions, *e.g.* with H_2SO_4[37], the rule of thumb that the phenyl substituent is about as effective as two alkyl groups in determining the relative importance of carbonium ion and other mechanisms (ref. 1, p. 325) can be applied to the $A_{AL}1$ reaction also. Only triphenylmethyl esters are hydrolyzed by the $A_{AL}1$ mechanism under all conditions. Benzhydryl compounds react predominantly with acyl–oxygen cleavage in dilute acid, and benzyl acetate behaves as a normal primary aliphatic ester even in moderately concentrated sulphuric acid[37]. Alkyl–oxygen cleavage is, however, readily facilitated by the introduction of mesomerically electron-donating substituents in these benzylic esters.

Triphenylmethyl acetate[67,84] and benzoate[75] are solvolyzed with alkyl–oxygen fission even under neutral conditions, and the reaction is strongly acid-catalyzed in each case. The hydrolysis of triphenylmethyl acetate in 80% dioxan–water has been studied by Bunton and Konasiewicz[67]. The reaction involves near-quantitative alkyl–oxygen fission in both acidic and initially neutral solution, as shown by the incorporation of ^{18}O from enriched H_2O into the triphenylmethyl alcohol produced, and by the lack of incorporation into the acetic acid. The rate of the acid-catalyzed reaction at 25°C is about 40% slower in 90% dioxan than in the more aqueous solvent.

Bunton *et al.*[85] and Harvey and Stimson[86] have also studied the hydrolysis of diphenylmethyl esters. These authors found that diphenylmethyl acetate is hydrolyzed in 0.06 and 0.12 *M* HCl in 70% dioxan at 99.5°C with acyl–oxygen cleavage[85], although this finding is complicated by acid-catalyzed exchange between the solvent and both products, acetic acid and diphenylmethyl alcohol. With 60% aqueous acetone, on the other hand, Harvey and Stimson found good evidence that both $A_{AC}2$ and $A_{AL}1$ mechanisms are involved[86]. Their evidence is of two kinds. Firstly, the rate of hydrolysis of diphenylmethyl acetate is similar at 100°C to that for the hydrolysis of diphenylmethyl 2,4,6-trimethylbenzoate, which is undoubtedly much less reactive in $A_{AC}2$ reactions. Secondly, the enthalpy and entropy of activation are typical of an $A_{AL}1$ reaction in the temperature region about 100°C, closely similar for example, to those for the trimethylbenzoate[86]. The data for the acid-catalyzed hydrolysis of diphenylmethyl acetate do not, however, give a linear Arrhenius plot. For 60% acetone the energy of activation, calculated from the slopes defined by pairs of points at different temperatures, gradually falls from

TABLE 10

ACTIVATION PARAMETERS FOR THE HYDROLYSIS OF DIPHENYLMETHYL ESTERS IN 60% AQUEOUS ACETONE
(From ref. 86)

Ester	*Temperature Range (°C)*	$\Delta H^{\ddagger}$ *(kcal·mole⁻¹)*	$\Delta S^{\ddagger}$ *(eu)*
2,4,6-trimethyl-benzoate[a]	97.1–109.2	29.7	11.4
Benzoate	84.5–108.6	29.8	8.2
Acetate	97.1–109.2	29.3	11.0
Acetate	85.6–97.1	26.5	1.4
Acetate	70.7–73.1	24.7	−1.4
Acetate	50.0–70.7	22.4	−8.2
Formate	85.0–97.1	17.8	−16.0

[a] In 80% aqueous acetone.

about 30 kcal·mole^{-1} to 23 kcal·mole^{-1}. Some of the data are summarized in Table 10.

These data for diphenylmethyl acetate resemble those for *t*-butyl formate, discussed above (p. 92) and are readily explained in terms of a progressive change from a predominant $A_{AC}2$ reaction at lower temperatures to predominant $A_{AL}1$ above about 90–100°C[86]. Although the solvent is different from the 70% aqueous dioxan used by Bunton *et al.*[85], the rates of hydrolysis under comparable conditions are closely similar, so it is not the case that alkyl–oxygen cleavage is intrinsically faster in 60% acetone. Since the rate of the $A_{AC}2$ reaction is not usually very sensitive to changes in the polarity of the solvent it seems likely that a significant amount of alkyl–oxygen cleavage does occur under Bunton's conditions also, and that both $A_{AL}1$ and $A_{AC}2$ mechanisms are involved in this region of temperature.

A very similar case, in which all the evidence is consistent, is the hydrolysis of 9-fluorenyl acetate (14) studied by Bunton *et al.*[87]. This ester is about three

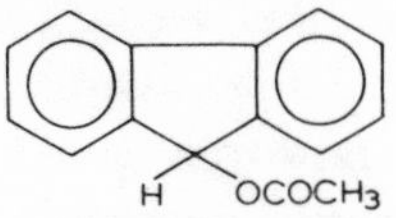

H OCOCH$_3$

(14)

times less reactive than diphenylmethyl acetate, and the results of hydrolysis in acidic dioxan–$H_2{}^{18}O$ indicate that the acyl–oxygen bond is cleaved. The kinetic data are also consistent with an $A_{AC}2$ mechanism: $\Delta H^{\ddagger} = 15.8$ kcal·mole^{-1} and $\Delta S^{\ddagger} = -28.4$ eu. These workers also established that the substitution of deuterium into the α-position of this ester, and into that of diphenylmethyl acetate, has no measurable effect on the rate of hydrolysis.

Harvey and Stimson's kinetic data[86] (Table 10) for diphenylmethyl formate leave little doubt that this ester is hydrolyzed by the $A_{AC}2$ mechanism, and this conclusion is supported by the more detailed evidence of Bunton *et al.*[85]. In this case an undetermined but significant amount of alkyl–oxygen fission occurs on hydrolysis in initially neutral 70% aqueous dioxan at 99.5°C, even though the formic acid produced catalyzes the reaction and thus increases the proportion of acyl–oxygen cleavage. But the acid-catalyzed hydrolysis proceeds with exclusive acyl–oxygen cleavage, in 70% aqueous dioxan (0.078 and 0.141 M in HCl) at 35°C. The activation parameters, $\Delta H^{\neq} = 15.5$ kcal·mole^{-1} and $\Delta S^{\neq} = -23.2$ eu, are typical for an $A_{AC}2$ reaction.

The introduction of a single *p*-methoxyl group into diphenylmethyl acetate is sufficient to change the mechanism of acid hydrolysis to $A_{AL}1$[88]. The observed levels of incorporation of ^{18}O from enriched H_2O into the alcohol and acetic acid produced are consistent with quantitative alkyl–oxygen cleavage, within the limits of experimental error, for hydrolysis in both acidic 60% aqueous dioxan and initially neutral 70% aqueous dioxan. The hydrolysis, as expected, is considerably faster than that of diphenylmethyl acetate[85]. The second-order rate coefficient for the acid-catalyzed hydrolysis of the *p*-methoxy compound being 1.44×10^{-3} l·mole^{-1}·sec^{-1} at 44.8°C, compared with 8.3×10^{-4} l·mole^{-1}·sec^{-1} at 99.5°C for the unsubstituted ester (in 70% dioxan).

Bunton and Hadwick[88] also measured the rate of hydrolysis of *p*-methoxydiphenylmethyl acetate at 25°C in aqueous 60% dioxan containing varying concentrations of perchloric acid, up to 1.3 M. Using the known values of H_0 in this system, the second-order rate coefficients were shown to follow H_0 rather than $[H_3O^+]$, the slope of the logarithmic plot against H_0 being -1.3. Other data consistent with the $A_{AL}1$ mechanism are the observed deuterium isotope effect, $k_D/k_H = 2$, in 60% aqueous dioxan, and the activation parameters, $\Delta H^{\neq} = 25$ kcal·mole^{-1} and $\Delta S^{\neq} = -0.6$ eu[88].

Bunton and Hadwick[89] have also studied the hydrolysis of diphenylmethyl trifluoroacetate. Alkyl–oxygen cleavage and attack on the carbonyl group are both expected to be facilitated by the introduction of the powerfully electron-withdrawing trifluoromethyl group. Both acyl–oxygen and alkyl–oxygen cleavage are faster than the acid-catalyzed oxygen exchange of diphenylmethanol, and tracer experiments using water enriched with $H_2{}^{18}O$ show that mixed bond fission occurs in both acidic and neutral solution. The proportion of each type of cleavage depends, as usual, on the temperature and on the solvent composition. In neutral 60% aqueous dioxan the amount of alkyl–oxygen cleavage varies from 6% at 25° to more than 37% at 72.9°C. With the same solvent containing 1 M $HClO_4$, the proportion is larger, 42% at 25° and 62% at 44.6°C[89].

The extent of the apparent acid catalysis of the hydrolysis of this ester, like that for the hydrolysis of other trifluoroacetates, is not large compared with that for other esters. Furthermore, trifluoroacetic acid, in common with trichloroacetic and oxalic acids, does not behave as a base in 100% sulphuric acid (p. 142). For this and other reasons it is doubtful whether the increase in the rate of hydrolysis of trifluoroacetate esters in acid solution is in fact due to acid catalysis. This problem is discussed in more detail for hydrolyses involving acyl–oxygen cleavage on p. 144. Only the particular properties of diphenylmethyl trifluoroacetate are discussed here.

The rates of hydrolysis of methyl and phenyl trifluoroacetates increase in dilute acid, but reach a maximum at about 1.5 *M* acid[89,90]. The similar behaviour of *o*-nitrophenyl hydrogen oxalate and its ethyl ester has been ascribed to specific salt effects by Bruice and Holmquist[91] (see p. 144). The decrease in rate at the higher acid concentrations is taken to result from the decreased activity of water at high ionic strength. This factor must affect specifically A-2 reactions, and so the different behaviour of diphenylmethyl trifluoroacetate can be explained, at least qualitatively, in terms of the unimolecular alkyl–oxygen cleavage of the neutral ester, formally the $B_{AL}1$ mechanism. Bruice and Holmquist[91] show that the specific effect of the hydrogen ion is greater than that of the lithium cation in the hydrolysis of *o*-nitrophenyl hydrogen oxalate in water, and Bunton and Hadwick's data[89] on the hydrolysis of diphenylmethyl trifluoroacetate in neutral 60% dioxan–water at 25°C show that the addition of lithium perchlorate causes a large increase in the observed first-order rate coefficient. Although it is not possible to disentangle the effects of added ions on the mixture of two mechanisms involved in these reactions, it is clear from the data that the effects of $LiClO_4$ and $HClO_4$ on the rate of alkyl–oxygen cleavage may be comparable (Table 11).

The observed activation parameters are consistent with the interpretation

TABLE 11

EFFECT OF ADDED IONS ON THE OVERALL RATES OF HYDROLYSIS OF DIPHENYLMETHYL TRIFLUOROACETATE IN 60% DIOXAN–WATER AT 25°C[89]

Salt (M)	*Alkyl–oxygen cleavage (%)*	$10^5 k_1$ *(sec^{-1})*
	6	9.14, 9.44
$LiClO_4$ (0.194)		14.4
$LiClO_4$ (0.820)		32.0
$HClO_4$ (0.197)		24.2
$HClO_4$ (0.784)		84.8
$HClO_4$ (0.983) + $LiClO_4$ (0.51)		160
$HClO_4$ (1.0)	42	*ca.* 120

in terms of the mixed $B_{AC}2$ and $B_{AL}1$ mechanisms, with the proportion of alkyl–oxygen cleavage increasing at higher temperatures and at higher ionic strength. ΔH^{+} for hydrolysis in neutral 60% dioxan–water increases from 16.6 kcal·mole^{-1} at 0–10° to 20.0 kcal·mole^{-1} at 30–40°C. For 0.936 M $HClO_4$ it is 16.2 kcal·mole^{-1}, with $\Delta S^{+} - 18.2$ eu, rising to 18.3 kcal·mole^{-1}, $\Delta S^{+} - 10.0$ eu for 1.69 M acid. These figures are consistent with a gradual change from $B_{AC}2$ to $B_{AL}1$, but the entropy of activation seems far too low, even in 1.69 M $HClO_4$, to suggest any very large proportion of reaction by the $A_{AL}1$ mechanism. ΔS^{+} lies close to + 10 eu for the hydrolysis of many different types of ester by this mechanism (Table 10) including that of diphenylmethyl acetate[86].

The true $A_{Al}1$ reactions described so far have been studied at fairly low acid concentrations, and becomes rapid in 2–3 M acid. Two types of reaction, thought to involve the $A_{AL}1$ mechanism, which require rather high concentrations of sulphuric acid, are the hydrolysis of secondary alkyl esters, such as isopropyl acetate, and that of benzyl acetate[37]. Data for these esters appear in Fig. 1 (p. 74), together with the results of Bunton *et al.*[56] for the hydrolysis of *t*-butyl acetate, also in sulphuric acid at 25°C. Yates and McClelland[37] found that their hydration number treatment, described below (p. 115) shows the same type of behaviour as is observed for aryl and primary alkyl acetates: reaction rates depend on the second power of the water activity at low to moderate acid concentration, but become much less sensitive to the activity of water in very strongly acidic solution. This behaviour is illustrated in Fig. 6. The slope of the plot of $\log k - \log \{h_0{}^m/(h_0{}^m + K^m_{SH^+})\}$ against $\log a_{H_2O}$ is expected to be a measure of the kinetic order in water (see p. 115), and is found to be in the region of 2.0 for isopropyl and benzyl acetates in weak to moderately concentrated acid[37]. In sufficiently strong acid, however, a change occurs, and a linear dependence with a smaller, negative slope is found. This slope is somewhat steeper, at about -0.6, than the slope of the similar plot in this region for esters hydrolyzed by the $A_{AC}1$ mechanism (see p. 82), and Yates and McClelland[37] suggest that this indicates that water is actually released in the rate-determining step of these $A_{AL}1$ reactions.

While this possibility cannot be ruled out, it is not sufficient to explain the behaviour of *t*-butyl acetate. The hydrolysis of this ester has been studied by many authors. Some recent results for aqueous sulphuric acid at 25°C are available from Bunton *et al.*[56], and these data are included in Fig. 6. The plot for this ester shows a very strongly negative slope, which can be shown to be in the region of -9 to -10 by a plot on expanded scales. This is an aberration from the behaviour expected by Yates and McClelland, and could arise from a breakdown of their approximations in the case of this ester, or from a factor not allowed for by them. Since all the other acetate esters used seem to behave

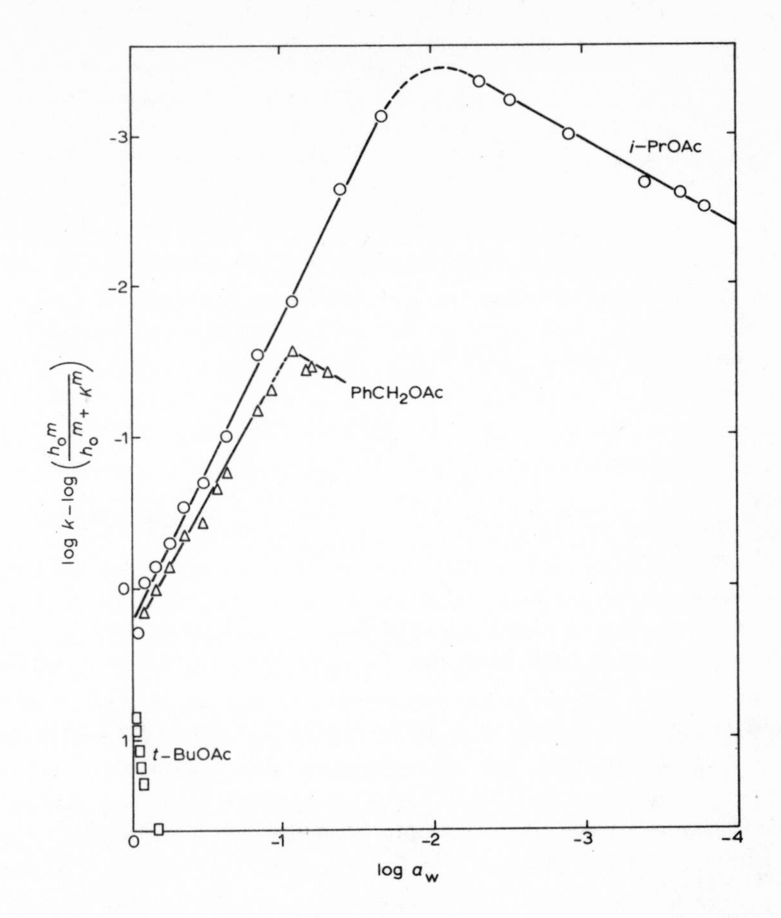

Fig. 6. Yates–McClelland plot of the rate coefficients for hydrolysis of protonated benzyl and isopropyl acetates in sulphuric acid solutions, against the activity of water. Data for *t*-butyl acetate (ref. 37) are also included.

very similarly to each other, it seems likely that the second alternative is correct. The observation is that the rate coefficient for the hydrolysis of the protonated form of *t*-butyl acetate increases very rapidly with increasing acid concentration, or with decreasing water activity, (this is shown very clearly by the results given by Bunton *et al.*[56]).

The simplest explanation is that this is an effect of increasing ionic strength. The $A_{AL}1$ mechanism is, of course, simply a S_N1 reaction involving a carboxylic acid as the leaving group, and it is well-known that increasing ionic strength strongly favours ionization reactions such as S_N1 (ref. 1, p. 366). The increase in the rate coefficient for the hydrolysis of the protonated ester is illustrated by the plot of Fig. 7, in which $\log k + mH_0$ is plotted against ionic strength (taking H_2SO_4 as a monobasic acid[56]).

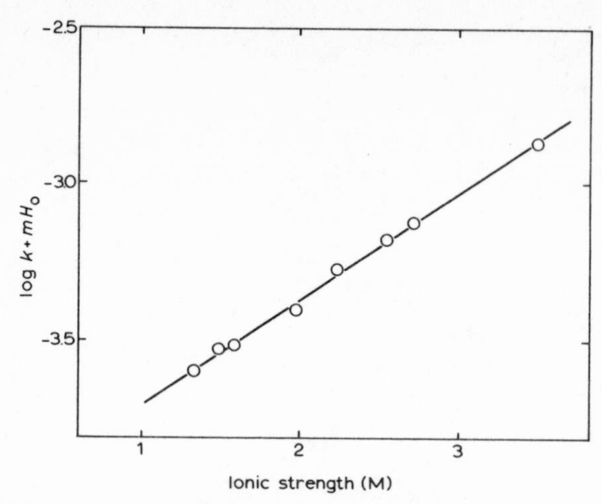

Fig. 7. The dependence on ionic strength of the rate coefficients for hydrolysis of protonated *t*-butyl acetate in aqueous sulphuric acid.

(*c*) *Enol esters*

There is some evidence that the acid-catalyzed hydrolysis of enol esters proceeds with alkyl–oxygen cleavage. In particular, Kiprianova and Rekasheva[92] found that ^{18}O from enriched water was not incorporated into the acetic and benzoic acids produced on hydrolysis of the vinyl esters. Landgrebe[93] studied several esters in 50% aqueous dioxan, and found near-zero entropies of activation. But Yrjänä[94] finds activation parameters typical for the acid-catalyzed hydrolysis of simple aliphatic esters. For vinyl acetate $\Delta H^{\neq} = 14.8$ kcal·mole^{-1} and $\Delta S^{\neq} = -26.7$ eu, and for 1-methylvinyl acetate $\Delta H^{\neq} = 17.8$ kcal·mole^{-1} and $\Delta S^{\neq} = -17.5$ eu[94]. Furthermore, Yrjänä calculates $\Delta S^{\neq}$ values of -20 to -25 eu from the data of Landgrebe also. Yrjänä's conclusion is that the normal values of the activation parameters, and of the deuterium solvent isotope effect ($k_D/k_H = 1.33$ for vinyl acetate, as it does also for ethyl formate[95]) are good evidence that the mechanism of acid-catalyzed hydrolysis is also normal, and thus $A_{AC}2$. This is in direct contradiction to the isotopic evidence, and only further work can resolve the discrepancy. However, it must be said that some other results of the isotopic study are surprising, particularly the conclusion[92] that there is no acyl–oxygen cleavage in the neutral solvolysis of vinyl acetate and benzoate, and only partial acyl–oxygen cleavage on alkaline hydrolysis.

Recently, Noyce and Pollack[306] have found kinetic evidence for a change-over from acyl to alkyl–oxygen cleavage in the hydrolysis of α-acetoxystyrenes. At low acidities (1 *M* H_2SO_4) compounds with electron-withdrawing substituents are hydrolyzed in a reaction which behaves as expected for the $A_{AC}2$ mechanism. The solvent deuterium isotope effect, $k_H/k_D = 0.75$, and the effect of substituents on the rate is small. As the acidity of the medium is increased the rate of hydrolysis increases, but the slope of the plot of log k_{hyd} *versus* H_0 decreases markedly at moderate acidities, before curving sharply upward at very high acid concentrations. For the *p*-nitro compound the deuterium isotope effect also changes, from 0.75 to 3.25 at high acid concentrations. This evidence is all consistent with a change of mechanism, from $A_{AC}2$ at low acid concentrations to alkyl–oxygen cleavage at high acidity, with rate-determining protonation of the double bond, *viz.*

$$X\text{-}C_6H_4\text{-}C(=CH_2)\text{-}O\text{-}C(=O)\text{-}CH_3 \underset{}{\overset{\text{slow, } H_3O^+}{\rightleftharpoons}} X\text{-}C_6H_4\text{-}\overset{+}{C}(CH_3)\text{-}O\text{-}C(=O)\text{-}CH_3 \xrightarrow[H_2O]{\text{fast}} ArCOCH_3 + CH_3COOH$$

This mechanism is favoured at high acid concentrations because the rate is proportional to $-H_0$, and unlike that of the $A_{AC}2$ reaction, does not fall off as the activity of water decreases. It is also favoured by electron-donating substituents X. Thus $k_H/k_D = 2.50$ for the hydrolysis of the *p*-methoxyphenyl ester at $H_0 = -0.07$, suggesting that alkyl–oxygen cleavage is dominant for this compound even at low acidity. The effect of substitution is also much greater for the reaction at high acidity, the rates being correlated by σ^+ and a ρ-value of -1.9[306]. Apart from this slightly smaller ρ-value, this behaviour is typical of reactions involving rate-determining protonation of the C–C double bond, and Noyce and Pollack's conclusions seem well-justified. These are that these enol esters behave like esters at low acidities, but like enol ethers at high acid concentrations or when they bear strongly electron-donating substituents.

2.2.3 *The $A_{AC}2$ mechanism*

The two unimolecular mechanisms for acid-catalyzed ester hydrolysis described above represent exceptional behaviour. They are generally observed only with compounds with narrowly defined structural characteristics, or under extreme conditions. The vast majority of esters are hydrolyzed under the vast majority of acidic conditions by the $A_{AC}2$ mechanism, and since ester hydrolysis has been a traditional proving ground for theories relating structure and reactivity, a wealth of data is available. Yet the detailed mechanism of the reaction is still in dispute.

This situation results from the combination of unfavourable circumstances summarized on pp. 57–58. The proper approach would seem to be to deal first with those areas which represent common ground between the various descriptions of the mechanism, before describing recent approaches towards a solution of the remaining problems.

(*a*) *Evidence for acyl–oxygen cleavage*

The first, and fundamental, piece of evidence necessary for a discussion of the detailed mechanism of any chemical change is the identification of the covalent bonds formed and broken; this may or may not be the same thing as the identification of the products of the reaction. In the case of ester hydrolysis or formation the alternatives involve the cleavage or formation of bonds from oxygen to the carbon atom of either an alkyl or an acyl group, and it is in principle, and generally also in practice, a simple matter to distinguish between these alternatives.

Since there are but two possibilities, acceptable evidence may be positive or negative. It is sufficient, for example, in order to establish acyl–oxygen fission, to show either that the acyl–oxygen bond is cleaved, or that the alkyl–oxygen bond is not, and both types of evidence have been used. Some of the evidence, for reactions involving alkyl–oxygen fission, has already been discussed in the sections dealing with the $A_{AL}1$ reaction (p. 87).

It is in the reversible reactions of ester hydrolysis and formation that ambiguity arises. For reactions of esters involving reagents other than water, and for those of carboxylic acids involving reagents other than alcohols, the nature of the products is sufficient evidence to identify the bonds formed and broken during the reaction. The alcoholysis of an ester, for example, must involve acyl–oxygen cleavage if a second ester is the product, since alkyl–oxygen fission leads to an ether and the free carboxylic acid, *viz.*

$$R'\text{—}C(OH)(\text{+})\text{—}O\text{—}R + :O(R'')H \longrightarrow R'\text{—}C(=O)OH + R\text{—}O\text{—}R''$$

$$R'\text{—}C(OH)(\text{+})\text{—}OR + :O(R'')H \longrightarrow R'\text{—}C(=O)OR'' + R\text{—}OH$$

Evidence that the alkyl–oxygen bond is not normally cleaved on acid-catalyzed ester hydrolysis is readily obtained, and has gradually accumulated in the literature. The alkyl–oxygen cleavage of an ester or of its conjugate acid involves a nucleophilic substitution at an sp^3-hybridised carbon atom, and will be subject to the usual stereochemical requirements of S_N reactions. If the process involves bimolecular attack by a water molecule inversion of the configuration at carbon will occur, and this result can readily be detected by using the ester of an optically active alcohol. On the other hand, if a unimolecular cleavage occurs, to form a carbonium ion, then racemization will be observed in the case of the ester of an optically active alcohol, while isomeric alcohols will be produced if the carbonium ion concerned is one of those which rearrange readily to more stable species.

An early demonstration of acyl–oxygen cleavage using an optically active ester was that of Holmberg[95], who showed that the malic acid produced on hydrolysis of O-acetylmalic acid retains the natural configuration. Ingold and Ingold[96] showed that the acid-catalyzed hydrolysis (and formation) of 1-methylallyl and 3-methylallyl acetates does not lead to the isomerisation which would be expected if the common cation were generated. And Quayle and Norton[97] found that neopentyl alcohol is esterified under acidic conditions without the rearrangement expected if the neopentyl carbonium ion is an intermediate.

Positive evidence that the acyl–oxygen bond is broken on ester hydrolysis and formation is available from experiments using $H_2{}^{18}O$. Hydrolysis in the enriched solvent results in a net incorporation of ^{18}O into the carboxylic acid produced if acyl–oxygen cleavage occurs, *viz.*

$$R'\text{—}C(=O)\text{—}OR \;+\; :{}^{18}OH_2 \;\underset{}{\overset{H^+}{\rightleftharpoons}}\; R'\text{—}C(=O)\text{—}{}^{18}OH \;+\; ROH$$

In the case of alkyl–oxygen fission, on the other hand, the carboxylic acid has initially normal isotopic constitution, but the alochol is now labelled.

$$R'\text{—}C(=O)\text{—}O\text{—}R \;+\; {}^{18}:OH_2 \;\overset{H^+}{\rightleftharpoons}\; R'\text{—}C(=O)\text{—}OH \;+\; R{}^{18}OH$$

The interpretation of the results of tracer experiments of this sort is sometimes complicated by ^{18}O-exchange reactions. ^{18}O from the solvent may be incorporated into the unreacted ester as hydrolysis proceeds (see below, p. 105), or into either or both of the products. The exchange reaction is significant with alcohols, such as triphenylmethyl alcohol, which give rise to relatively stable carbonium ions under acidic conditions (see, for example, refs. 67, 85), *viz.*

$$Ph_3C\cdot OH \;\overset{H^+}{\rightleftharpoons}\; Ph_3C\cdot\overset{+}{O}H_2$$

$$Ph_3C\cdot\overset{+}{O}H_2 \;\rightleftharpoons\; Ph_3C^+ + H_2O + H_2{}^{18}O$$

$$Ph_3C^+ + H_2O + H_2{}^{18}O \;\rightleftharpoons\; Ph_3C\cdot{}^{18}\overset{+}{O}H_2$$

$$Ph\,C\cdot{}^{18}OH \;\rightleftharpoons\; Ph_3C\cdot{}^{18}\overset{+}{O}H_2$$

It is also an important reaction of carboxylic acids under acidic conditions, (p. 128).

A considerable number of investigations using ^{18}O as a tracer confirm that, except under the special circumstances which favour alkyl–oxygen cleavage described above (pp. 86 and 100), acyl–oxygen fission is the normal route for acid-catalyzed ester hydrolysis and formation. Only a few leading references will be listed here. Some of the evidence is described in more detail when the kinetics of $A_{AC}2$ reactions are discussed below (pp. 106–117).

Ingold *et al.*[98] showed that no ^{18}O is incorporated into the methanol produced when methyl hydrogen succinate is hydrolyzed in $H_2{}^{18}O$, and Roberts and Urey[99] showed that isotopically normal water is produced in the HCl-catalyzed esterification of benzoic acid with ^{18}O-labelled methanol. Bunton *et al.* have established that acyl–oxygen cleavage occurs in the hydrolysis in acid of a number of esters, including diphenylmethyl formate[85], 9-fluorenyl acetate[87] and methyl[90] and phenyl[89] trifluoroacetates. Also Long and Friedman[100] found that the mass spectrum of the γ-hydroxybutyric acid produced from the acid hydrolysis of γ-butyrolactone is consistent with acyl–oxygen fission of that ester. Finally the well-established ^{18}O-exchange reaction of carboxylic acids under acidic conditions is further evidence supporting acyl–oxygen cleavage as the usual mechanism for the closely related reactions of ester hydrolysis and formation under these conditions.

(*b*) *Evidence for the two-step nature of the $A_{AC}2$ reaction*

Various kinds of evidence have been adduced to establish that bimolecular displacements at the carbonyl group involve tetrahedral addition intermediates[101], and in some cases such species are kinetically significant. It is clear, however, that the lifetime of any such intermediate in solution must be exceedingly brief[101]. The tetrahedral intermediate formed by the addition of a molecule of water to an ester, or of a molecule of an alcohol to a carboxylic acid, has the structure of a partially esterified *ortho*-acid, $R'C(OH)_2OR$. Its breakdown is expected to be highly sensitive to acid catalysis[102], much more sensitive, in particular, than esters normally are. For example, the second-order rate coefficient for the acid-catalyzed hydrolysis of ethyl orthobenzoate is 340 $l \cdot mole^{-1} \cdot sec^{-1}$ at 20°C[103], compared with a value of 3.9×10^{-4} $l \cdot mole^{-1} \cdot sec^{-1}$ at 99°C for the hydrolysis of ethyl benzoate[104], although the basicities of the two compounds are similar[103]. It is perhaps not surprising, therefore, that there has been no kinetic demonstration of the existence of a tetrahedral addition intermediate in acid-catalyzed ester hydrolysis or formation; the convincing evidence that is available comes from studies using ^{18}O as a tracer.

The first, and crucial, observations were made by Bender[104]. The experiments were designed to decide between the two alternative mechanisms for bimolecular ester hydrolysis which are both consistent with all the other information available. These alternatives are the direct displacement (i), which may be written for acid-catalyzed hydrolysis as follows

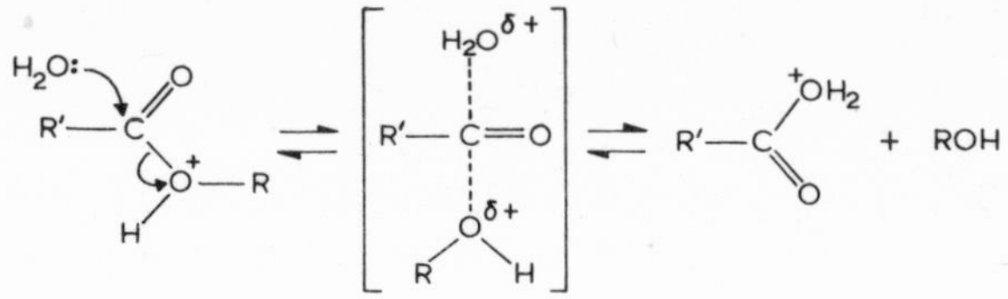

(i)

and the addition-elimination mechanism, (ii) in which an intermediate is formed. This is best written as

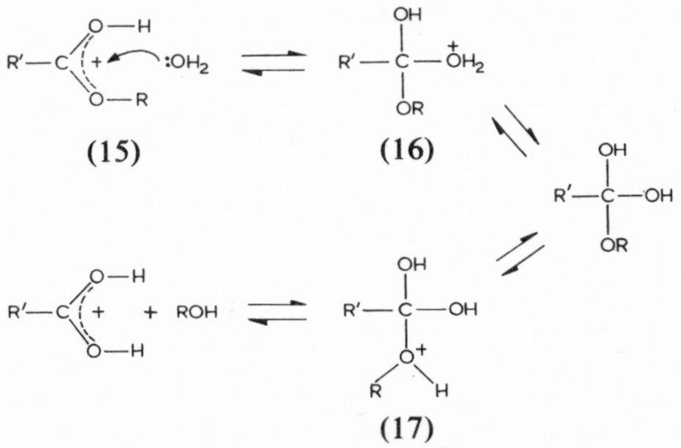

(15) (16) (17) (ii)

Consider first mechanism (ii). Whatever the detailed mechanism by which (16) is converted into (17), the proton-transfers involved are expected to be rapid compared with the rate at which (16) reverts to ester or (17) breaks down to products, so that the oxygen atom of the carbonyl group of (15) becomes chemically indistinguishable from that derived from the solvent as soon as addition occurs. Thus the carbonyl oxygen atom of any ester (15) regenerated from (16) by the reverse of the initial addition step has an equal chance of being the original carbonyl oxygen atom, or one derived from solvent. On the other hand, the carbonyl oxygen atom of the ester plays no part in mechanism (i), and will not exchange with the oxygen atoms of the solvent providing re-esterification is not important.

Bender[104] found that when ethyl benzoate, labelled with excess ^{18}O in the carbonyl group, is hydrolyzed at 99°C in isotopically normal aqueous 1 M acid, oxygen exchange between the unreacted ester and the solvent takes place, and the enrichment of the remaining ester decreases steadily as hydrolysis proceeds. This is precisely the result expected if hydrolysis involves a full intermediate and the addition elimination mechanism receives further support from the observation that hydrolysis and exchange proceed at similar rates, with a constant ratio, k_{hyd}/k_{ex} of 5.2 for ethylbenzoate. The reaction can thus be written as

References pp. 202–207

$$Ph{-}C({=}^{18}O){-}OEt \overset{H^+}{\rightleftharpoons} Ph{-}C^{+}({-}^{18}OH)({-}OEt) \overset{H_2O}{\rightleftharpoons} Ph{-}C({-}^{18}OH)({-}\overset{+}{O}H_2)({-}OEt) \rightleftharpoons Ph{-}C(OH)({-}OH)({-}OEt) + H^+$$

$$Ph{-}C({=}O){-}OEt + H^+ \rightleftharpoons Ph{-}C^{+}({-}OH)({-}OEt) + H_2{}^{18}O \rightleftharpoons Ph{-}C({-}^{18}OH_2^+)({-}OH)({-}OEt) \rightleftharpoons Ph{-}C(OH)({-}OH)({-}OEt) + H^+$$

$$Ph{-}C(OH)({-}OH)({-}OEt) \rightleftharpoons Ph{-}C({-}^{18}OH)({-}OH)({-}\overset{+}{O}(Et)H) \rightleftharpoons Ph{-}C^{+}({-}^{18}OH)({-}OH) + EtOH$$

$$Ph{-}C({=}^{18}O){-}OH + H^+ \rightleftharpoons Ph{-}C^{+}({-}^{18}OH)({-}OH)$$

The observation of ^{18}O-exchange between the carbonyl oxygen atom of an ester and solvent is evidence for the formation of a tetrahedral addition compound, but not proof that this actually lies on the reaction pathway. But the behaviour of the exchange and hydrolysis reactions is so similar that there can be little doubt that this is, in fact, the case. The evidence is most complete for alkaline hydrolysis (see p. 163). One further piece of evidence obtained under acidic conditions is the observation[47] that there is no exchange between the solvent and the carbonyl-^{18}O-labelled ester when methyl 2,4,6-trimethylbenzoate is hydrolyzed in 3.09, 5.78 and 11.5 M sulphuric acid. Other evidence makes it clear that this ester is hydrolyzed by the $A_{AC}1$ mechanism, and that no reversible addition of water is expected.

Finally, an important study by Lane[38] shows that the ^{18}O-exchange and hydrolysis reactions of ethyl acetate in aqueous sulphuric acid at 25°C depend in a closely similar way on the activity of water. Lane's plot of $\log k_{obs}/[BH^+]$ against $\log a_{H_2O}$ has a slope of 2.07 for hydrolysis, while the similar plot of the data for ^{18}O-exchange has a slope of 1.84. Lane interprets this as evidence that two molecules of water are involved in the transition state for each reaction. Since the two reactions are similar in detail it seems more than ever likely that they involve a common rate-determining step.

(c) *The kinetic order of the $A_{AC}2$ reaction*

In an acid-catalyzed solvolysis reaction three species are of primary kinetic importance, *viz.* the hydronium ion, the substrate and the solvent. The order with respect to the first two was established for the hydrolysis of simple esters in the nineteenth century. Ostwald[105] showed that the acid-catalyzed hydrolysis of methyl acetate is first-order with respect to ester concentration, and that the rate is proportional to the electrical conductances of the solutions of many different acids, but almost independent of the nature of the acid concerned. In other words, the reaction is first-order in hydronium ion concentration also. A similar demonstration by Goldschmidt[106] that the acid-catalyzed esterification of benzoic acid in ethanol is first-order in both ester and hydronium ion followed in 1895. Both results have been confirmed many times since; but they do not, of course, define the composition of the transition state for hydrolysis, because simple determinations of kinetic order cannot measure the dependence of the rate on the concentration of the solvent.

The record of the investigations designed to establish the kinetic dependence on water of the acid-catalyzed hydrolysis of esters is a history of our growing understanding of the physical chemistry of solutions. In an early attempt, Friedman and Elmore[107] followed the hydrolysis of methyl acetate at 25°C in acetone containing only a small concentration of water. They used constant concentrations of water and acid (5 ml of 9.42 M H_2SO_4 in 100 ml of acetone, giving a water concentration of about 1.6 M) and from about 2 to 7.5 M ester. They concluded that their results were consistent with a second-order rather than a first-order reaction, and consequently that the hydrolysis of methyl acetate is a reaction of the first order in water concentration. Since the activity of water in their solutions must have been low, and also sensitive to the changes in concentration as the reaction proceeded, this work cannot be regarded as being necessarily an accurate determination of the kinetic order.

A rather similar result has been obtained much more recently by Sadek *et al.*[108]. These authors carried out a detailed investigation of the hydrolysis of benzyl acetate in aqueous dioxan, varying the solvent composition, and also the temperature (from 25 to 45°C). In the low-acidity region used (0.05 M HCl) the ester is hydrolyzed by the $A_{AC}2$ mechanism, and the solvent effects are probably typical. The rate of hydrolysis decreases steadily as the proportion of dioxan in the medium is increased, as a result of a steady increase in $\Delta H^{\ddagger}$, which is only partially offset by an increasingly favourable entropy of activation. However, the increase in $\Delta S^{\ddagger}$ eventually levels out at about 70% w/w dioxan, as illustrated by the data for 30°C:

% dioxan (w/w)	15.45	30.60	40.71	50.40	60.48	70.21	80.33
$\Delta H^{\ddagger}$ *(kcal·mole⁻¹)*	14.15	15.52	15.88	16.30	16.68	16.80	16.80
$\Delta S^{\ddagger}$ *(eu)*	−37.71	−35.34	−33.93	−33.02	−31.57	−31.47	−30.96

Very similar results are obtained with ethanol–water mixtures[109]. Presumably the ester can only be protonated if some minimum number of water molecules is available to solvate it, and a point is reached where the minimum water content of the immediate solvation shell of the protonated ester is greater than that of the medium as a whole. Beyond this point, which apparently corresponds quite closely to an average solvent composition of two water molecules to one of dioxan, this minimum water content must be acquired at the expense of the solvent at large, and thus of a slightly less favourable entropy term; however, the enthalpy of activation should remain approximately constant as long as there is sufficient water in the solvent for this selective solvation[110] to be possible.

When the logarithm of the rate coefficient for hydrolysis is plotted against the logarithm of the water concentration, straight lines are obtained at all five different temperatures used. These have unit slope, up to about 70–80% dioxan, after which the rate becomes almost independent of the water content of the solvent. It is unlikely that this represents a changeover to an A-1 mechanism, since $A_{AC}1$ reactions require a very strongly acidic medium, and $A_{AL}1$ reactions are retarded by decreasing polarity of the medium. So hydrolysis by the $A_{AC}2$ mechanism must be presumed to be occurring throughout the range of solvent composition studied, and it seems likely, in view of the discussion above, that it is an over-simplification to claim that the unit slopes observed here indicate that the hydrolysis reaction is first-order with respect to solvent.

A later technique that was much used at one time in the investigation of hydrolysis mechanisms involves the application of the Zucker–Hammett hypothesis[68,111]. It is found that many acid-catalyzed reactions fall into one of two classes, when followed with acid concentration above 1 *M*. In some cases the logarithms of the observed first-order rate coefficients are proportional to Hammett's acidity function, $-H_0$, which is an independently obtained measure of the ability of the medium to protonate a neutral substrate, while in other cases it is proportional simply to the logarithm of the hydrogen ion concentration. Hammett found that the reactions for which $\log k_{obs}$ is proportional to $-H_0$ are generally those thought for other reasons to involve a rate-determining unimolecular reaction of the protonated substrate: an example is the $A_{AC}1$ hydrolysis of methyl mesitoate discussed above (p. 77). On the other hand, those reactions which followed $\log [H_3O^+]$ appeared to be those in which a molecule of solvent is involved in the transition state, as well as the protonated substrate, *e.g.* the acid-catalyzed hydrolysis of simple carboxylic esters[112]. The generalization arising from these observations is the Zucker–Hammett hypothesis, which says that it should be possible to distinguish between A-1 and A-2 mechanisms for hydrolysis according to whether $\log k_{obs}$ increases linearly with $-H_0$ or with $\log [H_3O^+]$. Much evidence consistent with the hypothesis is summarized by Long and Paul[112].

Apart from the exceptions noted in the discussion of the $A_{AL}1$ and $A_{AC}1$ mechanisms, the rates of hydrolysis of simple esters in solutions of strong acids usually follow $[H_3O^+]$ rather than h_0. Examples are the hydrolysis of methyl acetate in HCl up to 3.65 *M*[113], of methyl benzoate and the monoglyceryl esters of benzoic and *p*-methoxybenzoic acids in sulphuric and perchloric acid solutions[48], and of ethyl formate and ethyl acetate in HCl[44]. Although the fit is not exact in all cases, the data are in each instance more closely correlated by $[H_3O^+]$ than by h_0. An example of this approach, which will serve to illustrate kinetic behaviour typical of the $A_{AC}2$ reaction, is the careful study, by Long *et al.*[114,115] of the hydrolysis of γ-butyrolactone. Since the reaction is reversible in aqueous solution, this study also gives information about ester formation, although the conditions are not of course typical of those used in esterification.

Long and Friedman[100] established, by the use of ^{18}O-enriched water, that the hydrolysis involves acyl–oxygen fission. There is no significant uncatalyzed water reaction, of the type observed with esters activated in the acyl group, and the lactone appears to behave as a normal aliphatic ester[115]. At low electrolyte concentrations the hydrolysis is accurately first-order in both lactone and acid concentration, and salt effects are very small below an electrolyte concentration of 0.1 *M*. Data on the hydrolysis reaction can be used to calculate the rate of lactone formation also, since the equilibrium constants for the reversible reaction at different ionic strengths have been measured[114], *viz.*

$$\text{(}\gamma\text{-butyrolactone: } CH_2{-}C(=O){-}O{-}CH_2{-}CH_2\text{ ring)} + H_2O \overset{K}{\rightleftharpoons} HO(CH_2)_3COOH$$

The effects of added salts are shown in Fig. 8. Sodium chloride has a small positive effect on the hydrolysis rate, and sodium chloride and sodium perchlorate have a similar, rather larger, effect on the rate of lactone formation. This is the expected result, for many salts increase the protonating power of the medium as measured by Hammett's acidity function[116], and thus assist acid-catalyzed reactions. Sodium perchlorate, unusually, has a small negative effect on the hydrolysis rate. Qualitatively similar results have been found by Bunton *et al.*[56], who studied the effects of added salts on the acid-catalyzed hydrolysis of ethyl acetate. Added lithium and sodium chloride assist the $A_{AC}2$ hydrolysis of ethyl acetate, but the perchlorates have essentially no effect. In each case the effect is a little more positive than for γ-butyrolactone hydrolysis, and, in particular, chloride anions appear to assist $A_{AC}2$ hydrolysis more effectively than do the perchlorates.

Measurements with more concentrated solutions of hydrochloric and per-

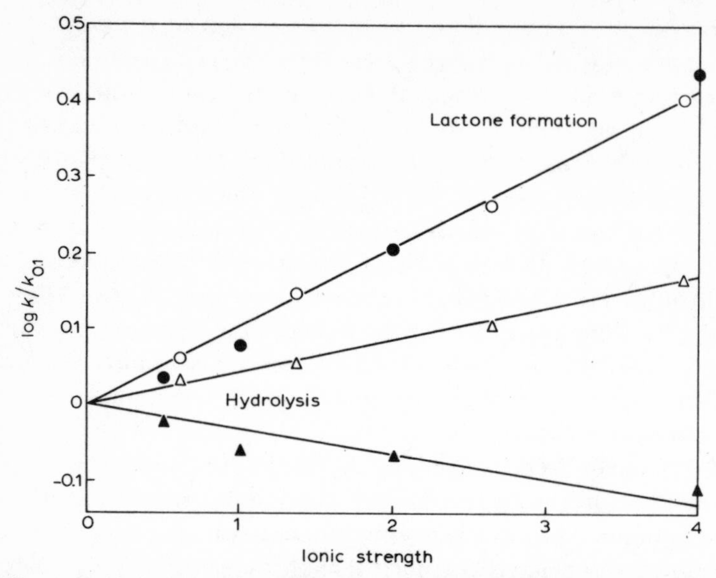

Fig. 8. Effects of added salts on the rate coefficients for the hydrolysis and formation of γ-butyrolactone. Open symbols represent points for NaCl; closed symbols for $NaClO_4$. Data from ref. 114.

chloric acids show that both forward and reverse reactions fall rather accurately into one or other of the categories of the Zucker–Hammett hypothesis. The hydrolysis of the lactone in perchloric acid at 5°C, and in HCl at 0°C, was followed. The plots of the logarithms of the observed first-order rate coefficients against log $[H_3O^+]$ are good straight lines, with slopes of 1.0 and 1.1, respectively. For lactone formation the corresponding plots are strongly curved, but the data are now linear with respect to $-H_0$, with slopes of 0.9 in both cases[112,115]. This is precisely the result expected from the Zucker–Hammett hypothesis, since a molecule of water must be involved in the hydrolysis reaction, but none is formally required for lactone formation.

Although the Zucker–Hammett hypothesis works rather well in most cases, and usually gives the correct answer in ester hydrolysis reactions, for example, it does lead on occasion to incorrect conclusions, and since it is now quite clear that one of its main underlying assumptions, that the protonation behaviour of all neutral substrates is quantitatively similar, is incorrect, it has been superseded by more accurate treatments. Unfortunately, none of these is as simple to operate as the Zucker–Hammett hypothesis, nor has any yet attained widespread use. But these improved treatments have resulted in a distinct advance in our understanding of ester hydrolysis in strong acid, and one recent approach will be discussed here in some detail.

In retrospect it is clear that the Zucker–Hammett hypothesis was regarded too uncritically too soon, and sometimes treated almost as a law which reactions were expected to obey. Criteria of mechanism are almost invariably based on empirical correlations, without firm theoretical foundations, and no mechanism can safely be excluded on the basis of only one or two diagnostic tests of this sort. In practice relatively few reactions fit exactly into one of the Zucker–Hammett categories, with $\log k_{obs}$ following $-H_0$ or $\log [H_3O^+]$ according to a good straight line of slope 1.0. In fact Bunnett[50,117,118] has based a more sensitive criterion of mechanism on just these divergences from what might be termed ideal Zucker–Hammett behaviour. He found that plots of $\log k_{obs} + H_0$ (or alternatively of $\log k_{obs} - \log [H_3O^+]$), *versus* the logarithm of the activity of water, are generally linear; and their slopes define a parameter w (or w^*) which has a numerical value in the region -2 to $+7$. These w-values are used by Bunnett as a new criterion of mechanism for acid-catalyzed reactions[50].

This treatment admits the activity of water as a major influence on reactions in acidic media, and relates divergences from ideal Zucker–Hammett behaviour to changes in the activity of water. This represents an advance on the conventional treatments, and together with a more recent modification by Bunnett and Olson[51,119], in which a new parameter (ϕ) is defined, provides an improved general procedure for the treatment of data for reactions in moderately concentrated acids.

However, recent work seems to indicate that no treatment is likely to be of general applicability if it is based on a single acidity function. Not only does H_0 not describe the protonation behaviour of all neutral substrates, but it cannot even be extrapolated from the substituted primary anilines, on which it is based, to such closely related compounds as the corresponding tertiary amines[120]. In fact, nearly every type of neutral substrate investigated seems to generate its own characteristic acidity function.

It would appear, therefore, that the minimum requirements for a valid treatment of the data for the hydrolysis of any given substrate are both an accurate knowledge of its protonation behaviour in the medium used, and due allowance for the variation in the activity of water. A few very recent investigations fulfil these requirements.

Lane[38] measured the rate of hydrolysis of ethyl acetate at 25°C in 11–79% sulphuric acid by both spectrophotometric and dilatometric techniques. He also measured spectrophotometrically the concentration of the conjugate acid of the ester in solution by measuring the absorptivity at 190 nm, and extrapolating to zero time, and by the change in the chemical shift of the acetyl protons. He found a linear relation between the ionization ratio and Hammett's acidity function

$$\log \left(\frac{[BH^+]}{[B]} \right) = 0.645\,(-6.93 - H_0)$$

References pp. 202–207

The half-protonation point was reached in 77% sulphuric acid.

If the rate-determining step of the reaction involves only the protonated ester and a water molecule or molecules, the dependence of the rate of hydrolysis of the protonated ester on the activity of water may be obtained by plotting $\log k_{obs}/[BH^+]$ *versus* $\log a_{H_2O}$. Lane's plot is reproduced as Fig. 9. The slope should give the order in water, and is close to 2 for hydrolysis and for ^{18}O-exchange also.

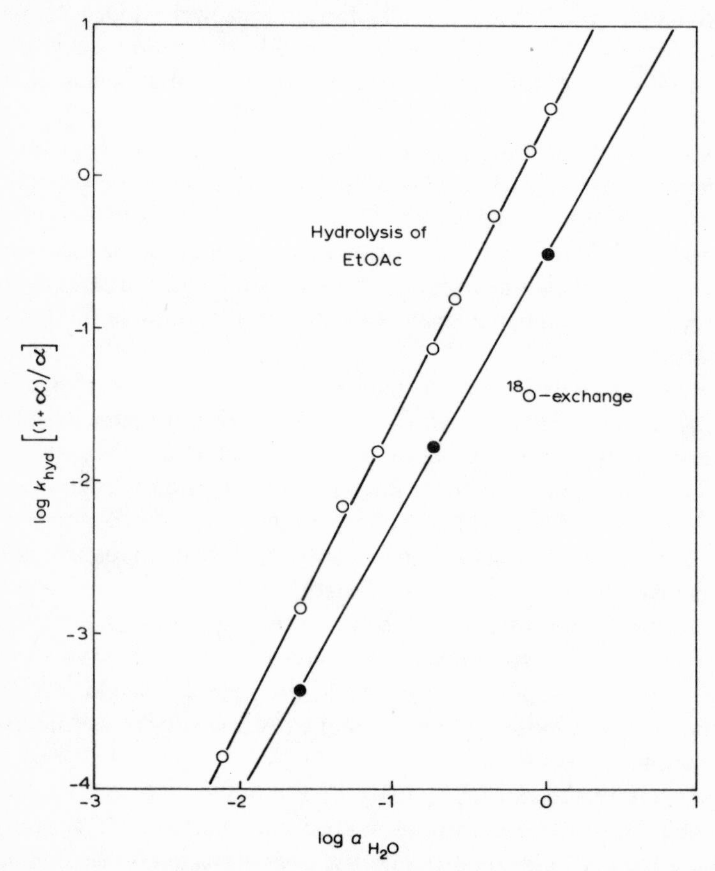

Fig. 9. Rate coefficients for reactions of protonated ethyl acetate. The open circles represent data for hydrolysis. Closed circles are for the oxygen-exchange reaction of the carbonyl-^{18}O labelled ester. Data from ref. 38.

Jaques[45] has confirmed this result for the hydrolysis reaction, and has extended the measurements to 100% sulphuric acid, as described previously. Above about 80% H_2SO_4 there is a sudden change of slope from close to 2.0 to about −0.16 (cf. Fig. 4, p. 83) corresponding to a presumed change of mechanism to $A_{AC}1$ at very low water activity.

This technique and these results have been extended in a very complete investigation by Yates and McClelland[37]. These authors have measured the rates of hydrolysis of eight representative acetate esters in 10–100% sulphuric acid: some of their results are plotted in Fig. 1, (p. 74.) Yates and McClelland also measured the protonation behaviour of three of the esters in the same medium, by the spectrophotometric method (p. 70). They found that the basicities of the three esters, methyl, *n*-propyl and isopropyl acetate, are closely similar, and that their protonation behaviour is described by the equation[37]

$$\log ([BH^+]/[B]) = m(-H_0 - pK)$$

For example, the data for *n*-propyl acetate fit the equation

$$\log ([BH^+]/[B]) = 0.62(-H_0 - 7.18)$$

the slope and pK are very similar for methyl acetate, at 0.63 and −7.25, respectively, and the spectrophotometric results indicate almost identical protonation behaviour for isopropyl acetate up to 90% H_2SO_4, at which concentration its hydrolysis becomes too rapid to obtain reliable zero-time absorptivities. These results, and the parallel behaviour of ethyl acetate, noted by Lane[38], suggest that the acid–base behaviour of at least aliphatic acetate esters depends very little on the nature of the alkyl group. In fact Yates and McClelland make the assumption that the protonation behaviour of all their acetate esters, including those that are too reactive to investigate directly, is described by the equation

$$\log ([BH^+]/[B]) = 0.62(-H_0 - 7.2)$$

In effect this equation defines an acidity function, $H_s = mH_0$, for the ionization of protonated acetate esters. It is an experimental observation that H_s is the same linear function of H_0 for several of the esters concerned, so that the approximations involved are greatly reduced from those implicit in the use of H_0 itself, while the treatment remains more general than the experimentally unattainable ideal of measuring the acidity function directly for every substrate. In practice the excellent results obtained with aliphatic esters are in contrast to the slightly less clear-cut picture for aryl esters (see below) and suggest that H_s describes the protonation behaviour of substituted phenyl acetates less than perfectly. However, there is little doubt that H_s will be a linear function of H_0 for any given ester, since this appears to be the case for the acidity functions defined for a wide variety of neutral substrates. Yates and McClelland[37] show that all available acidity functions measured in sulphuric acid, namely H_R, H_A[121], H_0[120] and H_I[122], are linear in H_0 over a very wide range of acidity.

If it is assumed that ester hydrolysis by the $A_{AC}2$ mechanism involves fast pre-equilibrium protonation of the substrate, followed by rate-determining attack of water on the conjugate acid of the ester, the mechanism can be written as

$$\mathrm{S(H_2O)_s + H^+(H_2O)_n \overset{fast}{\rightleftharpoons} SH^+(H_2O)_p + (s+n-p)H_2O}$$

$$\mathrm{SH^+(H_2O)_p + rH_2O \overset{slow}{\rightleftharpoons} SH^{\ddagger}(H_2O)_t}$$

$$\mathrm{S^{\ddagger}(H_2O)_t \longrightarrow products}$$

Thus, if the concentration of the protonated substrate is known accurately, the only kinetically significant hydration change is the addition of r molecules of water to form the transition state.

The rate expression is given by Bronsted's equation[123]

$$v = k_{obs}[\mathrm{S_T}] = ka_{\mathrm{SH^+}} a^r_{\mathrm{H_2O}} / f_{\ddagger}$$
$$= k[\mathrm{SH^+}] a^r_{\mathrm{H_2O}} f_{\mathrm{SH^+}} / f_{\ddagger}$$

where S_T is the stoichiometric ester concentration. The concentration of the protonated species is given by the relation

$$[\mathrm{SH^+}] = \frac{h_S}{(h_S + K_{\mathrm{SH^+}})} [\mathrm{S_T}]$$

so that

$$v = k_{obs}[\mathrm{S_T}] = k[\mathrm{S_T}] \cdot \frac{h_S}{(h_S + K_{\mathrm{SH^+}})} \cdot a^r_{\mathrm{H_2O}} f_{\mathrm{SH^+}} / f$$

and thus

$$k_{obs} = \frac{kh_S}{(h_S + K_{\mathrm{SH^+}})} \cdot a^r_{\mathrm{H_2O}} \cdot \frac{f_{\mathrm{SH^+}}}{f_{\ddagger}}$$

The hydrated protonated ester, $SH^+(H_2O)_p$, and the transition state, have the same charge, and are structurally closely similar, and it is assumed that the ratio of their activity coefficients is medium-independent[124]. Thus, by taking logarithms and rearranging we have

$$\log k_{obs} - \log \left\{ \frac{h_S}{(h_S + K\quad)} \right\} = r \log a_{\mathrm{H_2O}} + \text{constant} \qquad \text{(iii)}$$

The expression on the left hand side can be simplified at low acid concentrations, where the concentration of the protonation form is relatively small ($K_{SH^+} \gg h_S$) to

$$\log k_{obs} - \log (h_S/K_{SH^+})$$

so that under these conditions

$$\log k_{obs} + H_S = r \log a_{H_2O} + \text{constant} \qquad \text{(iv)}$$

The full expression (iii) must be used when a significant concentration of the conjugate acid of the substrate is present, but for sufficiently dilute acid either expression applies and they differ only by the dissociation constant, pK_{SH^+}.

(iii) and (iv) are the equations used by Yates and McClelland[37]. (iv) is seen to be of the same form as the equation derived by Bunnett, which defines his w-parameter[58,117], *viz.*

$$\log k_{obs} + H_0 = w \log a_w + \text{constant}$$

When $\log k_{obs} + mH_0$, or, more generally, $\log k_{obs} + h_0{}^m/(h_0{}^m + K^m_{SH^+})$, for the hydrolysis of various acetate esters is plotted against $\log a_{H_2O}$, the results shown in Figs. 4, 6, 10 and 11 are obtained. Figure 10 shows the plots for

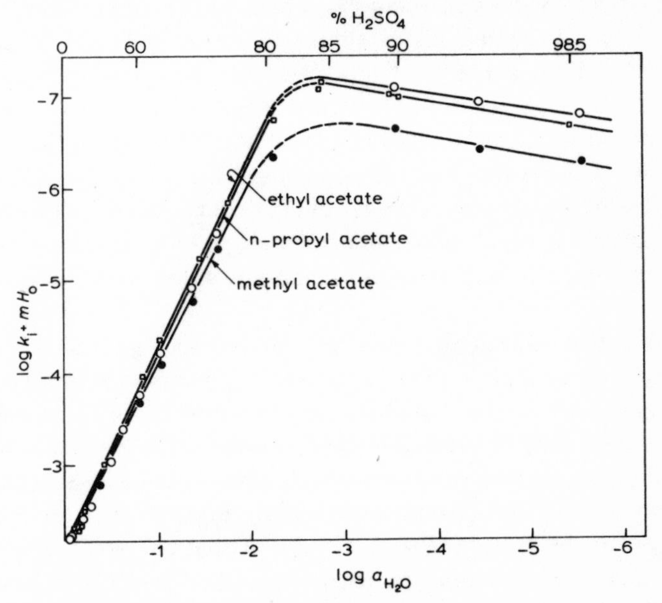

Fig. 10. Yates–McClelland plot for the hydrolysis of three protonated primary alkyl acetates in sulphuric acid at 25°C. Reproduced with permission from ref. 37.

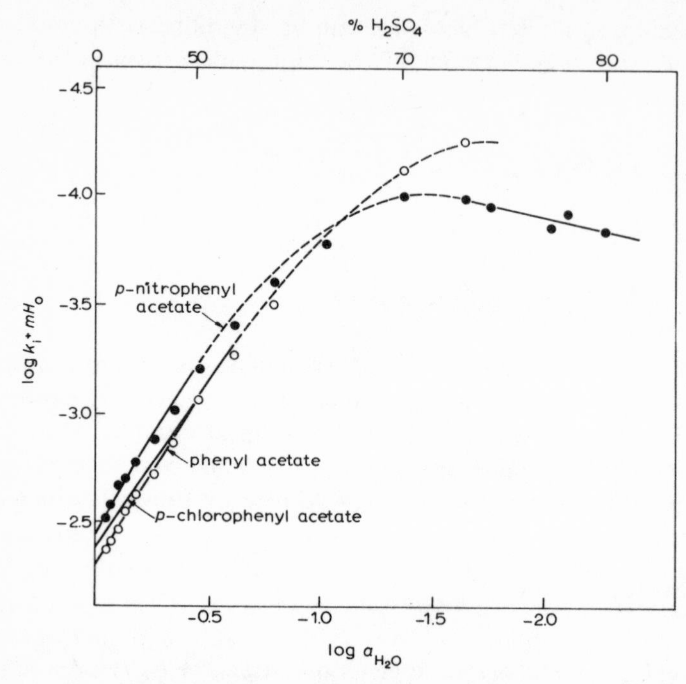

Fig. 11. Yates–McClelland plot for the hydrolysis of three aryl acetates in sulphuric acid at 25°C. Reproduced with permission from ref. 37.

methyl, ethyl and *n*-propyl acetates. These three esters give very similar results: excellent straight lines of almost identical slope ($r \simeq 2$) in the region 0–80% H_2SO_4, then a break at 80–85% acid, followed by a second linear region, where the slope has a much smaller, negative value ($r \simeq -0.2$).

The results of the same treatment of the data for a typical secondary alkyl ester, and for benzyl acetate, have been shown previously (Fig. 6, p. 99). Here, too, a good straight line ($r \simeq 2$) at lower acid concentrations is succeeded for each ester by a break at an intermediate acidity region, and a different linear dependence, with a smaller, negative, slope ($r \simeq -0.6$) in the high acidity region.

The results for three aryl esters are plotted in Fig. 11. Again an initial linear region with a slope, r, in the region of 2, gives way in more concentrated acid to a region where the slope appears to settle down to a small, negative value. The data are limited for concentrated acid because sulphonation becomes an increasingly important side-reaction of the less deactivated esters, but the broadness of the change-over region, together with the consequently reduced span of the truly linear plot, suggest that the protonation behaviour of the aryl esters is less than perfectly described by H_s. Nevertheless, the treatment gives consistent results for the whole series of esters studied. These results are summarized in Table 12.

TABLE 12

r-VALUES FOR ESTER HYDROLYSIS IN SULPHURIC ACID[37]

Acetate	*r Value*	*Concentration range (% H_2SO_4)*	*r Value*	*Concentration range (% H_2SO_4)*
Methyl	1.92	0–80	−0.2	> 80
Ethyl	2.10	0–80	−0.18	> 85
n-Propyl	2.06	0–80	−0.2	> 85
iso-Propyl	2.11	0–75	−0.57	> 80
sec-Butyl	2.18	0–70	−0.66	> 75
Benzyl	1.9	0–60	~ −0.5	> 65
Phenyl	1.6	0–50	<0	> 70
p-Nitrophenyl	1.6	0–50	~ −0.2	> 70
p-Chlorophenyl	1.5	0–55		

The corresponding treatment for data for amide hydrolysis gives *r* values approaching 3, and the fact that low integral values are obtained in each case is in agreement with the interpretation of *r* as the kinetic order of the reaction with respect to water. The results for ester hydrolysis, where an *r* value close to 2.0 is found for all but the most reactive (*t*-butyl) ester, at low to medium acid concentrations, are encouragingly consistent with a common mechanism in this acidity region. All the other evidence is firm for the majority of these esters, that this mechanism is that designated $A_{AC}2$. So the conclusion to be drawn from the treatment of Yates and McClelland[37] is that the $A_{AC}2$ reaction is of the second order with respect to the activity of water.

(*d*) *The mechanism of the $A_{AC}2$ reaction*

It is well-established that $A_{AC}2$ reactions are typically of the first order in both ester and acid concentration, and on the basis of the recent results just described, it is a strong presumption that they are of the second order in water. There is further support for this latter observation from a number of sources. Long *et al.*[125], have measured the rates of hydrolysis of ethyl formate and methyl acetate in dilute solutions of HCl in mixtures of H_2O and D_2O. They find that the variation of the rate of hydrolysis with the fraction of deuterium present is most accurately accounted for if at least five, rather than the minimum of three, exchangeable protons are present in the transition state for the reaction. Similarly, Bunnett's *w* parameter for the $A_{AC}2$ hydrolysis of a number of esters does not fall in the region characteristic of simple A-2 reactions, but has values between 4 and 5[118], usually interpreted as indicating that water is involved as a proton-transfer agent. The same result is obtained on the basis of Bunnett's ϕ-parameter[51,119] calculated from the same data. Finally, Laidler and Landskroener[126] found reasonable agreement with a theory of the effect of solvent on reaction rates when they used a model for ester hydrolysis involving two molecules of water, one as the hydronium ion, in the transition state.

If the determination of kinetic order with respect to solvent is correct, then the order is known with respect to all the reactants involved in the $A_{AC}2$ reaction, and it is possible to specify the composition of the transition state completely. The four species, ester molecule, proton, and two solvent molecules, will be present in solution as three chemical species, but it is not clear whether the proton is primarily involved in the reaction as a hydronium ion or as the conjugate acid of the ester. Furthermore, it is unlikely that a termolecular reaction is involved: elementary processes of this complexity are known, but rare. They are generally significant only where powerful reagents, such as the strong acids, are not available; and they are characterized by very large negative entropies of activation, of the order of −40 to −50 eu (for some suggested examples, see p. 157). Entropies of activation for the $A_{AC}2$ reaction (Table 13) fall typically in the range −20 to −30 eu.

TABLE 13

ENTHALPIES AND ENTROPIES OF ACTIVATION FOR $A_{AC}2$ REACTIONS

Ester	*Conditions*	$\Delta H^{\ddagger}$ *(eu)*	$\Delta S^{\ddagger}$ *(eu)*	*Ref.*
$HCOOCH(C_6H_5)_2$	70% dioxan–water	15.5	−23.2	85
C_6H_5COOH	$HClO_4$	15.3	−30	49
$HCOOC_2H_5$	HCl	14.7	−20.5	78
$HCOOCH(CH_3)_2$	HCl	14.8	−20.3	78
$HCOOC(CH_3)_3$	HCl	14.2	−21.2	78
$(CH_3)_3N^+CH_2CH_2COOC_2H_5$	HCl	16.0	−25.5 ± 4	127
$CH_3COOC_2H_5$	60% ethanol–water	15.5	−26.1	128
$C_6H_5COOC_2H_5$	60% ethanol–water	18.9	−26.4	128

Although this argument is indicative rather than conclusive, it is firmly in line with general experience, and if accepted simplifies the picture very considerably, since it eliminates all mechanisms not involving an intermediate addition compound, as well as those invoking various forms of general species catalysis of the formation of a tetrahedral intermediate. Since this restriction falls conveniently on a group of mechanisms not favoured by the present writer, due weight will be attached to the entropy of activation argument in the following discussion.

The case for a two-step mechanism for the $A_{AC}2$ reaction involving a tetrahedral addition intermediate, has already been discussed (p. 104), and has been widely accepted. It is not possible, however, to exclude completely the possibility that a concerted nucleophilic displacement is involved, and it is possible to write such mechanisms involving transition states of the correct composition. All such mechanisms necessarily involve a termolecular collision and are therefore not readily reconciled with the observed entropies of activation.

If the $A_{AC}2$ reaction does involve a direct displacement, then it is a simple matter to show that there is only a single reasonable mechanism that is consistent both with the observed kinetics and with the chemical evidence. The crucial observation is that hydrolysis and the exchange of ^{18}O from the enriched solvent into the carbonyl group of the unreacted ester proceed side-by-side at comparable rates. This makes it clear that the two oxygen atoms of the carboxyl group must be in similar chemical environments in the species under attack, which must, therefore, if a concerted displacement is involved, be the conjugate acid of the ester, *viz.*

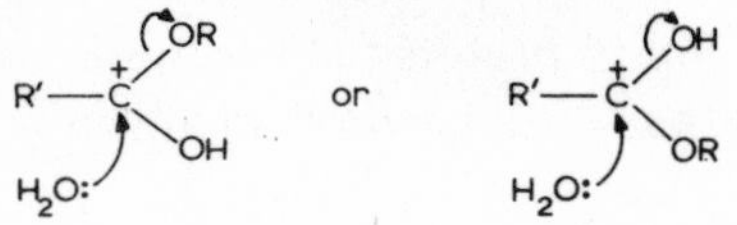

R′—C⁺(OR)(OH) ← H_2O: or R′—C⁺(OH)(OR) ← H_2O:

Secondly, it is not chemically reasonable to expect a neutral water molecule to displace a hydroxide or alkoxide ion from carbon under any conditions, so that it is mandatory that the leaving group should be at least partially protonated in the transition state. Since the extra proton from the medium has already been used, this can only be done by implicating a proton of the attacking water molecule which becomes acidic as bond formation develops. The role of the remaining water molecule must then be to relay this proton from the developing acid to the incipient basic site, as shown (18)

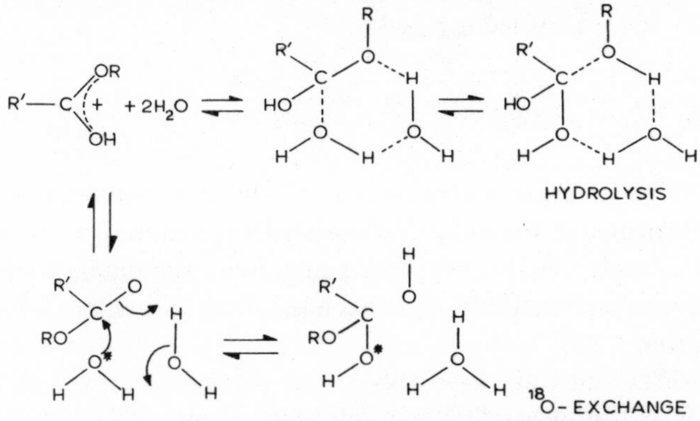

HYDROLYSIS

^{18}O- EXCHANGE

(18)

This mechanism has been suggested by Lane[38] as the best rationalization of the available evidence. It has the merit that the second water molecule plays a simple, integral part in the transition state, and is consistent with as many of the observed facts as is any other mechanism. In particular it is a symmetrical mechanism, applicable equally to acid-catalyzed hydrolysis and ester formation. It does, however, involve a termolecular collision, and this fact is at variance with the observed entropy of activation, as discussed above.

Mechanisms involving general species catalysis of the formation of a tetrahedral intermediate seem unlikely. General acid catalysis of the addition of water by the hydronium ion, *viz.*

$$H_2\ddot{O}: + (R')(RO)C{=}O + H{-}\overset{+}{O}H_2 \rightleftharpoons H_2\overset{+}{O}{-}C(R')(OR){-}OH + H_2O \tag{19}$$

can be ruled out, since it involves the addition of water to what can be considered as a partially protonated carbonyl group, a process which must be slower than the addition to the fully protonated group. Since the dependence of the rate of reaction of primary alkyl esters on the activity of water is the same in regions where significant amounts of the protonated ester are present as it is in dilute acid, where the concentration of protonated species is negligible, it is not likely that proton transfer to the neutral ester is involved in the rate-determining step.

Perhaps the most acceptable of this group of mechanisms is the general base-catalyzed attack of water by water on the protonated ester, *viz.*

$$H_2O: + H{-}O(H) + R'C(RO)\overset{+}{\cdots}(OH) \rightleftharpoons H_3O^+ + HO{-}C(R')(OR){-}OH \tag{20}$$

An immediate objection to this scheme is that it is difficult to believe that general base catalysis, particularly by so weak a base as water, should be necessary for the addition of water to a protonated ester. In other words, the entropically more favourable bimolecular addition

$$H_2O: + R'C(RO)\overset{+}{\cdots}(OH) \rightleftharpoons H_2\overset{+}{O}{-}C(R')(OR){-}OH \tag{21}$$

would be expected to be a faster reaction. It is possible, however, that the protonated orthoester formed in this way is so much more strongly acidic than H_3O^+ that the more complex path (20), which generates the stable hydroxonium ion, does compete significantly, and mechanism (20) remains a possibility for the addition step.

A vital fact not considered fully thus far is that the similar rates of ^{18}O-exchange and hydrolysis found for several typical esters (p. 129) prove that the forward and back reactions of any tetrahedral intermediate must proceed at similar rates, and thus that neither its formation nor its breakdown can alone be rate-determining. Both processes are kinetically important, and both transition states must be considered in any detailed discussion of mechanism. Both presumably have the same composition, since there is no sign of any change in kinetic order as the concentration of the reactants are varied, except where there is thought to be an actual change of mechanism. The first step, the formation of the tetrahedral intermediate, has been considered briefly. Since its breakdown is of comparable kinetic importance, this second step must now be considered.

The tetrahedral intermediate formed by the acid-catalyzed addition of water to an ester is an alkyl or aryl dihydrogen orthoester. The hydrolysis of fully esterified orthoesters has been studied in some detail, and the present position is summarized in an excellent review by Cordes[102]. The hydrolysis of most orthoesters in aqueous solution is subject to general acid catalysis by weak acids, such as carboxylic acids, and the catalytic coefficients are related by the Bronsted equation, with large values of α. Significantly, the point for hydronium ion catalysis falls on the Bronsted plot (see, for example, ref. 129), and it seems likely that hydronium ion catalysis of orthoester hydrolysis may also involve a mechanism that is formally general acid catalysis[102]. This viewpoint is strongly supported by a study by Bunton and Dewolfe[103], in which estimates of the basicity of orthoesters are used to calculate the rate coefficients required to account for the observed hydrolysis rates by the A-1 mechanism. This exercise leads to the conclusion that the carbon–oxygen bond cleavage (22) of the hypothetical conjugate acids of orthoesters must occur at rates comparable to those for proton transfers from strong acids to strong bases[103], *viz.*

$$R'C(OR)_3 + H_3O^+ \underset{k_2}{\overset{k_1}{\rightleftharpoons}} R'C(OR)_2\overset{+}{O}(H)R + H_2O$$

$$R'C(OR)_2\overset{+}{O}(H)R \xrightarrow{k_3} R'C^+(OR)_2 + ROH \qquad (22)$$

It seems much more likely that the transfer of the proton is actually involved in the rate-limiting step for the hydrolysis of carboxylic orthoesters. This is consistent with the observed catalytic constants. If Bunton and Dewolfe's estimate of $K_a \simeq 10^7$ is accepted for the dissociation constant of the conjugate acid of an orthoester, and if the rate coefficient for the loss of the proton, k_2, is of the order of 10^{11} sec^{-1}, then k_1 will be about 10^4 l·mole^{-1}·sec^{-1}, close to the value observed, for example, for the catalytic coefficient for the acid-catalyzed hydrolysis of ethyl orthoacetate at 20°C[103].

It seems unlikely, however, that the transfer of the proton to the orthoester will itself be rate-determining, since this requires that the reverse of this reaction, the diffusion-controlled transfer of a proton from a strong acid, be slower than the cleavage of the carbon–oxygen bond. The most reasonable conclusion is that the proton transfer and the carbon–oxygen bond cleavage are concerted, and that the hydrolysis is catalyzed by the hydronium ion acting as a general acid, *viz.*

References pp. 202–207

$$R'C(OR)_2\text{—}O(R)\cdots H\text{—}\overset{+}{O}H_2 \rightleftharpoons [R'\text{—}C(OR)_2\text{---}O(R)\text{---}H\text{---}\overset{+}{O}H_2] \rightleftharpoons R'\text{—}\overset{+}{C}(OR)_2 + ROH + H_2O \qquad (23)$$

In view of the relatively high values of α, and the insensitivity of the rate of hydrolysis to the nature of the R′ group[102], the best description of the transition state has the proton transfer well advanced, but a relatively small amount of carbon–oxygen bond breaking.

It seems probable that this picture is relevant to ester hydrolysis, and the above conclusions make good sense chemically. But it must be realised that the comparison of trialkyl and monoalkyl orthoesters involves a not-inconsiderable extrapolation. The latter are presumably intermediates in the hydrolysis of trialkyl orthoesters, and their hydrolysis is clearly considerably faster than that of the fully esterified compounds, since the loss of the first alkyl group is rate-determining. In particular, concerted elimination mechanisms, illustrated in (24)

$$H_2O{:}\ H\text{—}O\text{—}C(R)(OR)\text{—}\overset{+}{O}(R)H \longrightarrow H_3O^+ + O{=}C(R')(OR) + ROH \qquad (24)$$

become possible for the intermediate hydrolysis products. However, since conjugate acid of the trialkyl orthoester seems to play no part in its hydrolysis, it seems unlikely that that of the more reactive monoalkyl orthoester will do so. And in any case, mechanisms of the type of (24) do not lead to useful conclusions for ester hydrolysis. It will be assumed, therefore, that the tetrahedral addition intermediates formed during acid-catalyzed ester hydrolysis behave qualitatively as do fully esterified orthoesters.

The acid-catalyzed hydrolysis of orthoesters is very much faster than that of esters. The second-order rate coefficient for the hydrolysis of ethyl acetate is of the order of 10^{-4} l·mole^{-1}·sec^{-1} at 25°C, whereas that for the hydrolysis of ethyl orthoacetate[103] is of the order of 10^{4} l·mole^{-1}·sec^{-1}, and that for the breakdown of a monoalkyl orthoester must be faster still. If the breakdown of the tetrahedral intermediate is partially rate-determining in acid-catalyzed ester hydrolysis, therefore, its concentration must be very small; that is, the equilibrium for its formation must be highly unfavourable. This would account for the failure of attempts to demonstrate its existence directly, and is an important requirement of any mechanism involving an intermediate.

If it is accepted that the breakdown of the tetrahedral intermediate is likely to be subject to general acid catalysis by the hydronium ion, a consistent picture of the mechanism of the $A_{AC}2$ reaction can be elaborated. Since the entering and leaving groups, H_2O and ROH, are so similar, at least in case of the lower aliphatic alcohols, it can be assumed with some confidence that the mechanism will be symmetrical: in other words, if the loss of ROH from the tetrahedral intermediate is subject to general acid catalysis, then so also will be the loss of water which is the reverse of the first step. Lane's data[38], demonstrating the parallel behaviour of the ^{18}O-exchange and hydrolysis reactions of ethyl acetate, provide evidence for this contention. But if the reverse of the addition of water is subject to general acid catalysis, then the addition itself, which must proceed by way of the same transition state (25) must be general base catalyzed, *viz.*

$$H_2O\colon + H{-}O(H) + [HO{\cdots}C(R'){\cdots}OR]^{+} \rightleftharpoons \overset{\delta+}{H_2O}{---}H{---}O(H){---}C(R')(\overset{\delta+}{OH}){\cdots}OR \rightleftharpoons H_2\overset{+}{O}{-}H + HO{-}C(R')(OH){-}OR \quad (25)$$

In the discussion of the general base catalyzed addition step above (p. 120) the objection was raised that it was difficult to believe that general base catalysis would be necessary for the addition of water to so reactive a species as a protonated ester. An answer to this objection is implicit in the discussion above of the mechanism of hydrolysis of orthoesters. It appears that the protonated orthoester, which would be the initial product of the simple addition of a molecule of water to a protonated ester, is too reactive a species to exist in aqueous solution, and that carbon–oxygen bond-cleavage is concerted with the transfer of the proton to the orthoester. The formation of a protonated orthoester by the addition of a molecule of water to the conjugate acid of an ester will be even less likely, and it seems entirely reasonable, therefore, that the formation of the neutral orthoester, by a general base catalyzed process, should be the favoured mechanism.

The full mechanism based on these conclusions can be written as

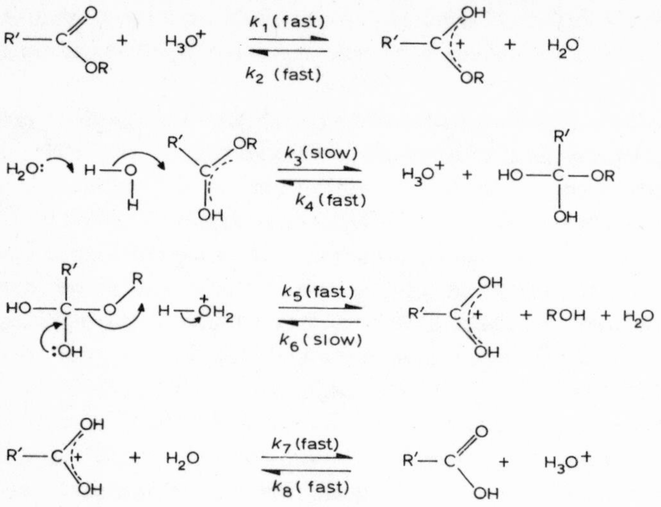

Scheme 1. Mechanism ($A_{AC}2$) for acid-catalyzed ester hydrolysis.

Steps k_4 and k_5 are labelled fast, to emphasize that the rate coefficients are high for these steps. They are, of course, partially rate-determining for ester formation and hydrolysis, respectively, because of the unfavourable equilibrium constants, k_3/k_4 and k_6/k_5, for the formation of the tetrahedral intermediate. The actual magnitude of the equilibrium constant for the formation of the tetrahedral intermediate from the protonated ester is not easy to estimate with any certainty. Taking K_a for the protonated ester as 10^7, and noting that the acid-catalyzed hydrolysis of the tetrahedral intermediate is probably more than 10^8 times faster than the overall rate of ester hydrolysis (p. 122) k_3/k_4 is likely to be of the order of 10^{-2} or less. The proton transfers to and from the starting ester and product acid are expected to be fast compared with the rate-determining steps. If K_a for the dissociation of the protonated ester or acid is of the order of 10^7, and the rate coefficient for proton loss, k_2 and k_7, is about 10^{10}–10^{11} sec^{-1} in each case, then k_1 and k_8 will be of the order of 10^3–10^4 $l \cdot mole^{-1} \cdot sec^{-1}$, much greater than the observed rate coefficients of ester hydrolysis and formation.

The general rate expressions for the hydrolysis and formation of an ester by this mechanism are complicated, and even the mechanism as written is a simplification, since ROH and ROH_2^+ will act as general base and general acid, respectively, just as H_2O and H_3O^+ do. In a solution containing only the reactants the initial rate of disappearance of ester, E, can be shown, by the use of the steady-state approximation for the tetrahedral intermediate, to be[130]

$$-\frac{d[E]}{dt} = \frac{k_3 k_5 [EH^+][H_2O]^2}{k_4 + k_5}$$

$$= \frac{K'_w k_3 k_5 [E][H_3O^+][H_2O]}{K_a(k_4 + k_5)}$$

(where K'_w is the dissociation constant of the hydronium ion). Similarly, the initial rate of appearance of ester, in a solution containing the carboxylic acid, A, and alcohol, is given by

$$\frac{d[E]}{dt} = \frac{k_4 k_6 [AH^+][ROH][H_2O]}{k_4 + k_5}$$

The mechanism depicted in Scheme 1 on p. 124 is consistent with most of the available evidence. It explains the kinetic order of two in water, and the high w and ϕ parameters found by Bunnett. Indeed Bunnett[118] suggested general acid catalysis of the breakdown of the tetrahedral intermediate as an explanation of the observed value of w for ester hydrolysis by the $A_{AC}2$ mechanism. It is consistent with the solvent deuterium isotope effect (*e.g.* $k_D/k_H = 1.57$ for the hydrolysis of ethyl formate[125]), since the k_D/k_H in the region of 3, expected for simple A-1 reactions, is reduced by the isotope effect ($k_D < k_H$) on the steps involving general species catalysis. The single piece of evidence not easily reconciled with the mechanism is the entropy of activation, which is typically in the region -20 to -30 eu for $A_{AC}2$ ester hydrolysis, as discussed above, and seems rather high for what is, in effect, a termolecular reaction. However, the mechanism is too complicated to predict the entropy of activation with any confidence, since the observed value contains contributions for the protonation equilibrium (for which $\Delta S°$ may be positive, as discussed on p. 92) as well as for the two rate-determining steps of the reaction. Thus, the conclusion stands that the mechanism of the $A_{AC}2$ reaction is best represented, in the light of the evidence presently available, by Scheme 1 shown on p. 124.

2.3 ESTER FORMATION

Ester formation is formally the reverse of ester hydrolysis, *viz.*

$$R'-C(=O)OH + ROH \rightleftharpoons R'-C(=O)OR + H_2O$$

Thus, to define the mechanism of either reaction under given experimental conditions is to define the mechanisms of both, since the transition state or states of lowest energy are necessarily the same in either direction. In practice, however, ester hydrolysis and formation are not carried out under the same conditions. Hydrolysis is carried out in water, but esterification reactions are carried out in the alcohol concerned. So Scheme 1 written for ester hydrolysis by the $A_{AC}2$ mechanism (p. 124) must be modified, as well as reversed, before it can be applied to ester formation, as shown below (Scheme 2). It should be emphasized that the evidence for Scheme 2 for ester formation is indirect, and the details of the mechanism are derived entirely from the mechanism shown in Scheme 1 (p. 124) for ester hydrolysis. Nevertheless, authors in the field almost invariably accept or assume mechanisms for ester formation which are simpler versions of Scheme 2.

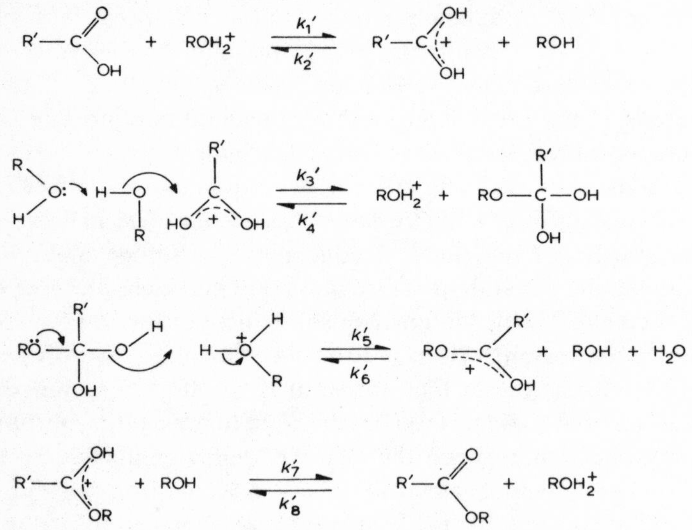

$$R'{-}C(=O)OH + ROH_2^+ \underset{k_2'}{\overset{k_1'}{\rightleftharpoons}} R'{-}\overset{+}{C}(OH)OH + ROH$$

$$ROH + H{-}O(R) + R'\overset{+}{C}(OH)_2 \underset{k_4'}{\overset{k_3'}{\rightleftharpoons}} ROH_2^+ + RO{-}C(R')(OH){-}OH$$

$$R\ddot{O}{-}C(R')(OH){-}OH + H{-}\overset{+}{O}(H)R \underset{k_6'}{\overset{k_5'}{\rightleftharpoons}} RO{-}\overset{+}{C}(R'){-}OH + ROH + H_2O$$

$$R'{-}\overset{+}{C}(OH)OR + ROH \underset{k_8'}{\overset{k_7'}{\rightleftharpoons}} R'{-}C(=O)OR + ROH_2^+$$

Scheme 2. Mechanism for acid-catalyzed ester formation.

2.3.1 Reactions involving alkyl–oxygen bond formation

The reverse of the $A_{AL}1$ cleavage of an ester is the reaction of a neutral carboxylic acid with a carbonium ion, *viz.*

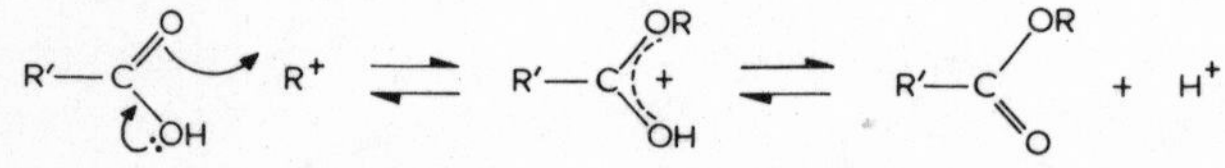

$$R'{-}C(=O)OH + R^+ \rightleftharpoons R'{-}\overset{+}{C}(OR)OH \rightleftharpoons R'{-}C(=O)OR + H^+$$

This is a well-known reaction, of some synthetic importance, used in particular for the preparation of *t*-butyl esters[131]. The *t*-butyl group is, of course, highly labile under acidic conditions, and, this property makes it a valuable protecting group for amino-acid derivatives in peptide synthesis. *t*-Butyl esters, including those of amino acids and their N-acyl derivatives[132] are most conveniently prepared by the strong-acid-catalyzed reaction of the carboxylic acid with a large excess of liquid isobutene. These conditions are necessarily very different from those under which the $A_{AL}1$ cleavage of esters of this type is observed; the mechanism has not been studied, but there seems little doubt that the *t*-butyl cation is involved, *viz.*

$$(CH_3)_2C{=}CH_2 \underset{}{\overset{H^+}{\rightleftharpoons}} (CH_3)_3C^+ + R'{-}C(=O)OH \rightleftharpoons R'{-}C(=O)OC(CH_3)_3 + H^+$$

Only carbonium ions and alkylating agents of similar high reactivity, such as trialkyloxonium cations[133], will attack the very weakly nucleophilic carboxylic acid group at a useful rate under normal conditions. Ester formation by alkylation of the carboxyl oxygen atom normally involves the carboxylate anion, *viz.*

$$R'{-}CO_2^- + R{-}X \longrightarrow R'{-}C(=O)OR + X^-$$

This is formally the reverse of the $B_{AL}1$ cleavage of an ester, and is the only one-stage mechanism for ester formation available for the ionized carboxyl group. Numerous methods are, of course, available which involve initial electrophilic attack on the carboxylate group, followed by a displacement at the carbonyl carbon atom of the intermediate formed, which is often an anhydride. An example[134] is the esterification of carboxylic acids in the presence of *p*-toluenesulphonyl chloride in pyridine, *viz.*

$$R'{-}CO_2^- + ArSO_2Cl \xrightarrow{\text{pyridine}} R'{-}C(=O){-}O{-}SO_2Ar + Cl^-$$

$$R'{-}C(=O){-}O{-}SO_2Ar + ROH \xrightarrow{\text{pyridine}} R'{-}C(=O)OR + ArSO_3^- + H^+$$

Simple alkylations of carboxylate anions, by such alkylating agents as alkyl halides, dialkyl sulphates, and quaternary ammonium compounds, are generally straightforward S_N2 reactions, and will not be discussed here.

One other type of esterification process which formally involves the carboxylic acid is the reaction with diazoalkanes. Diazomethane in particular is widely used for the synthesis of methyl esters, *viz.*

$$R'{-}C(=O)OH + CH_2N_2 = R'{-}C(=O)OCH_3 + N_2$$

The mechanism of this type of esterification reaction has been much studied in recent years, using diphenyldiazomethane. In alcoholic solvents the mechanism of the reaction involves a rate-determining proton transfer from the acid to the carbon atom of the diazoalkane, to form a benzhydryldiazonium–carboxylate ion-pair[135], *viz.*

$$(C_6H_5)_2C{=}N_2 + R'{-}C(=O)OH \longrightarrow (C_6H_5)_2CH{-}\overset{+}{N}_2 \quad {}^{-}O{\cdots}C(R'){\cdots}O$$

This then collapses, possibly by way of the benzhydryl–carboxylate ion-pair, to the benzhydryl ester. Since, however, it is clear that ester formation plays no part in the slow step of these reactions, they will not be discussed in detail here.

2.3.2 Reactions involving acyl–oxygen bond formation

The great majority of ester-forming reactions involve acyl–oxygen cleavage, and in particular acyl–oxygen cleavage by the $A_{AC}2$ mechanism. The acylium ion route is convenient for the preparation of esters from a few sterically hindered acids, the best-known example being Newman's synthesis of methyl mesitoate, in which a solution of mesitoic acid in 100% sulphuric acid is poured into methanol[17]. But although Deno *et al.*[20] have shown that this method works for mesitoic acid in considerably less concentrated sulphuric acid, and may be extended to a range of acids by working with oleum, it is not likely to supersede the usual methods for normal substrates.

Ester formation under $A_{AC}2$ conditions has been widely studied, and the field was particularly active in the early years of this century. The reaction is clearly closely similar to ester hydrolysis, but precise comparisons are rarely possible because our understanding of reactions in alcohols and mixed aqueous solvents is less well-developed than that of those in water.

Two types of esterification reaction that can be studied with water as solvent are lactone formation, in which the alcohol is part of the same molecule as the acid, and the ^{18}O-exchange reaction of carboxylic acids, which makes it possible to examine A-2 reactions of carboxylic acids under the conditions used for ester hydrolysis. Work in both these fields confirms the similarities between ester hydrolysis and formation. The hydrolysis and formation of γ-butyrolactone have already been discussed (p. 109). We deal here with the ^{18}O-exchange reactions of carboxylic acids.

Roberts and Urey[99] were the first to demonstrate the similarities between ester hydrolysis and formation, and the ^{18}O-exchange reaction of carboxylic acids. Not only are all three reactions of the first order in carboxylic acid or ester and the hydronium ion, but the rates of all three are closely similar under similar conditions. More recent work, especially by Bender, Bunton, and O'Connor, and their respective collaborators, have extended and confirmed these conclusions.

Bender *et al.*[136] have measured the rate of incorporation of ^{18}O from enriched water into several substituted benzoic acids. The catalyst was 0.07 *M* HCl, and the solvent 33% dioxan–water. The rate coefficients for exchange at 80°C are given in Table 14, which also contains a comparison of these rate coefficients with those for the hydrolysis of the corresponding ethyl esters, measured by Timm and Hinshelwood[128] in 60% acetone and 60% ethanol. As noted earlier by Roberts and Urey, the absolute rates are very similar for the two reactions. Also, as expected, the exchange rate of benzoic acid, with two equivalent oxygen atoms, is almost exactly twice as fast as that of ethyl benzoate, with only one: (k_{hyd}/k_{exch} is 5.2 for the ester).

TABLE 14

THE KINETICS OF THE OXYGEN-EXCHANGE REACTIONS OF SUBSTITUTED BENZOIC ACIDS IN 33% DIOXAN–WATER AT 80.0±0.1°C

(Data from ref. 136)

Acid	*HCl* *(M)*	$10^5 k_{obs}$ *(sec^{-1})*	$10^4 k_{H^+}$ *(l·mole^{-1}·sec^{-1})*	k_{hyd}/k_{exch} [b]
Benzoic[a]	0.0708	1.14±0.10	1.61	2.5
m-Chlorobenzoic	0.0639	1.13±0.04	1.77	
p-Chlorobenzoic	0.0639	0.70±0.1	1.1	3.4
p-Methoxybenzoic	0.0639	0.64±0.02	1.00	2.7
p-Methylbenzoic	0.69	1.03±0.3	1.5	2.2
2,4,6-Trimethylbenzoic	0.69	0.00102±0.00015	0.00148	

[a] Rate coefficient is 2.37×10^{-4} l·mol^{-1}·sec^{-1} for exchange in H_2O[137].

[b] Ratio of rate coefficients for exchange in 33% dioxan–water and the hydrolysis of the corresponding ethyl benzoates in 60% ethanol or acetone–water (mean of the two values) at 80°C in the presence of 0.05 *M* HCl.

One of the characteristics of the acid-catalyzed hydrolysis of esters, that is shared by ester formation also, is that substituent effects on the rate coefficients are small, and not simply related to σ values (see below, p. 131). The data in Table 14 show that this is also true for the ^{18}O-exchange reaction of substituted benzoic acids. This is borne out by the relative constancy of the ratio k_{hyd}/k_{exch} for the different substituted acids: it was not possible to obtain a meaningful ρ value from the data of Table 14, because of the small number of points and the large amount of scatter evident on the Hammett plot. Mesitoic acid is highly unreactive, compared with the *m*- and *p*-substituted esters used, as is its methyl ester towards alkaline hydrolysis[138], and presumably reacts by the seriously hindered $A_{AC}2$ route.

O'Connor and Bunton and their coworkers have measured the rates of ^{18}O-exchange of a number of aliphatic carboxylic acids in acidic, basic and neutral solution. In Table 15 are summarized activation parameters for the reaction with a number of acids measured in this work. Here, of course, *only* acyl–oxygen cleavage is measured, and comparison with Table 13 (p. 118) shows at once how similar are these figures to those for typical ester hydrolysis reactions involving the $A_{AC}2$ mechanism. The very different activation parameters for mesitoic acid, which is known to exchange by the $A_{AC}1$ mechanism, confirm the assignment of the $A_{AC}2$ mechanism for the other exchange reactions.

TABLE 15

THERMODYNAMIC PARAMETERS FOR THE ACID-CATALYZED ^{18}O-EXCHANGE OF CARBOXYLIC ACIDS IN WATER

Acid	$\Delta H^{\ddagger}$ *(kcal·mole⁻¹)*	$\Delta S^{\ddagger}$ *(eu)*	*Ref.*
CH_3COOH	14.3	−21	137
$(CH_3)_3COOH$	14.3	−28	139
CCl_3COOH	14.4	−27	140
CF_3COOH	13.4	−26	140
C_6H_5COOH	15.3	−30	49
$H_3N^+CH_2COOH$	16.3	−34	141
HOOCCOOH	16.3	−18	140
2,4,6-Trimethylbenzoic acid	~ 32	+9	49

The acid-catalyzed esterification of carboxylic acids in alcohol solvents has been studied by many authors over many years. The reaction goes at a convenient rate, in methanol, for example, at much lower temperatures than are required for acid-catalyzed ester hydrolysis, and for this reason has often been preferred to the hydrolysis reaction for investigations of structure-reactivity relationships, and of steric effects in particular. For purposes of this sort, where the relative rates of reactions are of primary importance, a detailed understanding of the properties of the medium is less vital, and large amounts of useful data have been collected. Consequently the information to be gleaned from an analysis of the results is primarily about the effect of structure on reactivity, and the work will be discussed under this heading. The slow step of the reaction has generally been assumed to be the attack of the alcohol molecule on the protonated ester, and the observed effects of changing structure on reactivity in the reaction can give information about the transition state if examined in the light of this assumption.

2.4 STRUCTURE AND REACTIVITY

It has been recognized for many years that acid-catalyzed ester hydrolysis and formation are reactions which show very low sensitivities to polar effects. This is qualitatively predictable on the basis of the mechanistic schemes shown on pp. 124 and 126. If the addition of a molecule of solvent (water or alcohol) to the protonated ester or acid, and probably also the acid-catalyzed breakdown of the tetrahedral intermediate, are rate-determining, the response to substitution of the overall reaction will be determined by the individual sensitivities of all these separate processes, and by their relative importance in determining the rate. Now ρ for the protonation of substituted benzoic acids is -1.09 (see p. 70), and ρ for the acid-catalyzed hydrolysis of trimethylorthobenzoates, $ArC(OCH_3)_3$, and thus probably also for the similar decomposition of the tetrahedral intermediate in ester hydrolysis, is about -2[129] (there is some doubt about the correctness of this figure, as explained by Cordes[102]). The addition of water to the protonated ester, on the other hand (closely similar to the reverse of this latter reaction), is expected to be characterized by a fairly large positive ρ value. Although this last parameter is difficult to estimate, it is clear that the opposing electronic requirements of this reaction and of the other two processes contributing to the determination of the rate of the overall reaction make a large ρ value unlikely.

2.4.1 The effects of substitution in the acyl group

In fact ρ for the acid-catalyzed hydrolysis and formation of benzoate esters is small, and since the fit to the Hammett equation is characteristically rather poor for these reactions, ρ is often taken as zero. Some typical values are listed in Table 16. Too much weight should not be attached to the absolute values of the reaction constants given in the table. The data do not fit the Hammett equation accurately, and it is known that much of the earlier work was subject to considerable error, because no allowance was made for the slow reaction of the catalyst with the solvent. This could affect the observed rate coefficients in three ways—by reducing the catalyst concentration, by decreasing the acid content as measured by titration, and by producing water, *e.g.*

$$CH_3OH + HCl = CH_3Cl + H_2O$$

This would have both a solvent effect and a mass law effect on the rate of ester formation. The error is systematic, since it is most serious for the slower ester formation reactions, and consequently the ρ value calculated by Jaffé[144] from the data of Hartman and Borders[142] is not accurate. Later workers allowed for this side-reaction[146] or used aromatic sulphonic acids rather than HCl as the catalyst[145,147]. However, whatever the exact ρ values, it is quite clear that the polar effects of substituents on acid-catalyzed ester hydrolysis and formation are small.

TABLE 16

HAMMETT REACTION CONSTANTS, ρ, FOR ACID-CATALYZED ESTER HYDROLYSIS AND FORMATION

Acid	*Conditions*	*Temperature (°C)*	ρ	*Ref.*
Ester formation				
ArCOOH	HCl, methanol	25	−0.23	142, 143, 144
ArCOOH	HCl, methanol	60	−0.21	142, 143, 144
ArCOOH	*p*-TsOH, methanol	60	−0.57	145
ArCOOH	HCl, cyclohexanol	65	+0.42	146
$ArCH_2COOH$	*p*-TsOH, methanol	40	−0.31	147
ArCH=CHCOOH	87.8% ethanol	30	−0.21	148, 149
o-substituted ArC≡CCOOH	HCl, methanol	35	−0.22	148, 150
m,p-substituted ArC≡CCOOH	HCl, methanol	35	−0.44	148, 150
Ester hydrolysis				
$ArCOOC_2H_5$	60% ethanol	100	0.14	128, 144
$ArCOOC_2H_5$	60% acetone	100	0.11	128, 144
$ArCOOCH_3$	80% methanol	100.8	~0	151

The same conclusion has been reached for the corresponding reactions of aliphatic esters. For example, the introduction of a chlorine atom into the α position of ethyl acetate, to give ethyl chloroacetate, causes only a small decrease in the rate of acid-catalyzed hydrolysis in aqueous acetone, little more than the effect of introducing a methyl group into the same position to give ethyl propionate[152]. Consequently the relative rates of acid-catalyzed ester hydrolysis or formation reactions are determined almost exclusively by non-polar effects, of which the size of the groups in the vicinity of the reacting centre appears to be decisive. Taft[152] has defined the steric substituent constant E_s of a group R′, in an aliphatic compound as the (logarithmic) measure of the effect of the group, compared with that of the methyl group of ethyl acetate, on the rate coefficient for the acid-catalyzed hydrolysis of R′COOEt, *viz.*

$$\log (k/k_0)_A = E_s$$

Qualitatively, the E_s values obtained in this way are clearly related to the size of the groups concerned (Table 17), and in practice the steric factor can often be accounted for quantitatively by the use of the equation

$$\log (k/k_0) = \rho^*\sigma^* + \delta E_s \qquad (5)$$

where δ is a measure of the sensitivity of the reaction to steric effects. These figures are derived from data similar to those reproduced below (Table 21, p. 135) for acid-catalyzed ester hydrolysis and formation, and provide a

TABLE 17

STERIC SUBSTITUENT CONSTANTS, E_s, FOR ALIPHATIC SUBSTITUENTS, R′, IN $R'COOC_2H_5$ AT 25°C[152]

Substituent R′	*E_s for R′*
H	1.24
CH_3	0
CH_2CH_3	−0.07
$(CH_2)_2CH_3$	−0.36
$(CH_2)_3CH_3$	−0.39
$(CH_2)_4CH_3$	−0.40
$(CH_2)_7CH_3$	−0.33
$(CH_3)_2CH$	−0.47
$(C_2H_5)_2CH$	−1.98
$(CH_3)_3C$	−1.54
$(C_2H_5)_3C$	−3.8
Cyclopentyl	−0.51
Cyclohexyl	−0.79
FCH_2	−0.24
$ClCH_2$	−0.24
$BrCH_2$	−0.27
ICH_2	−0.34
$C_6H_5CH_2$	−0.38
$(C_6H_5)_2CH$	−1.76
$(CH_3)_3CCH_2$	−1.74
CF_3	−1.16
CCl_3	−2.06
CBr_3	−2.43
CHF_2	−0.67
$CHCl_2$	−1.54
$CHBr_2$	−1.86

TABLE 18

KINETIC DATA FOR THE ACID-CATALYZED HYDROLYSIS AND FORMATION OF SUBSTITUTED BENZOATE ESTERS

X	*k_{obs} ($l \cdot mole^{-1} \cdot sec^{-1}$)*	*$\Delta H^{\ddagger}$ ($kcal \cdot mole^{-1}$)*	*$\Delta S^{\ddagger}$ (eu)*
X-$C_6H_4COOC_2H_5$ in 60% ethanol–water at 100°C, catalyzed by 0.05 M $C_6H_5SO_3H$[128]			
H	9.0×10^{-5}	18.9	−26.4
p-CH_3	7.8×10^{-5}	19.0	−26.4
p-Cl	7.9×10^{-5}	18.3	−28.4
p-Br	8.3×10^{-5}	18.2	−28.4
p-CH_3O	6.2×10^{-5}	19.8	−24.7
p-HO	4.5×10^{-5}	19.7	−25.7
m-NO_2	8.8×10^{-5}	18.1	−28.6
p-NO_2	10.6×10^{-5}	16.8	−31.8
o-NO_2	0.51×10^{-5}	19.4	−30.4
X-$C_6H_4COOCH_3$ in 80% methanol–water at 100.8°C[151]			
H	1.88×10^{-4}		
m-CH_3	1.39×10^{-4}		
m-Cl	1.53×10^{-4}		
m-NO_2	1.29×10^{-4}		
Esterification of X-C_6H_4COOH in methanol at 60°C, catalyzed by p-TsOH[153]			
H	2.90×10^{-3}	15.4	−24.1
m-CH_3	2.90×10^{-3}		
m-Cl	1.80×10^{-3}		
m-NO_2	1.14×10^{-3}		

straightforward explanation for relative reactivities in most cases.

The data in Table 18 illustrate the small effects of polar substituents on the rates of the acid-catalyzed hydrolysis and formation of substituted benzoate esters. The rate of hydrolysis of the derivative with the most strongly electron-attracting substituent in the *para* position is faster than that of the compound with the most strongly electron-donating group, but by a factor of barely two. On the other hand, the *p*-nitro compound is hydrolyzed more than twenty times as fast as the ester with the nitro group in the *ortho* position. It is quite clear that this must be a steric effect: with four groups attached to the central carbon atom, the tetrahedral intermediate, and thus also, to a somewhat lesser extent, the transition state for hydrolysis, is more crowded than the trigonally hybridized ground state, with the result that the addition of substituents sufficiently close to the carbonyl carbon atom causes an increase in non-bonded interactions in the transition state, and thus slows the reaction.

TABLE 19

KINETIC DATA FOR ESTER FORMATION FROM METHYL-SUBSTITUTED BENZOIC ACIDS IN METHANOL AT 25°C[153]

Acid	$k_{obs} \times 10^{-4}$ ($l \cdot mole^{-1} \cdot sec^{-1}$)	$\Delta H^{\ddagger}$ ($kcal \cdot mole^{-1}$)	$\Delta S^{\ddagger}$ (*eu*)
Benzoic	2.04	14.0	−28.4
3-Methylbenzoic	2.17	13.6	−32.4
4-Methylbenzoic	1.97	13.2	−29.8
3,4-Dimethylbenzoic	1.71	14.4	−27.5
3,5-Dimethylbenzoic	2.09	14.3	−27.5
2-Methylbenzoic	0.64	13.8	−31.2
2,3-Dimethylbenzoic	0.73	14.2	−30.0
2,4-Dimethylbenzoic	0.59	14.8	−28.3
2,5-Dimethylbenzoic	0.57	14.9	−28.1
2,6-Dimethylbenzoic	Too slow to measure		

This point is well-illustrated by the data of Table 19, which show the effect of methyl substituents on the rate coefficients for methyl ester formation from benzoic acid. The compounds fall naturally into three classes. Those with no *ortho* substituent react 3–4 times as fast as those which have one *ortho*-methyl group, while 2,6-dimethylbenzoic acid, the only compound with two *ortho*-substituents, did not give the ester at a measurable rate.

Ortho-substituents in general reduce the rates of acid-catalyzed ester formation and hydrolysis. Some typical data are collected in Table 20. In each series the rate coefficient decreases as the size of the *ortho* substituent increases. Only the smallest substituent of all, the fluorine atom, does not have a marked effect.

TABLE 20

KINETIC DATA FOR THE ACID-CATALYZED HYDROLYSIS AND FORMATION OF *ortho*-SUBSTITUTED BENZOATE ESTERS

	k_{obs} ($l \cdot mole^{-1} \cdot sec^{-1}$) for		
		Esterification of $X\text{-}C_6H_4COOH$	
Substituent (*X*)	*Hydrolysis of* $X\text{-}C_6H_4COOC_2H_5$ *at 100.8°C*[151]	*at 60°C in methanol*[145]	*at 55°C in cyclohexanol*[146]
H	18.8×10^{-5}	29.0×10^{-4}	15.6×10^{-6}
o-CH_3	3.68×10^{-5}	8.95×10^{-4}	7.67×10^{-6}
o-C_2H_5	2.05×10^{-5}	5.37×10^{-4}	
o-$CH(CH_3)_2$		4.97×10^{-4}	
o-F	18.5×10^{-5}	25.7×10^{-4}	23.5×10^{-6}
o-Cl	6.12×10^{-5}	10.6×10^{-4}	13.6×10^{-6}
o-Br	4.78×10^{-5}	7.48×10^{-4}	
o-I	2.77×10^{-5}	5.77×10^{-4}	6.6×10^{-6}
o-NO_2	0.51×10^{-5} [a]		2.87×10^{-6}

[a] Datum from ref. 128.

TABLE 21

RATE COEFFICIENTS FOR THE HYDROLYSIS AND FORMATION OF ALIPHATIC ESTERS

Acid R'COOH	k_{obs} *for hydrolysis of R'COOEt at 24.8°C in 70% acetone–water* ($l \cdot mole^{-1} \cdot sec^{-1}$)	*Ref.*	k_{obs} *for the formation of R'COOMe in methanolic HCl at 25°C* ($l \cdot mole^{-1} \cdot sec^{-1}$)	*Ref.*
HCOOH (in aq. HCl)	3.36×10^{-3}	78	0.83	157
CH_3COOH	4.47×10^{-5}	154	5.93×10^{-2}	157
CH_3CH_2COOH	3.70×10^{-5}	154	5.73×10^{-2}	157
$CH_3(CH_2)_2COOH$	1.96×10^{-5}	154	2.90×10^{-2}	157
$CH_3(CH_2)_3COOH$	1.79×10^{-5}	154	3.08×10^{-2}	157
$CH_3(CH_2)_4COOH$	1.77×10^{-5}	154	2.92×10^{-2}	157
$CH_3(CH_2)_5COOH$	1.64×10^{-5}	154		
$CH_3(CH_2)_6COOH$	1.55×10^{-5}	154		
$(CH_3)_2CHCOOH$ (at 25°C)	1.35×10^{-5}	155	1.95×10^{-2}	158
$(CH_3)_3CCOOH$	1.28×10^{-6}	154	1.94×10^{-3}	159
$C_2H_5(CH_3)CHCOOH$	5.72×10^{-6}	154		
$(C_2H_5)_2CHCOOH$	1.17×10^{-6}	154	5.02×10^{-4}	158
$C_6H_{11}COOH$	7.4×10^{-6}	155	1.18×10^{-2}	158
	3.0×10^{-6}	156		
$(n\text{-}C_3H_7)_2CHCOOH$			4.25×10^{-4}	158
$(n\text{-}C_4H_9)_2CHCOOH$			3.87×10^{-4}	158
$(C_6H_5)_2CHCOOH$			1.87×10^{-3}	158
$C_6H_5CH_2COOH$	9.34×10^{-6}	156	2.62×10^{-2}	160
$C_6H_5(CH_2)_2COOH$	8.05×10^{-6}	156	2.67×10^{-2}	158
$C_6H_5(CH_2)_3COOH$	8.00×10^{-6}	156	2.67×10^{-2}	158
$C_6H_5(CH_2)_4COOH$	8.61×10^{-6}	156	2.79×10^{-2}	158
$C_6H_5(CH_3)CHCOOH$	1.71×10^{-7}	156		
$C_6H_5(C_2H_5)CHCOOH$	6.69×10^{-8}	156		
$(C_6H_5)_2CHCOOH$	2.96×10^{-8}	156		
$C_6H_{11}CH_2COOH$			6.62×10^{-3}	158
$C_6H_{11}(CH_2)_2COOH$			2.81×10^{-2}	158
$C_6H_{11}(CH_2)_3COOH$			2.60×10^{-2}	158
FCH_2COOH	2.35×10^{-5}	161		
$F_2CHCOOH$	9.5×10^{-5}	161		

More extensive data are available for aliphatic carboxylic acids and esters, where the number of substituion positions is not limited to two. A selection of these data is included in Table 21, for ester formation in methanol and for the hydrolysis of methyl esters. From attempts to interpret the data given in this table have evolved many of our current ideas about steric effects. They are the source of Taft's steric substituent constants E_s[152], and illustrate Newman's rule of six[162], which emphasizes the large contribution of the steric factor of atoms in the six position (counting from the carbonyl oxygen as atom number one). Hancock *et al.*[163–165] has introduced a hyperconjugation term[163,164], and the change in the six-number[165], as independent variables in eq. (5) (p. 132).

Recent work has begun to explore the limitations of the applicability of Taft's steric substituent constants. Chapman *et al.*[145,151] and Bowden[166] showed that E_s^o values, derived for *ortho* substituents in benzoic acids and esters by Taft's procedure, are not independent of polar effects. The E_s^o values obtained differ for basic and acidic hydrolysis, and even for esterification by different alcohols (Table 22). These authors concluded[145] that data for ester formation are not suitable for the determination of E_s^o values. One reason would be that ρ values for ester formation, though not large, are not negligible (Table 16, p. 132). But more recently Charton[250,251] has produced evidence to suggest that the effects of *ortho* substituents are purely electrical.

Charton used statistical methods to investigate the contributions to Taft's E_s values of the size of the groups concerned and their polar effects. As a measure of group size he took the van der Waal's radius, r_v, and he used σ_I and σ_R as measures of inductive and mesomeric effects. Correlations of steric substituent constants were then carried out[250] using the equation (for group x)

$$E_s(x) = \alpha\sigma_I(x) + \beta\sigma_R(x) + \psi r_v(x) + h$$

TABLE 22

VARIATION OF THE STERIC SUBSTITUENT CONSTANT WITH CONDITIONS IN REACTIONS OF *ortho*-SUBSTITUTED BENZOIC ACID DERIVATIVES

	$-E_s$			
Substituent	a	b	c	d
H	0	0	0	0
o-CH_3	0.49	0.709	0.56	0.35
o-C_2H_5		0.963	0.84	
o-$(CH_3)_2CH$			0.86	
o-$(CH_3)_3C$			3.7	
o-F	0.00	0.008	0.14	−0.13
o-Cl	0.31	0.488	0.49	0.12
o-Br	0.49	0.595	0.64	0.26
o-I	0.69	0.833	0.84	0.37

[a] According to Taft[152].

[b] From data for the hydrolysis of *ortho*-substituted methyl benzoates at 100.8°C in 80% methanol–water[151].

[c] From data for the esterification of *ortho*-substituted benzoic acids in methanol, measured at 60°C and extrapolated to 100.8°C[145].

[d] From data for the esterification of *ortho*-substituted benzoic acids in cyclohexanol at 60°C, calculated from the data of ref. 158[145].

For E_s values, obtained for aliphatic compounds, ψ was clearly the most significant coefficient. E_s values are a linear function of the van der Waal's radii, and are correlated by the equation[301]

$$E_s(x) = \psi r_v(x) + h$$

They are thus independent of group polarity, and are genuine measures of steric effects.

The *ortho* substituent constants, E_s^o, on the other hand, gave no significant correlation with van der Waal's radii. The effects of groups in the *ortho* position are fully accounted for in terms of σ_I and σ_R, with the mesomeric term much the more important. Thus the steric effects of groups in the *ortho* position are constant, negligible, or non-existent[251]. This result was confirmed by an analysis of the data for acid-catalyzed ester hydrolysis and formation, including those of Tables 20 and 22.

These conclusions seem to have the following implications. There must be negligible steric hindrance to the formation of the near-tetrahedral transition state from a single *ortho* substituent. Since all *ortho* substituents, with the exception of fluorine, and oxygen and nitrogen-linked groups, cause a considerable rotation of the plane of the carboxyl group out of the plane of the ring[252] in the ground state, the approach of the nucleophile from one side of the carboxyl group is probably quite severely hindered. This would explain why the introduction of an *ortho* substituent reduces the reactivity of the carboxyl group, but to an extent that, in general, shows no correlation with the size of the substituent. A very large *ortho* substituent, however, would be expected to hinder the formation of the transition state also, and it is significant that the effect of the *ortho-t*-butyl group is very much greater than that of other alkyl substituents[145].

2.4.2 The effects of substitution in the leaving group

The data of Tommila and Hinshelwood[217] (Table 23), for the acid-catalyzed hydrolysis of alkyl and substituted-phenyl acetates show that although electron-withdrawing substituents do consistently reduce k_{H^+}, on the whole acid-catalyzed ester hydrolysis is little affected by changes in polarity of the leaving group. The steric effects of the group R of R′COOR are also small, because of the interposition of the oxygen atom between the alkyl and the carbonyl groups. Thus the rate coefficients for the hydrolysis of ethyl, isopropyl and *t*-butyl formates in dilute aqueous HCl at 25°C are 3.36×10^{-3}, 2.92×10^{-3}, and 5.98×10^{-3} l·mole^{-1}·sec^{-1}, respectively[78]. Although the last figure almost certainly contains a contribution from alkyl–oxygen cleavage,

TABLE 23

KINETIC DATA FOR THE ACID-CATALYZED HYDROLYSIS OF ACETATE ESTERS IN 60% ACETONE–WATER AT 25°C

Ester	k_{hyd} *(l·mole⁻¹·sec⁻¹)*	$\Delta H^{\ddagger}$ *(kcal·mole⁻¹)*	$\Delta S^{\ddagger}$ *(eu)*	*Ref.*
$CH_3COOCH_2CH_3$[a]	4.47×10^{-5}	15.6	−26.1	154
$CH_3COOCH_2C_6H_5$	3.08×10^{-5}	15.8	−25.9	217
$CH_3COOCH_2C_6H_4$-*p*-CH_3	3.16×10^{-5}	15.6	−26.6	217
$CH_3COOCH_2C_6H_4$-*p*-NO_2	2.76×10^{-5}	15.6	−26.6	217
$CH_3COOC_6H_5$	2.77×10^{-5}	16.6	−23.3	217
$CH_3COOC_6H_4$-*p*-CH_3	2.93×10^{-5}	16.6	−23.3	217
$CH_3COOC_6H_4$-*p*-COOH	2.37×10^{-5}	16.5	−24.3	217
$CH_3COOC_6H_4$-*m*-NO_2	1.95×10^{-5}	16.1	−25.8	217

[a] At 24.8°C in 70% acetone–water.

the effects are clearly much smaller than those found for substitution at the acyl position (Table 21).

These two sets of data are for compounds which would not be expected to show a high sensitivity to steric effects, since the acyl substituent is methyl or hydrogen, the least sterically demanding groups. An indication that steric hindrance can become important if both the leaving group and the acyl group are large, comes from the results of Harfenist and Baltzly[167], for acid-catalyzed alcoholysis reactions of various β-naphthyl esters at 25°C (Table 24). These

TABLE 24

RELATIVE RATES[a] OF HCl-CATALYZED ALCOHOLYSIS REACTIONS OF β-NAPHTHYL ESTERS IN ALCOHOLS AT 25°C[167]

	Ester		
Alcohol	CH_3COOAr $(k_{rel}/913)$	$(CH_3)_2CHCOOAr$ $(k_{rel}/50.4)$	$(CH_3)_3CCOOAr$ k_{rel}
CH_3OH	69.6	183	455
CH_3CH_2OH	16.8	28.6	
$CH_3CH_2CH_2OH$	16.2	24.4	35.3
$(CH_3)_2CHOH$	1.0	1.0	1.0

[a] k_{rel} is relative to β-naphthyl trimethylacetate.

data are semi-quantitative only: the HCl concentration was varied widely (from $6\times10^{-3}\,M$ to $4\,M$) to attain convenient rates, and second-order rate coefficients were calculated on the basis of the single acid concentration used; also, of course, the solvent is different for each set of alcoholysis reactions. However, the trends are unmistakeable; in the reaction

$$R'-C(=O)-O-C_{10}H_7 + ROH = R'-C(=O)-OR + C_{10}H_7OH$$

reactivity becomes progressively more sensitive to the steric requirements of the alcohol, ROH, as the size of the acyl substituent, R′, is increased.

2.4.3 The mechanism of steric effects

For acid-catalyzed ester formation and hydrolysis the transition state (or states) is more crowded than the initial state, because the central atom is close to tetrahedral, and also more highly ordered, because two molecules have combined to form one. The introduction of a bulky substituent into either R or R′ in the reaction

$$R'-C(=O)-OR + H_2O \underset{}{\overset{(H^+)}{\rightleftharpoons}} R'-C(OH)_2-OR \overset{(H^+)}{\rightleftharpoons} R'-C(=O)-OH + ROH$$

can affect the free energy of activation, and thus the rate, in several ways. Experimentally it is a relatively simple matter to separate effects on the enthalpy of activation from those on the entropy term. In practice a group which has a significant effect on one also alters the other, and generally the effects are partially compensating. A selection from the data available is given in Table 25.

TABLE 25

ACTIVATION PARAMETERS FOR ACID-CATALYZED ESTER HYDROLYSIS AND FORMATION[a]

Aliphatic						
	Hydrolysis of RCOOEt			*Formation of RCOOMe in methanol*		
Acid RCOOH	$\Delta H^{\ddagger}$ (*kcal·mole*$^{-1}$)	$\Delta S^{\ddagger}$ (*eu*)	*Ref.*	$\Delta H^{\ddagger}$ (*kcal·mole*$^{-1}$)	$\Delta S^{\ddagger}$ (*eu*)	*Ref.*
HCOOH	14.7	−20.5	78	9.4	−27.5	157
CH_3COOH	15.6	−26.1	154	9.4	−32.5	157
CH_3CH_2COOH	15.5	−26.4	154	9.4	−33.9	157
$CH_3(CH_2)_4COOH$	15.6	−27.9	154	9.3	−33.9	157
$(CH_3)_2CHCOOH$	15.8	−27.9	155	9.2	−35.4	160
$(CH_3)_3CCOOH$	18.5	−23.6	155	10.8	−34.6	159
$(C_2H_5)_2CHCOOH$				11.8	−34.4	159

TABLE 25 (CONTINUED)

Aliphatic

Acid RCOOH	Hydrolysis of RCOOEt $\Delta H^{\ddagger}$ (kcal·mole^{-1})	$\Delta S^{\ddagger}$ (eu)	Ref.	Formation of RCOOMe in methanol $\Delta H^{\ddagger}$ (kcal·mole^{-1})	$\Delta S^{\ddagger}$ (eu)	Ref.
$(n\text{-}C_3H_7)_2CHCOOH$				11.8	−34.3	159
$(n\text{-}C_4H_9)_2CHCOOH$				12.3	−32.8	159
C_6H_5COOH	18.9	−26.4	128	15.4	−24.1	145
				14.0	−28.4	153
$C_6H_5CH_2COOH$	15.5	−28.8	155	9.3	−34.5	160
$C_6H_5(CH_2)_2COOH$	15.7	−28.4	156	9.0	−35.4	158
$C_6H_5(CH_2)_4COOH$	16.0	−27.2	156	9.3	−34.3	158
$C_6H_5CH(CH_3)COOH$	16.3	−29.4	156			
$C_6H_5CH(C_2H_5)COOH$	16.5	−30.6	156			
$(C_6H_5)_2CHCOOH$	17.1	−30.2	156			
$C_6H_{11}COOH$	16.3	−28.3	156	9.4	−35.7	159
$C_6H_{11}CH_2COOH$				9.9	−35.3	159
$C_6H_{11}(CH_2)_3COOH$				9.5	−33.8	159
$(CH_3)_3N^+CH_2CH_2COOH$	16.0	−25.4±4	127			
$ClCH_2COOH$	14.7	−29.1	128			
$Cl_2CHCOOH$	13.6	−33.9	128			
Cl_3CCOOH	11.5	−37.9	128			

Aromatic

Substituent (X) in X-C_6H_4COOH	Hydrolysis of ArCOOEt[128] $\Delta H^{\ddagger}$ (kcal·mole^{-1})	$\Delta S^{\ddagger}$ (eu)	Formation of ArCOOMe in methanol[145] $\Delta H^{\ddagger}$ (kcal·mole^{-1})	$\Delta S^{\ddagger}$ (eu)	Formation of ArCOOC_6H_{11} in cyclohexanol[146] $\Delta H^{\ddagger}$ (kcal·mole^{-1})	$\Delta S^{\ddagger}$ (eu)
H	18.9	−26.4	15.4	−24.1	18.9	
m-CH_3			13.2	−29.8[153]	19.4	−21.3
p-CH_3	19.0	−26.4	13.8	−31.2[153]	17.9	−26.3
p-Cl	18.3	−28.4			17.7	−27.1
p-Br	18.2	−28.4				
p-OH	19.7	−25.7				
p-OCH_3	19.9	−24.7			19.6	−22.4
m-NO_2	18.1	−28.6			17.7	−25.8
p-NO_2	16.8	−31.8			18.0	−26.5
o-NO_2	19.4	−30.4			16.8	−28.9
o-F			14.3	−27.7	16.9	−28.3
o-Cl			14.9	−27.7	17.5	−27.6
o-Br			14.3	−30.2	18.0	−26.5
o-I			13.6	−32.8	17.0	−30.3
o-CH_3			14.4	−29.6	17.8	−27.8
o-C_2H_5			14.2	−31.1		
o-$CH(CH_3)_2$			14.2	−31.2		

[a] Conditions were as follows: refs. 154, 155, 156, 70% acetone–water; ref. 78, aqueous HCl; ref. 128, 60% ethanol–water.

A bulky substituent close to the reaction centre may increase the non-bonded compression energy as the transition state is formed: this will cause an increase in $\Delta H^{\ddagger}$. It will also hinder the close approach of solvent molecules to the reaction centre, thus reducing the maximum amount of stabilization possible (steric inhibition of solvation). This will result in a further increase in $\Delta H^{\ddagger}$, but since decreased solvation means less ordering of solvent molecules about the transition state, there is a compensating increase in $\Delta S^{\ddagger}$. Another effect of the bulky substituent may be to block certain vibrational and rotational degrees of freedom more in the (more crowded) transition state than in the initial state, and so to reduce $\Delta S^{\ddagger}$. These are the most important of the simple effects of a bulky substituent and can be used to explain most of the relationships of Table 25.

Formate esters are formed and hydrolyzed faster by one to two orders of magnitude than acetates and higher homologues. This cannot be a result of the different polar effects of the hydrogen atom and the methyl group because this sort of difference is reflected primarily in $\Delta H^{\ddagger}$, and the enthalpy of activation is the same for the esterification of formic and acetic acids in methanol, and not different for hydrolysis of the ethyl esters. The difference in rates is almost entirely accounted by the more favourable entropy of activation for the reaction of the formic acid derivative, and is thus due to the difference in size of the hydrogen atom and the methyl group. It is unlikely that non-bonded compression is an important factor, since both groups are small, and the size of the alkyl substituent does not affect the enthalpy term until it is as large as *t*-butyl. The rates of reaction of acids R′COOH and esters R′COOR decrease with increasing chain length of primary alkyl groups R′, until this decrease levels out. The enthalpy of activation is essentially independent of the length of the chain, and the effect is entirely due to changes in the entropy term. This no doubt represents the reduction of the number of degrees of freedom in the transition state with increasing size of R′. It is relevant that when the degrees of freedom are already limited in the initial state, as they are, for example, in cyclohexane carboxylic acids, the entropy term is normal.

When non-bonded compression does eventually appear, its effects on both $\Delta H^{\ddagger}$ and $\Delta S^{\ddagger}$ are unmistakeable. A sharp increase in $\Delta H^{\ddagger}$, as is observed, for example, for the hydrolysis of ethyl trimethylacetate, is accompanied by a decrease in $\Delta S^{\ddagger}$, as expected if steric hindrance of solvation is occurring.

Steric effects on the reactivity of benzoic acid derivatives are a little more complicated. The first significant point is that the aromatic carboxylic acids and esters are often some two orders of magnitude less reactive than the corresponding aliphatic compounds. Chapman *et al.*[147], have explained this difference in terms of three co-operative factors: (*i*) the stabilization of the initial state by delocalization in the case of the aromatic compounds, (*ii*) inductive electron-withdrawal by the ring, which is significant in esterification reactions which have significant ρ values, and (*iii*) the steric effects of the hydrogen atoms in the *ortho* position of benzoic acid derivatives. This last factor is unlikely to be large, because the difference in reactivity is entirely due to a higher enthalpy of activation for the reactions of benzoic acids and ester; the entropies of activation are similar for the hydrolysis and esterification reactions of both aliphatic and aromatic compounds.

Substituents in the *ortho* position, almost without exception, reduce the reactivity of benzoic acid derivatives: a significant rate increase is a *prima facie* case for intramolecular catalysis (see p. 201). This effect is associated in the case of the esterification of *ortho*-substited benzoic acids with a somewhat lower enthalpy of activation, and a greater decrease in the entropy of activation. The first factor may be due to a secondary steric effect[145], whereby the energy of the initial state is raised because the carboxyl group is twisted out of the plane of the ring, to reduce non-bonded interaction with the *ortho*-substituent (see p. 178), and delocalization energy is lost. The lower entropy term is presumably accounted for by the bulk effect of the substituent.

The data for acid-catalyzed ester formation in cyclohexanol are doubly interesting. The activation parameters are closely similar to those for the acid-catalyzed hydrolysis of the corresponding ethyl esters. The enthalpy of activation is considerably higher than for esterification in methanol; this is probably a result of steric inhibition of solvation, as well as non-bonded compression in the transition state, as suggested by the entropies of activation, which are also significantly higher than with methanol, especially for compounds without *ortho* substituents which presumably have more transition state solvation to lose.

2.4.4 Reactions of very weakly basic acids and esters

Several acids with very strongly electron-withdrawing substituents are not protonated to a significant extent in 100% sulphuric acid[4]. These include trichloroacetic and trifluoroacetic acids, and perhaps also oxalic acid. Both Hantszch[4] and Wiles[168] found *i*-values for the latter acid somewhat greater than 1.0, but the data are not precise because oxalic acid is decomposed in sulphuric acid. The hydrolysis of the esters of these compounds is catalyzed very weakly, if at all, by acid[169]. A considerable amount of accurate data is now available for the reactions of oxalate and trifluoroacetate esters, and some firm conclusions can be drawn.

Oxalate and trifluoroacetate esters are reactive compounds, which are hydrolyzed in aqueous acid at rates which are not a great deal faster than the rates in neutral solution. Moffat and Hunt[170], for example, studied the hydrolysis of a series of alkyl trifluoroacetates in 70% acetone–water, and found that the acid released had no observable effect on the rate coefficient for hydrolysis up to at least 60% of reaction (although other authors do claim to have observed autocatalysis). When a strong mineral acid is added complex behaviour is observed. Typically the observed first-order rate coefficient shows an initial increase, but after reaching a maximum value, it actually decreases as the acid concentration is increased further[89–91,171]. This behaviour is illustrated in Fig. 12 for the hydrolysis of *o*-nitrophenyl oxalate[171] and methyl[90] and phenyl[89] trifluoroacetates. This pattern of behaviour is not to be confused with that of simple aliphatic esters (Fig. 1, p. 74), where a maximum is observed in the region of 70% H_2SO_4. The maxima here occur in dilute solution, between 1 and 2 *M* acid.

All these reactions involve acyl–oxygen cleavage. This has been demon-

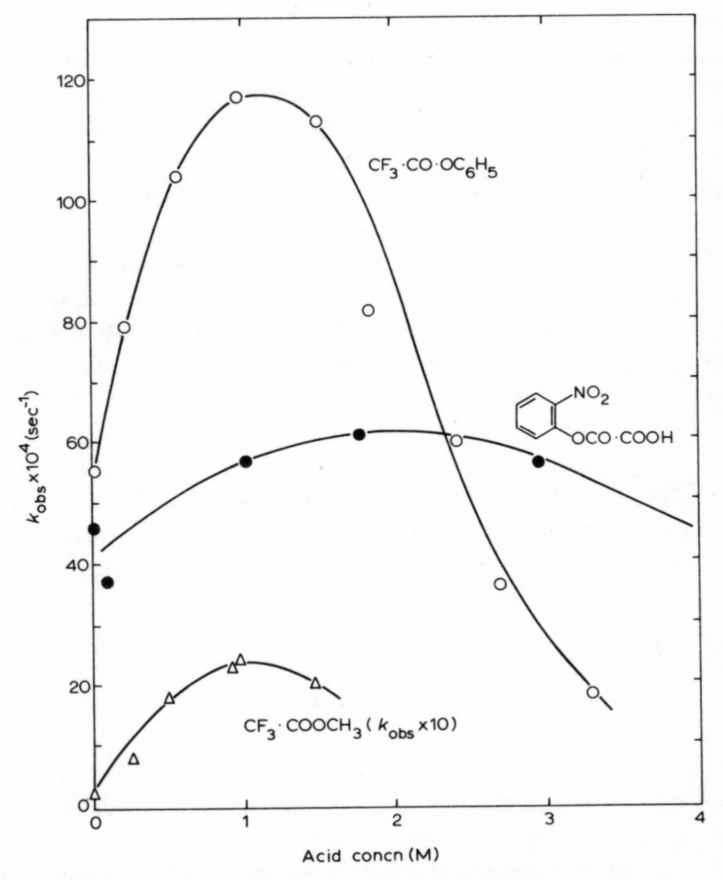

Fig. 12. First-order rate coefficients for the hydrolysis of weakly basic esters, as a function of acid concentration. Data for phenyl and methyl trifluoroacetates in $HClO_4$ in 70% dioxan–water at 0°C (from refs. 89 and 90), and for *o*-nitrophenyl hydrogen oxalate in aqueous HCl at 25.5°C (from ref. 171; corrected for the reaction of the anion using $pK_a = 0.35$).

strated for the esterification of optically active 2-octanol by trifluoroacetic acid[172], and by ^{18}O incorporation experiments for methyl[90] and phenyl trifluoroacetates[89]. It may be assumed for the oxalate ester, since aromatic nucleophilic displacement on *o*-nitrophenyl compounds is much slower than the reactions considered here. The acid hydrolysis of diphenylmethyl trifluoroacetate, on the other hand, involves both alkyl and acyl–oxygen cleavage[89].

For the three reactions represented in Fig. 12 the maximum rate of hydrolysis in acid represents only a modest acceleration, compared with the rate in initially neutral solution. Bunton and Hadwick[89,90] explained the maximum for methyl and phenyl trifluoroacetate in terms of negative salt effects on both acid-catalyzed and neutral reactions. Consistent with this interpretation, it was demonstrated directly that the rate of neutral hydrolysis is decreased by added salts. The effect of added salt should be to decrease the activity of water, and perhaps also to salt in the ester.

Bruice and Holmquist[91] have extended the investigation of Bender and Chow[171] of the hydrolysis of *o*-nitrophenyl hydrogen oxalate, in particular by making measurements in HCl solutions of constant ionic strength, up to 8 *M*. The pH–rate profile is complicated by the ionization to the anion, and it was shown that when *o*-nitrophenyl ethyl oxalate was used, the initial increase in rate with increasing acid concentration disappeared, and in the absence of added salt only the decrease over the range 1–8 *M* acid was observed. Bruice and Holmquist[91] attribute this decrease to the decrease in the activity of water caused by the increasing salt concentration: a Bunnett plot of ($\log k_{obs} - H_0$) against $\log a_{H_2O}$ gave slopes, w, of +7.4 and +9.7 for *o*-nitrophenyl hydrogen oxalate and its ethyl ester, respectively[91]. (A similar plot using the data of Bunton and Hadwick[89,90] for the hydrolysis of phenyl trifluoroacetate shows an even larger positive slope.) Large positive values of w are usually ascribed to reactions in which water acts as a proton transfer agent in the rate-determining step, and are consistent with the accepted mechanism for neutral hydrolysis, in which the attack of one molecule of water is subject to general base catalysis by one or more others (see p. 158).

$$H_2O\colon \curvearrowright H{-}O(H) \curvearrowright C(R')(=O)(OR) \longrightarrow$$

The addition of LiCl to the solutions of *o*-nitrophenyl hydrogen and ethyl oxalates at pH 2.0 causes a progressive decrease in the rate of hydrolysis of each ester, and the rate depression, about tenfold over the range 1.0–6.0 *M* LiCl, is very similar to that observed when HCl is added[91]. This makes it clear that the decrease in rate is no more than an effect of changing ionic strength. When the ionic strength is maintained constant, at 8.0, with LiCl, a rate increase is observed as the concentration of hydronium ion is increased[91]. Bruice and Holmquist ascribe this "small and non-proportional increase in rate with increasing hydronium ion concentration" to the slightly different effects of H_3O^+ and Li^+ on the activities of water and of the ester.

The above is quite good evidence that acid catalysis of the hydrolysis of *o*-nitrophenyl oxalate esters is negligible, up to 8 *M* HCl. The question remains, whether the initial increase in the rate of hydrolysis of trifluoroacetate esters, up to about 1.5 *M* acid, is also a specific salt effect on the neutral hydrolysis reaction, or represents a genuine, though very weakly, acid-catalyzed reaction. Since inductive electron-withdrawal by the CF_3 group, as measured by σ_I[173] is slightly greater than by COOH or COOEt, it is unlikely that trifluoroacetate esters are inherently more sensitive to acid catalysis, but it is possible that significant catalyzed reactions of the nitrophenyl oxalates are concealed by the faster neutral hydrolysis.

To summarize, it seems reasonable to conclude that, relative to the water reaction, acid-catalyzed hydrolysis is insignificant for *o*-nitrophenyl oxalates, is possibly significant for phenyl trifluoroacetate, and is probably significant for methyl trifluoroacetate. It is doubtless no coincidence that this order is that of increasingly poor leaving group, and this suggests that breakdown of the tetrahedral intermediate is the step most affected by acid-catalysis in these compounds. However, the data of Bruice and Holmquist[91] rule out the further possibility that unsymmetrical partitioning of the tetrahedral intermediate is the explanation of the rate depression in acid. This is so for the superficially similar case of ethyl trifluorothiolacetate[174], in which the breakdown of the tetrahedral intermediate to regenerate starting materials is catalyzed, though the formation of products is not, *viz*.

$$CF_3{-}C(=O){-}SEt \underset{k_2[H_3O^+]}{\overset{k_1[H_2O]}{\rightleftharpoons}} CF_3{-}C(OH)(SEt){-}OH \xrightarrow{k_3} CF_3COOH + EtSH$$

Bender and Heck[175], using carbonyl-^{18}O-labelled ester, have shown that the extent of ^{18}O-exchange accompanying the hydrolysis of ethyl trifluorothiolacetate is that predicted on the basis of this unsymmetrical mechanism, but that k_{exch}/k_{hyd} for carbonyl-^{18}O-labelled ethyl trifluoroacetate is independent of pH, and thus that the partitioning of the tetrahedral intermediate is symmetrical. Their value of k_{exch}/k_{hyd} is 0.6 ± 1, confirming that the breakdown of the tetrahedral intermediate is partially rate-determining. Bender and Heck write two kinetically equivalent mechanisms for the hydrolysis of ethyl trifluoroacetate in moderately concentrated acid solutions, of which the following accounts most simply for all the observed facts.

$$CF_3COOEt + H_2O \underset{k_2[H_3O^+]}{\overset{k_1[H_2O]}{\rightleftharpoons}} CF_3-\overset{OH}{\underset{OEt}{C}}-O^- \xrightarrow{k_3[H_3O^+]} CF_3COOH + EtOH$$

As the leaving group is varied from ethanol to phenol to *o*-nitrophenol, k_3 will increase, relative to k_2, and a neutral, water-catalyzed elimination of the leaving group may become important. This will result in the breakdown of the tetrahedral intermediate becoming relatively faster than its formation, and thus also in a decrease in the effectiveness of acid catalysis, since the formation of the tetrahedral intermediate is not significantly acid-catalyzed.

3. Nucleophilic and general catalysis

The reactions discussed in this section are those of ester hydrolysis and formation under neutral or basic conditions, designated by the letter B in Ingold's classification (ref. 1, p. 753). This terminology will not be much used in this review, because the vast majority of reactions fall under the same heading of $B_{AC}2$. As discussed previously, unimolecular acyl–oxygen fission of esters requires very strongly acidic conditions, so that the $B_{AC}1$ mechanism can be ruled out. Also the reactions of esters which involve alkyl–oxygen fission under neutral or basic conditions are best regarded simply as S_N1 and S_N2 reactions, in which the leaving group happens to be a carboxylate anion. Reactions of this type are discussed here only briefly.

3.1 REACTIONS INVOLVING ALKYL–OXYGEN CLEAVAGE

3.1.1 Ester hydrolysis

Many of the esters which are hydrolyzed by the $A_{AL}1$ mechanism in acid are also hydrolyzed with alkyl–oxygen fission under neutral conditions[60,67,74,75,84,85,88,89]. These reactions have the high enthalpies and entropies of activation characteristic of unimolecular reactions, and involve the ionization of (usually) tertiary alkyl esters, to the carbonium ion and a carboxylate anion in the rate-determining step, *viz.*

$$R'-C(=O)-O-CR_3 \underset{}{\overset{\text{slow}}{\rightleftharpoons}} R'-CO_2^- \quad \overset{+}{C}R_3$$

$$\downarrow \text{fast, } H_2O$$

$$R'COOH + R_3COH$$

These are best regarded as S_N1 reactions, in which the leaving group is the carboxylate anion. The point is brought out very well by Moffat and Hunt's comparison[79] of the solvolyses of *t*-alkyl trifluoroacetates in 70% aqueous acetone with the reactions of the corresponding *t*-alkyl halides (Table 26). The activation parameters, and the fate of the carbonium ion, as measured by the percentage of olefin formed by the parallel E1 reaction, are closely similar for the two types of substrate.

TABLE 26

COMPARISON OF THE HYDROLYSIS OF *t*-ALKYL FLUOROACETATES WITH THE SOLVOLYSIS OF *t*-ALKYL HALIDES AT 25°C

(Ref. 79)

Compound	*Relative rate*	*Elimination (%)*	$\Delta H^{\ddagger}$ *(kcal·mole⁻¹)*	$\Delta S^{\ddagger}$ *(eu)*	*Solvent*
$(CH_3)_3C\cdot OOCF_3$	0.47	15	23.8	−4.3	70% Acetone
$(CH_3)_3C\cdot Cl$	0.60	17	22.6	−6.2	80% EtOH
$(CH_3)_3C\cdot Cl$			22.3	0.6	50% EtOH
$C_2H_5(CH_3)_2C\cdot OOCF_3$	1.0	25	23.6	−3.4	70% Acetone
$C_2H_5(CH_3)_2C\cdot Cl$	1.0	33	22.3	−6.2	80% EtOH
$C_3H_7(CH_3)_2C\cdot OOCF_3$	0.93	33	23.7	−3.2	70% Acetone
$C_3H_7(CH_3)_2C\cdot Cl$	0.95	33			80% EtOH

Another well-established class of ester reactions which involve alkyl–oxygen cleavage belongs to the large category of S_N2 reactions.

$$Y^- + R{-}O{-}C(=O){-}R' \longrightarrow RY + R'COO^-$$

Only a few points of interest which are specifically properties of the S_N2 reactions of esters will be discussed here.

Since nucleophilic reactivity towards sp^3-hybridized carbon, in S_N2 reactions, differs markedly from reactivity towards the sp^2-hybridized carbon of the carbonyl group, it is a relatively simple matter to design reactions of esters which involve bimolecular attack on the alkyl carbon. Primary alkyl esters, and especially the methyl compounds, will be most susceptible to S_N2 reactions. For example, a useful (though neglected) procedure for dealkylating methyl esters, including severely sterically hindered compounds, is to heat them with lithium iodide in pyridine or 2,4,6-trimethylpyridine[176,177]. The reaction allows a more than usually precise examination of the dependence on structure of reactivity in the S_N2 reaction, since a wide range of leaving groups of a single class are available. A relevant study is that by Hammett and Fluger[178] on the reaction of trimethylamine with substituted methyl benzoates in methanol at 100°C

$$(CH_3)_3N: + CH_3{-}OCOAr \longrightarrow (CH_3)_4N^+ \quad ArCOO^-$$

This was very early work, and not enough data for *meta*- and *para*-substituted compounds were obtained for an accurate determination of ρ, but the plot of $\log k$ *versus* pK_a of ArCOOH (both at 100°C) for *o*- and *p*-substituted methyl benzoates gave a good straight line of slope 0.67. This result, together with that of Eschenmoser *et al.*[177], emphasizes that steric hindrance at the carbonyl group has little or no effect on the displacement of the ether oxygen of the ester.

The S_N2 reaction of alkyl carboxylates can also be isolated under conditions where attack at the carbonyl group is the faster reaction, if the nucleophile concerned is identical with the leaving group. Under these conditions displacement at the carbonyl carbon leads to no net reaction (unless of course the alcohol is labelled[179]), *viz.*

$$R'{-}C(=O)OR + RO^- \overset{ROH}{\rightleftharpoons} R{-}C(OR)_2{-}O^-$$

and other, slower, reactions can be observed. Using this technique Bunnett *et al.*[180] were able to isolate dimethyl ether in 74% yield from the reaction of methoxide with methyl benzoate in methanol at 100°C.

$$Ph{-}C(OCH_3)_2{-}O^- \rightleftharpoons Ph{-}C(=O){-}O{-}CH_3 \;\; {}^{\ominus}OCH_3 \longrightarrow Ph{-}COO^- + CH_3OCH_3$$

Oxygen nucleophiles are normally more reactive towards sp^2- than sp^3-hybridized carbon, and consequently bimolecular alkyl–oxygen fission, identified (for example) by using optically active alkyl carboxylates, is not normally observed in ester solvolysis reactions. The single exception is the neutral hydrolysis or solvolysis of β-lactones, and proves the rule very convincingly. Olson and Miller[52] showed that optically active β-butyrolactone is hydrolyzed to opposite enantiomers of β-hydroxybutyric acid in water and alkali, an observation similar to that made earlier by Ingold *et al.*[181] using β-malolactonate. Olson and Miller showed, using ^{18}O as a tracer, that the neutral hydrolysis involves alkyl–oxygen cleavage, *viz.*

$$H_2{}^{18}O + \underset{CH_2-C=O}{\overset{CH_3}{CH}{-}O} \longrightarrow CH_3{-}CH(^{18}OH){-}CH_2{-}COOH$$

Similarly, Bartlett and Rylander[182] have shown that the methanolysis of β-propiolactone leads to β-methoxypropionic acid. Long and Purchase[54] have studied the neutral hydrolysis of β-propiolactone, and find that it reacts four times faster than β-butyrolactone, as expected for the S_N2 (but not the S_N1) mechanism; these workers measured the activation parameters for the neutral reaction, and found $\Delta H^{\ddagger} = 18.9$ kcal·mole^{-1} and $\Delta S^{\ddagger} = -14.7$ eu. All this evidence is consistent with the neutral hydrolysis of β-propiolactone being a unique case of the $B_{AL}2$ reaction. A final, decisive piece of evidence, is the identification of the corresponding unimolecular reaction of a β-lactone, by Liang and Bartlett[83]. β-Isovalerolactone is hydrolyzed in alkali more slowly than β-butyrolactone, but its neutral hydrolysis is about 100 times faster (at 25°C). This is the behaviour expected if the mechanism has changed to a rate-determining ionization

$$(CH_3)_2C{-}O{-}C(=O){-}CH_2 \text{ (ring)} \xrightarrow{\text{slow}} (CH_3)_2C^+{-}CH_2{-}CO_2^-$$

Confirmation that a carbonium ion is an intermediate can be found in the observation that the major product of neutral solvolysis is isobutene, resulting from its decarboxylation[83], *viz.*

$$(CH_3)_2C^+{-}CH_2{-}CO_2^- \longrightarrow (CH_3)_2C{=}CH_2 + CO_2$$

3.1.2 Ester formation

Esters are formed in nucleophilic substitution reactions in which the nucleophile is a carboxylate anion. The anions of carboxylic acids are relatively weak nucleophiles towards sp^3-hybridized carbon. Swain's nucleophilic constant, n, for acetate ion is 2.7[183], slightly smaller than that for chloride. Thus acetate is selectively alkylated by alkyl halides in aqueous solution, *e.g.*

$$CH_3{-}CO_2^- + CH_3Br \longrightarrow CH_3COOCH_3 + Br$$

Typical synthetic procedures include the reaction of alkyl halides with the silver salts of carboxylic acids, the reaction of carboxylate anions in alkali with an excess of a dialkyl sulphate, (especially dimethyl sulphate), and heating tertiary[184] or quaternary ammonium salts of carboxylic acids. These routes are particularly valuable for the preparation of esters of seriously sterically hindered acids. For example, Fuson *et al.*[185] made the methyl ester of 2,4,6-triethylbenzoic acid by heating the tetramethyl ammonium salt to 200–250°C, *viz.*

$$2,6\text{-}(C_2H_5)_2\text{-}4\text{-}C_2H_5C_6H_2\text{-}CO_2^{-} + CH_3\text{-}\overset{+}{N}(CH_3)_3 \longrightarrow 2,6\text{-}(C_2H_5)_2\text{-}4\text{-}C_2H_5C_6H_2\text{-}COOCH_3 + (CH_3)_3N$$

3.2 REACTIONS INVOLVING ACYL–OXYGEN CLEAVAGE

The category $B_{AC}2$ (ref. 1, p. 753) referred originally to alkaline hydrolysis, but included the then relatively small class of hydrolytic reactions proceeding in neutral solution. It is our picture of this class of reactions which has changed most dramatically in the past fifteen years, with the discovery of a multitude of catalytic processes which can assist hydrolysis in neutral solutions. Alkaline hydrolysis itself can be regarded as just one example, albeit a special case, of a large class of reactions which involve nucleophilic attack on esters

$$R'\text{—}C(=O)OR + Y^- \longrightarrow R'\text{—}C(=O)Y + RO^-$$

Catalysis of hydrolysis is observed if the reaction with the nucleophile is faster than hydrolysis under the reaction conditions, as long as the initial product is also more reactive than the starting material, *viz.*

$$R'\text{—}C(=O)OR + Y^- \longrightarrow R'\text{—}C(=O)Y + ROH$$

$$R'\text{—}C(=O)Y + H_2O \longrightarrow R'\text{—}C(=O)O^- + Y^-$$

The slow step of this process is often the initial nucleophilic displacement, and although the alkaline hydrolysis reaction ($Y^- = HO^-$) is unique in that only a single displacement is involved, the step concerned is the same as the rate-determining step for the reaction with many other nucleophiles. Kinetically, therefore, the reactions are comparable. A second important type of catalysis by nucleophilic reagents does not involve direct attack by the nucleophile on the ester. For example, the hydrolysis of ethyl dichloroacetate is catalyzed by aniline, and no anilide is produced in the reaction[186], *viz.*

$$(PhNH_2) + Cl_2CH{\cdot}COOC_2H_5 + H_2O = Cl_2CH{\cdot}COO^- + C_2H_5OH + H^+ (+ PhNH_2)$$

Reactions of this sort usually involve the nucleophile as a general base. They are not immediately distinguishable from nucleophilic reactions, since both may be first-order in both ester and nucleophile. The mechanism of general base catalysis of hydrolysis has been the subject of some discussion. Most authors agree in preferring a mechanism in which the nucleophile assists the attack of a water molecule on the carbonyl group, by removing a proton from it in the rate-determining step of the reaction, *viz.*

$$Y^- \; H{-}O(H) \; + \; R'C(=O)OR \rightleftharpoons YH + HO{-}C(R')(O^-){-}OR \longrightarrow R'COO^- + ROH$$

This is a termolecular process and would be expected to be entropically highly unfavourable, as indeed reactions of this sort generally are. A considerable amount of data is now available, (some of it is given in the appropriate sections below) and intermolecular reactions thought for other reasons to involve general base catalysis commonly have entropies of activation in the region −40 to −50 eu.

An alternative kinetically indistinguishable mechanism for general base catalysis involves general acid catalysis of the attack of hydroxide ion on the carbonyl group, *viz.*

$$HO^- + R'{-}C(=O)OR + H{-}Y \xrightarrow{\text{slow}} R'{-}C(OH)_2{-}OR + Y^- \xrightarrow{\text{fast}} R'{-}COO^- + ROH$$

This mechanism has been rejected by a number of authors in specific cases, and at present seems unlikely to be of wide importance.

Finally, different mechanisms must obtain in those cases where the breakdown of the tetrahedral intermediate is the slow step of the reaction. Two kinetically equivalent mechanisms are possible in this case also. Here, too, the catalyst may be involved as a general base, in which circumstances it would catalyze the elimination of the leaving group by the E-2 mechanism, *viz.*

$$Y^- \; H{-}O{-}C(R')(OH){-}OR \longrightarrow YH + R'{-}C(=O)OH + RO^-$$

or it may be involved as its conjugate acid, assisting the breakdown of the *anion* of the tetrahedral intermediate.

$$^-O{-}C(R')(OH){-}O{-}R \; + \; H{-}Y \longrightarrow R'{-}C(=O)OH + ROH + Y^-$$

This second mechanism is the reverse of that for general base-catalyzed addition, and inasmuch as the latter is the probable mechanism for the formation of the tetrahedral intermediate, it will represent the preferred transition state for its breakdown, at any rate for reactions that are close to symmetrical. An unsymmetrical mechanism is most likely where the leaving group is markedly different from HO^-. Normally RO^- in esters cannot be a much poorer leaving group than HO^-, but it can readily be a much better leaving group; in this case the breakdown of the tetrahedral intermediate will not be rate-determining. Thus catalysis of the breakdown of the tetrahedral intermediate is only likely to be significant for symmetrical reactions under neutral or alkaline conditions.

A detailed comparison of nucleophilic and general base catalysis of the reactions of carboxylic acid derivatives is already available in the excellent review by Johnson[101], and the subject will not be discussed further here as such. A good deal of progress has been made in distinguishing nucleophilic from general base catalysis, and some of the work involved is discussed below, but the borderline is not clear for all reactions, and it seems most useful to discuss nucleophilic and general base catalysis together for different classes of nucleophile. A few general points here, however, may help to set the problem in perspective.

The single most important factor governing the relative importance of nucleophilic and general base catalysis for a given reaction is the partitioning of the tetrahedral intermediate formed by the addition of the nucleophile to the carbonyl group, *viz.*

$$Y^- + R'(RO)C{=}O \underset{k_2}{\overset{k_1}{\rightleftharpoons}} Y{-}C(R')(OR){-}O^- \xrightarrow{k_3} R'(Y)C{=}O + RO^-$$

If Y^- is so much better a leaving group than RO^- that the tetrahedral intermediate breaks down essentially exclusively to regenerate the starting materials ($k_2 \gg k_3$), then nucleophilic catalysis cannot occur. (A measure of the leaving ability of a group is its basicity, usually measured as the pK_a of its conjugate acid, although direct comparisons are only reliable when they are between nucleophiles of the same structural type, such as substituted phenolate anions.)

As discussed above, hydroxide ion is one of the poorest leaving groups, so alkaline hydrolysis can always be classed as nucleophilic catalysis. Thus, even if a given nucleophile cannot catalyze the hydrolysis of an ester directly, it will be able to do so indirectly if it can assist the addition of water. And this is what a nucleophile does when it acts as a general base (or as a general acid).

As might be expected, the addition of water, catalyzed by general acids and bases, is most readily observed under conditions where the uncatalyzed reaction is significant. This water reaction is important only for reactive esters, which are those with electron-withdrawing substituents in either the acyl group or the leaving group, and is discussed below (p. 153). Since general base catalysis is expected to be most important for esters with leaving groups much more strongly basic than the nucleophile, it is thus observed most readily for esters which have poor leaving groups but are activated by electron-withdrawing substituents on the acyl group.

3.2.1 Neutral hydrolysis

The general rate expression for the hydrolysis of esters, and of many other species, is

$$-\frac{\mathrm{d}[\mathrm{R'COOR}]}{\mathrm{d}t} = [\mathrm{R'COOR}](k_0 + k_{\mathrm{H}^+}[\mathrm{H_3O^+}] + k_{\mathrm{OH}^-}[\mathrm{OH^-}])$$

In dilute aqueous solutions in which pH control is maintained, hydrolysis follows first-order kinetics, and the observed rate coefficient is given by the equation

$$k_{\mathrm{obs}} = k_0 + k_{\mathrm{H}^+}[\mathrm{H_3O^+}] + k_{\mathrm{OH}^-}[\mathrm{OH^-}]$$

Usually k_{H^+} and k_{OH^-} can be determined at low and high pH, respectively, where only one form of catalysis is significant. For unreactive esters k_0 is small, and can be neglected. The pH–rate profile, that is the plot of $\log k_{\mathrm{obs}}$ *versus* pH, then consists of two straight lines, of slope -1.0 in the region of acid catalysis, and $+1.0$ in the alkaline region, which intersect at the rate minimum. This behaviour is illustrated by curve A of Fig. 13, which is the pH–rate profile for

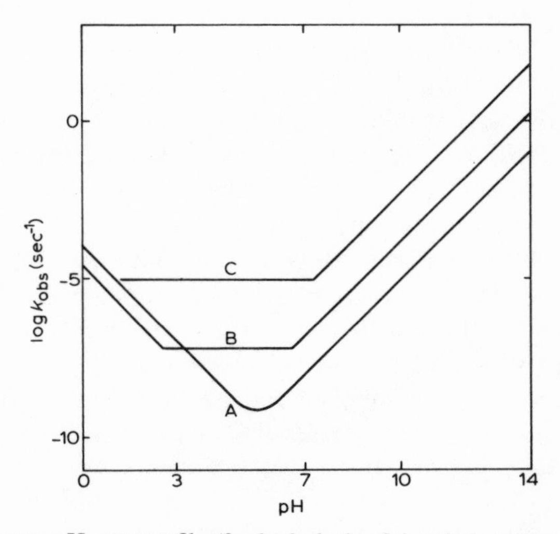

Fig. 13. Approximate pH–rate profiles for hydrolysis of A, ethyl acetate; B, phenyl acetate; and C, 2,4-dinitrophenyl acetate, all at 25°C. Data from Skrabal[188], Kirsch and Jencks[300] and Tommila and Hinshelwood[217].

ethyl acetate in aqueous solution at 25°C. Skrabal and Zahorka[187], have estimated that about 27% of this minimum rate is accounted for by the neutral hydrolysis reaction. At the minimum the rates of the acid- and hydroxide-catalyzed reactions are equal, and so

$$k_{H^+}[H_3O^+]_{min} = k_{OH^-}[HO^-]_{min}$$
$$= k_{OH^-}\cdot K_W/[H_3O^+]_{min}$$

The hydronium ion concentration is thus given by

$$[H_3O^+]_{min} = (k_{OH^-}\cdot K_W/k_{H^+})^{1/2}$$

The rate coefficient for the catalyzed reaction at the minimum is exactly twice that for the acid- or base-catalyzed process, so the rate coefficient at the minimum in the pH–rate profile, given by the sum of the neutral and catalyzed reactions, is[188]

$$k_{min} = k_0 + 2k_{H^+}(k_{OH^-}\cdot K_W/k_{H^+})^{1/2}$$
$$= k_0 + 2(k_{H^+}\cdot k_{OH^-}\cdot K_W)^{1/2}$$

As Skrabal and Zahorka[187] have pointed out, the neutral solvolysis reaction is readily detected if $k_0 \gg 2(k_{H^+}\cdot k_{OH^-}\cdot K_W)^{1/2}$ and is impossible to measure where $k_0 \ll 2(k_{H^+}\cdot k_{OH^-}\cdot K_W)^{1/2}$. For ethyl acetate at 25°C they obtained $k_0/2(k_{H^+}\cdot k_{OH^-}\cdot K_W)^{1/2} = 0.36$.

More reactive esters give pH–rate profiles of type B, illustrated for the hydrolysis of phenyl acetate. The presence of electron-withdrawing substituents in either the acyl or the leaving group of the ester affects the three reactions concerned to different extents. The acid-catalyzed reaction is affected very little by substitution: ρ for the acid-catalyzed hydrolysis of substituted ethyl benzoates is close to zero, and electron-withdrawing substituents in the leaving group cause a consistent but relatively small decrease in reactivity (p. 137). The same substituents have a much larger, positive effect on the rate of alkaline hydrolysis: ρ for the alkaline hydrolysis of substituted alkyl benzoates falls in the region 2–2.5[189], and a plot of $\log k_{OH^-}$ against the pK_a of the leaving group for a large number of acetate esters has a slope of -0.26[190]. Accordingly, the relative importance of alkaline hydrolysis increases with the reactivity of the ester, and the pH at which the minimum in k_{obs} is observed falls to lower values.

But for a sufficiently reactive ester the minimum is no longer observed experimentally, because it is obscured by a broad pH-independent region, marking the appearance of the neutral hydrolysis or water reaction. This reaction is the most sensitive of the three to the effects of substituents, at least to those in the leaving group. (No data are available for the neutral hydrolysis of substituted benzoates.) The slope of the linear plot of $\log k_{hyd}$ against the pK_a of the leaving group is 0.38 (Fig. 14) for a series of substituted

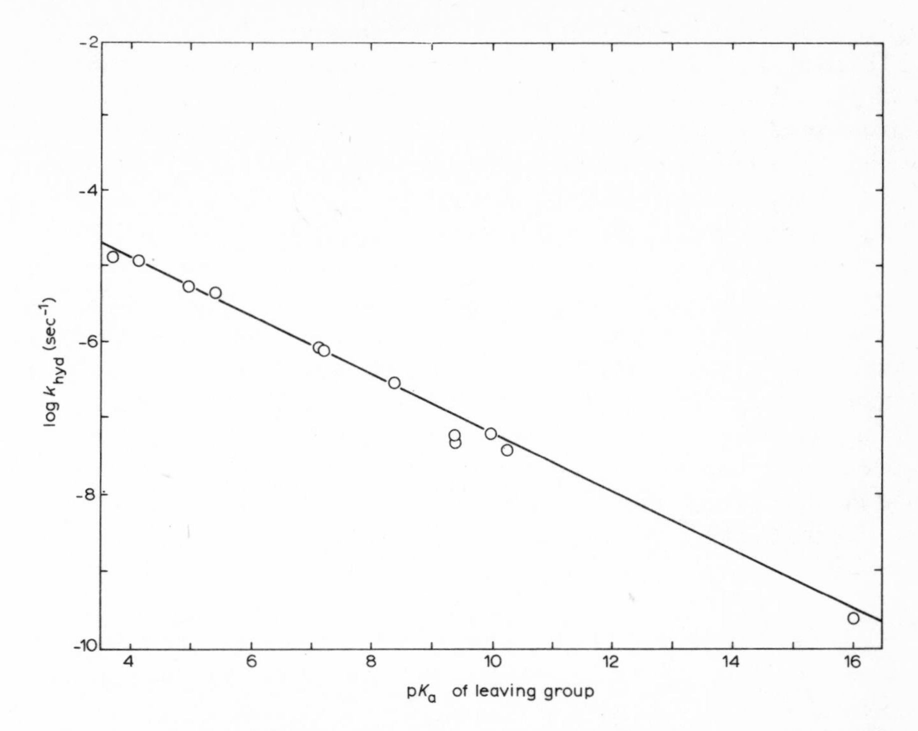

Fig. 14. Linear free energy relationships between the rate coefficient for neutral hydrolysis at 25°C, and the pK_a of the conjugate acid of the leaving group, for a series of acetate esters. Data for substituted phenyl acetates (ref. 191), and for ethyl acetate (ref. 187).

phenyl acetates studied by Gold *et al.*[191]. Also included on this plot is the value[187], for the neutral hydrolysis of ethyl acetate. Assuming that the linear free energy relationship holds good for ethyl acetate as well as for substituted phenyl acetates, it predicts a half-life of 75 years for the neutral hydrolysis of ethyl acetate at 25°C, in what must be considered remarkably good agreement with Skrabal's figure of 89 years, measured in the 1920's.

This sensitivity to substitution of neutral hydrolysis means that the pH-independent reaction gradually becomes more important than the hydroxide reaction at the high pH end of the region, and becomes much more rapidly more important than acid-catalyzed hydrolysis at low pH. Thus from Fig. 13, the acid-catalyzed reaction can be seen to be significant for the hydrolysis of ethyl acetate between pH 4 and 5, and for phenyl acetate about pH 2: but for 2,4-dinitrophenyl acetate the acid-catalyzed reaction is not detectable at pH 1, and is presumably important only in relatively strong acid. It seems certain that this fast neutral hydrolysis is at any rate a partial explanation for the low efficiency of acid catalysis in the hydrolysis of very weakly basic esters, such as the trifluoroacetates and oxalates, in moderately concentrated acid (see p. 145).

(*a*) *The mechanism of the neutral hydrolysis reaction*

The neutral hydrolysis of esters has been studied *per se* by only a handful of authors, and the only review of the subject is in Finnish[192]. The kinetic work has involved mainly the hydrolysis of haloacetate esters, although data have been collected for a sufficiently wide range of compounds to establish what is typical behaviour for reactions of this type.

Apart from ethyl acetate, the least reactive ester studied is N,O-diacetyl serinamide, which is hydrolyzed in a pH-independent reaction between pH 7 and 8 with a rate coefficient[193] of 2.66×10^{-5} sec^{-1}. Salmi and Suonpää[194] and Palomaa *et al.*[195], have measured the rates of neutral hydrolysis of a number of chloroacetate esters, and this work has been extended more recently by Euranto and Cleve[196–198], who have measured the activation parameters for the hydrolysis of several compounds. Moffat and Hunt[199] have obtained the same data for the hydrolysis of a variety of alkyl and aryl trifluoroacetates, and the data for substituted phenyl acetates[191] have been plotted in Fig. 14. Most of the available data are collected in Table 27.

TABLE 27

RATE COEFFICIENTS AND ACTIVATION PARAMETERS FOR THE NEUTRAL HYDROLYSIS OF ESTERS IN WATER AT 25°C

Ester	k_{hyd} *(sec⁻¹)*	$\Delta H^{\ddagger}$ *(kcal·mole⁻¹)*	$\Delta S^{\ddagger}$ *(eu)*	*Ref.*
$HCOOCH_2Cl$	8.5×10^{-5}	11.4	−39	197
$HCOOCHClCH_2Cl$	8.5×10^{-5}	11.4	−38.8	197
$HCOOCHClCH_2Cl$[a]	9.8×10^{-6}	12.9	−38	198
$CH_3COOC_2H_5$	2.47×10^{-10}			187
$CH_3COOCHClCCl_3$[a]	2.2×10^{-7}	14.1	−41.8	198
$CH_3COOC_6H_4CH_3$-*p*	3.9×10^{-8}			191
$CH_3COOC_6H_5$	6.6×10^{-8}			191
$CH_3COOC_6H_4Cl$-*p*	6.3×10^{-8}			191
$CH_3COOC_6H_4NO_2$-*m*	3.02×10^{-7}			191
$CH_3COOC_6H_4NH_2$-*p*	7.64×10^{-7}			191
$CH_3COOC_6H_4NO_2$-*p*	8.46×10^{-7}			191
$CH_3COOC_6H_3(NO_2)_2$-3,4	4.60×10^{-6}			191
$CH_3COOC_6H_3(NO_2)_2$-2,3	5.62×10^{-6}			191
$CH_3COOC_6H_3(NO_2)_2$-2,4	1.18×10^{-5}			191
$CH_3COOC_6H_3(NO_2)_2$-2,6	1.42×10^{-5}			191
$ClCH_2COOC_2H_5$	1.08×10^{-8}			186
$ClCH_2COOCH_2Cl$	1.08×10^{-6}	11.8	−37	197
$ClCH_2COOC_6H_4NO_2$-*p*	6.0×10^{-6}			200
$Cl_2CHCOOCH_3$	1.56×10^{-5}	11.4	−42	197
$Cl_2CHCOOC_2H_5$	5.51×10^{-6}			194
$Cl_2CHCOO(CH_2)_2OCH_3$	1.14×10^{-6}			194
$Cl_2CHCOO(CH_2)_2Cl$	3.67×10^{-5}			194
$Cl_2CHCOOCH_2Cl$[a]	2.83×10^{-4}			197
$Cl_2CHCOOCH_2CCl_3$[b]	3.1×10^{-5}			194
$Cl_2CHCOOC_6H_5$	2.90×10^{-3}			200

TABLE 27 (CONTINUED)

Ester	k_{hyd} *(sec⁻¹)*	$\Delta H^{\ddagger}$ *(kcal·mole⁻¹)*	$\Delta S^{\ddagger}$ *(eu)*	*Ref.*
$F_2CHCOOC_2H_5$	5.7×10^{-5}			186
$Cl_3CCOOCH_3$	7.95×10^{-4}			186
$Cl_3CCOOCH_3$[b]	2.67×10^{-5}			195
$Cl_3CCOOC_2H_5$	2.8×10^{-5}			186
$Cl_3CCOOC_2H_5$[b]	5.55×10^{-6}			195
$Cl_3CCOOC_3H_7$-*n*[b]	3.48×10^{-6}			195
$Cl_3CCOOC_3H_7$-*i*[b]	8.38×10^{-7}			195
$Cl_3CCOOC_4H_9$-*n*[b]	3.50×10^{-6}			195
$Cl_3CCOO(CH_2)_2Cl$[b]	4.42×10^{-5}			195
$Cl_3CCOO(CH_2)_3Cl$[b]	1.06×10^{-5}			195
$Cl_3CCOO(CH_2)_2OH$[b]	1.83×10^{-5}			195
$Cl_3CCOO(CH_2)_2OCH_3$[b]	1.53×10^{-5}			195
$Cl_3CCOO(CH_2)_2OCH_3$[b]	7.00×10^{-6}			195
$CF_3COOC_2H_5$[a]	2.57×10^{-4}			199
$CF_3COOC_2H_5$[c]	3.99×10^{-5}			199
$CF_3COO(CH_2)_2CH_3$[c]	5.67×10^{-6}	9.21	−51.7	199
$CF_3COO(CH_2)_3CH_3$[c]	3.50×10^{-6}	10.19	−49.1	199
$CF_3COO(CH_2)_4CH_3$[c]	2.56×10^{-6}	9.20	−51.7	199
$CF_3COO(CH_2)_5CH_3$[c]	2.59×10^{-6}	9.86	−51.1	199
$CF_3COOC_6H_5$[c]	1.67×10^{-2}	6.26	−45.9	199
$CF_3COOC_6H_5CH_3$-*p*[c]	8.39×10^{-3}	6.23	−47.2	199
$CF_3COOC_6H_5CH_3$-*m*[c]	1.00×10^{-2}	6.62	−45.6	199
$CF_3COOC_6H_5CH_3$-*o*[c]	4.55×10^{-3}	7.00	−45.8	199

[a] In 50% acetone–water.
[b] In 50% dioxan–water.
[c] In 70% acetone–water.

The important generalizations which can be drawn from the data in Table 27 are that the entropy of activation is typically in or close to the region −40 to −50 eu, and that changes in reactivity due to substituents appear primarily in the enthalpy term. $\Delta S^{\ddagger}$ is particularly low for alkyl trifluoroacetates, and partially compensates, as so often, for particularly low values of $\Delta H^{\ddagger}$. The rates of these reactions can be accounted for by a combination of polar and steric effects, and will not be discussed in detail.

There is a considerable deuterium solvent isotope effect for reactions of this sort. Jencks and Carriuolo[186] found values of k_H/k_D of about 5 for the hydrolysis of ethyl dichloroacetate at 25°C and ionic strength 1.0, and 2.1 for ethyl difluoroacetate. A value in the region of 2 seems to be typical. For example, Gold *et al.*[201] found $k_H/k_D = 1.8$ for the neutral hydrolysis of methyl trifluoracetate at 25°C, compared with 2.8 for the corresponding reaction of acetic anhydride. The volume of activation has not been measured for the neutral hydrolysis of an ester, but is expected to be strongly negative, by analogy with the value of -22 ± 1 cm³·mole⁻¹ for the water reaction of acetic anhydride[202]. The rate of hydrolysis is markedly reduced by the addition of organic solvents, and at high salt concentrations[186].

All the facts point consistently to a considerable involvement of water in the transition state, and there is widespread agreement that two molecules of water are concerned in bond making and breaking. Euranto[192] has calculated, using an equation derived by Long *et al.*[125], that four exchangeable hydrogen atoms are present in the transition state. This is consistent with general base catalysis of the attack of water by a second molecule of water, or the kinetically equivalent general acid catalysis of the attack of hydroxide ion by the hydronium ion. Authors in this field agree in preferring the former mechanism[186,192], which has the further advantage of being equivalent to the mechanism considered most likely for the attack of water on protonated esters in the $A_{AC}2$ reaction (see p. 123). Since a protonated ester may be regarded as a reactive ester bearing a strongly electron-withdrawing substituent on the carbonyl oxygen atom it is not unreasonable to suppose that they will react with water similarly to esters with strongly electron-withdrawing substituents in the other two positions. Further confirmation is the observation of Jencks and Carriuolo[186] that the second-order rate coefficient for water catalysis of the hydrolysis of acyl-activated esters falls on the Bronsted plot correlating the data for general bases of widely different structural types.

It seems safe to conclude, therefore, that the neutral hydrolysis of esters, like the hydrolysis of the conjugate acids, involves a molecule of water acting as a general base to assist the addition of a second molecule of water to the ester carbonyl group (Scheme 3).

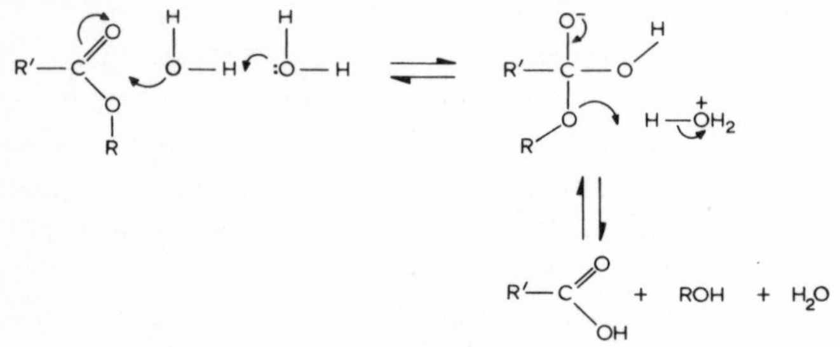

Scheme 3.

For alkyl esters the reaction is close to symmetrical, and it is likely that the breakdown of the tetrahedral intermediate is partially rate-determining, and will therefore be general acid-catalyzed. For aryl esters breakdown to products will be faster, and the formation of the tetrahedral intermediate should determine the rate. This might account for the more favourable entropies of activation found by Moffat and Hunt[199] for the hydrolysis of aryl trifluoroacetates: if the single transition state (of the addition step) occurs early, the loss of degrees of freedom compared with the initial state will also be less complete.

3.2.2 *Ester formation in neutral solution*

It is generally assumed that the ionized carboxyl group $R'COO^-$ is not susceptible to nucleophilic attack, and essentially all the available evidence supports this assumption. A solitary exception is discussed below (p. 160). Thus ester formation with acyl–oxygen cleavage, and oxygen exchange with the solvent, are not expected under alkaline conditions. There is no reason, however, why they should not occur in neutral solution, where a significant proportion of the undissociated acid (R'COOH) is present, and both reactions are well established under these conditions.

Reactivity in these reactions should be similar to that in ester hydrolysis reactions under neutral conditions, and is therefore expected to be most important for acids with strongly electron-withdrawing acyl substituents. Traynham[172] has shown, for example, that 2-octyl esters of trifluoroacetic and perfluorobutyric acids can be prepared by heating the alcohol with a slight excess of the acid. By using the optically active alcohol, which was regenerated without change of configuration on alkaline hydrolysis of the ester produced, it was shown that quantitative acyl–oxygen cleavage is involved in the esterification reaction. It seems very likely that this is an uncatalyzed reaction of the carboxylic acid.

Relevant kinetic measurements on the ^{18}O-exchange reactions of carboxylic acids in enriched water have been made by Llewellyn and O'Connor[139–141,203]. The exchange reaction between water and acetic acid could not be detected even at 123°C[139], but the uncatalyzed reaction is a major contributor to the exchange reaction of trichloroacetic and trifluoroacetic acids[203] and oxalic acid[140], and can be detected for the glycine cation, $H_3N^+CH_2COOH$[141]. The data are summarized in Table 28. They have been corrected where appropri-

TABLE 28

KINETIC DATA FOR THE ^{18}O EXCHANGE REACTION OF CARBOXYLIC ACIDS IN ENRICHED WATER AT 25°C, IONIC STRENGTH 4.0

Acid	k_{obs} (sec^{-1})	*Temp.* (°C)	$\Delta H^{\ddagger}$ ($kcal \cdot mole^{-1}$)	$\Delta S^{\ddagger}$ (*eu*)	*Ref.*
CH_3COOH	Not detected				139
$H_3N^+CH_2COOH$[a]	7.81×10^{-6}	100	18	−41	141
CCl_3COOH	7.86×10^{-5}	25	13.5	−26	203
CF_3COOH	7.66×10^{-4}	25.5	12.5	−35	203
HOOC·COOH	3.42×10^{-6}	25	12.5	−38	140

[a] At ionic strength 1.0 *M*.

ate to refer to the reaction of water with the undissociated carboxylic acid group. This seems the likely reaction in each case, in view of the conclusions reached above on the relative importance of these neutral and acid-catalyzed hydrolysis reactions of the corresponding esters (see pp. 142 and 156). Apart from the figures for trifluoroacetic acid, in which case $\Delta H^{\ddagger}$ and $\Delta S^{\ddagger}$ are surprisingly high, the rates and activation parameters are similar to those for the hydrolysis of the corresponding ethyl esters (Table 27), and it seems clear that the reactions must be very similar.

One very remarkable result obtained by Llewellyn and O'Connor[203] is that the anions of trifluoroacetic and trichloroacetic acid, also undergo base-catalyzed ^{18}O-exchange reactions. The reaction was not observed for the other acids examined, and the authors only comment is the suggestion that this reaction becomes significant because of the strong inductive effect of the three halogen atoms. These are not slow reactions. The second-order rate coefficients for the reactions of hydroxide ion with trifluoroacetate and trichloroacetate (7.39×10^{-4} and 3.11×10^{-4} l·mole^{-1}·sec^{-1}, respectively) are some two orders of magnitude greater than those for the reaction of water with the undissociated acids. For comparison, hydroxide attacks the ethyl esters some 10^8 times faster[204].

There seems no reason to suppose that the tetrahedral intermediate formed by the addition of a hydroxide ion to the carboxylic acid should be so stabilized by the trihalomethyl group that its breakdown should be a slow hydroxide-catalyzed reaction, *viz.*

$$CX_3COOH + {}^{18}OH^- \rightleftharpoons CX_3\text{—}C(O^-)(OH)\text{—}{}^{18}O\text{—}H \;\; OH^- \longrightarrow CX_3\text{—}C(O)({}^{18}O)^- + HO^- + H_2O$$

So it seems inescapable that the reaction, if it occurs, must involve the addition of a hydroxide ion to the carboxylate anion, *viz.*

$$CX_3\text{—}C(O)(O)^- + H{}^{18}O^- \rightleftharpoons CX_3\text{—}C(O^-)(O^-)\text{—}{}^{18}OH$$

This is a most unexpected reaction, and if confirmed represents an important discovery*.

However, the reaction products were not, apparently, identified[203]. Trifluoroacetate and trichloroacetate were extracted and precipitated from the acidified solution as their silver and mercurous salts, respectively, and the carbon dioxide produced by pyrolysis of these was analyzed in the mass spectrometer. So the occurrence of side-reactions leading to other carboxylate salts cannot be completely ruled out. Just one possibility is the halodecarboxylation of the anion, followed by hydrolysis of the dihalocarbene or haloform produced, to formate, *viz.*

---

* Nucleophilic attack on the trifluoroacetate anion is not completely without precedent. The reactions with Grignard reagents, for example, is a useful route to trifluoromethyl ketones[307], but it seems reasonable to suppose that the metal cation plays an important role in such reactions.

$$CX_3-CO_2^- \xrightarrow{-CO_2} CX_3^- \underset{}{\overset{H_2O}{\rightleftharpoons}} CHX_3 + OH^-$$

$$CX_3^- \xrightarrow{-X^-} :CX_2 \qquad CX_3-CO_2^- \xrightarrow{-CO_2,\ -X^-} :CX_2 \xrightarrow[H_2O]{HO^-,} HCOO^- + 2HX$$

3.2.3 Catalysis by oxyanions

The hydrolysis of the more reactive carboxylic esters is catalyzed by a wide range of oxyanions. The mechanism proposed for the neutral hydrolysis of esters on p. 158 involves two molecules of water, one as a nucleophile and one as a general base. In principle an oxyanion or other nucleophile can replace either of these molecules, and both general base and nucleophilic catalysis of ester hydrolysis are well-known. The detailed mechanism of nucleophilic catalysis depends, to some extent, on the type of anion concerned, but the differences occur at a relatively late stage in the reaction, and the similarities are sufficient to allow generalizations about oxyanion reactions as a class. Some of the differences are not normally kinetically significant, and are best mentioned briefly at this point.

It is not strictly accurate to describe alkaline hydrolysis as hydroxide-catalyzed, because hydroxide is consumed during ester hydrolysis, *viz.*

$$R'-C(=O)OR + HO^- \longrightarrow R'-CO_2^- + ROH$$

but the distinction is of no practical significance under conditions of controlled pH, where first-order kinetics are observed. For nucleophiles other than hydroxide ion it may be important to identify the detailed mechanism of nucleophilic catalysis. At least three possibilities can be distinguished. The initial product of nucleophilic substitution may be more reactive than the starting ester because:

(*a*) The nucleophile is itself rapidly displaced from the intermediate, as in catalysis by tertiary amines

$$R'-C(=O)OR + Y^- \rightleftharpoons R'-C(=O)Y\ (+RO^-) \overset{H_2O}{\rightleftharpoons} R'-COOH + Y^-\ (+ROH)$$

(*b*) The intermediate has a second electrophilic centre which is attacked more rapidly than the carbonyl group of the ester, as in many reactions catalyzed by carboxylate anions,

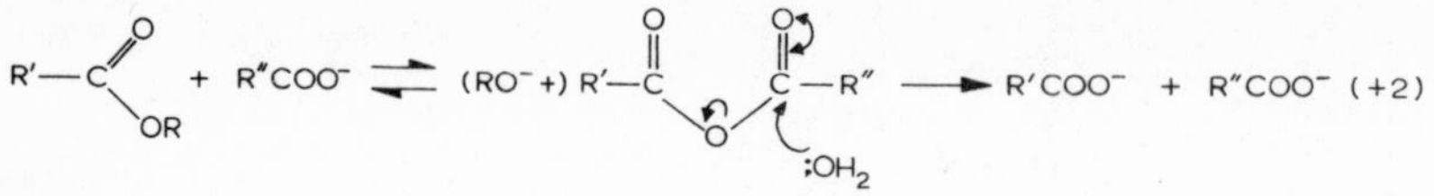

$$R'COOR + R''COO^- \rightleftharpoons (RO^- +)\ R'C(=O)\text{–}O\text{–}C(=O)R'' \xrightarrow{:OH_2} R'COO^- + R''COO^-\ (+2)$$

(*c*) The intermediate decomposes in a rapid unimolecular process, as is probably the case for catalysis by carbonate (and, in some cases, by phosphate), *viz.*

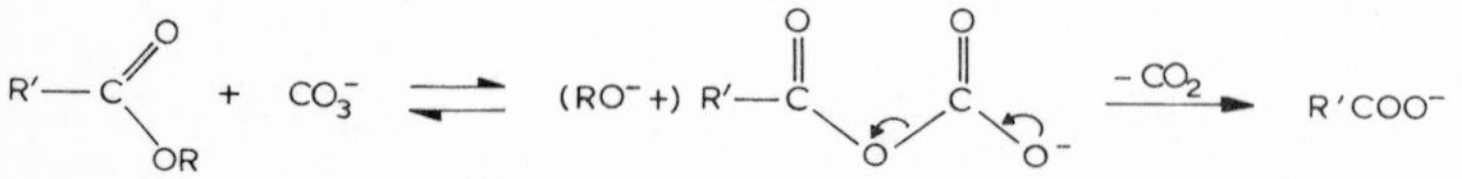

$$R'COOR + CO_3^= \rightleftharpoons (RO^- +)\ R'C(=O)\text{–}O\text{–}C(=O)O^- \xrightarrow{-CO_2} R'COO^-$$

Most nucleophiles are both more basic and more reactive towards sp^2-hybridized carbon than is water. Thus catalysis is readily observed with those esters which undergo hydrolysis at a significant rate in neutral solution, that is, with esters activated by electron withdrawing substituents in either the acyl or the leaving group. Hydroxide ion, however, is basic enough to attack any ester, and since the mechanism of alkaline hydrolysis appears to be relatively simple this reaction will be discussed first.

(*a*) *Catalysis by hydroxide ion*

The broad outline of the mechanism of catalysis of ester hydrolysis by hydroxide ion is not in doubt. The reaction is well known to involve acyl–oxygen cleavage, and seems invariably to be of the second order, being first order in both ester and hydroxide anion. General base catalysis in the usual sense is not a possibility: the partial removal of a proton from water cannot generate a species more reactive than hydroxide ion, so direct nucleophilic attack must be involved. (However, if it is accepted that the high ionic mobility of the hydroxide ion in water is explained by a Grotthus-type mechanism

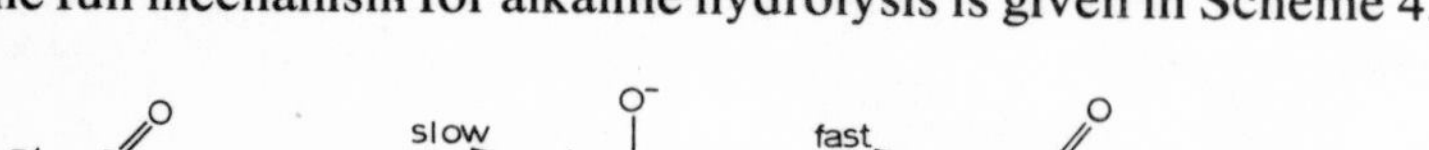

$$HO^- \cdots H\text{–}O(H) \cdots H\text{–}O(H) \cdots H\text{–}O(H) \rightleftharpoons HO\text{–}H \cdots O(H)\text{–}H \cdots O(H)\text{–}H \cdots OH^-$$

then it must be conceded there are some formal similarities to the general base-catalyzed attack of water, especially if one accepts the contention of Zimmerman and Rudolph[205] that proton transfers are very much faster than the electronic changes involved in covalent bond formation.)

The full mechanism for alkaline hydrolysis is given in Scheme 4.

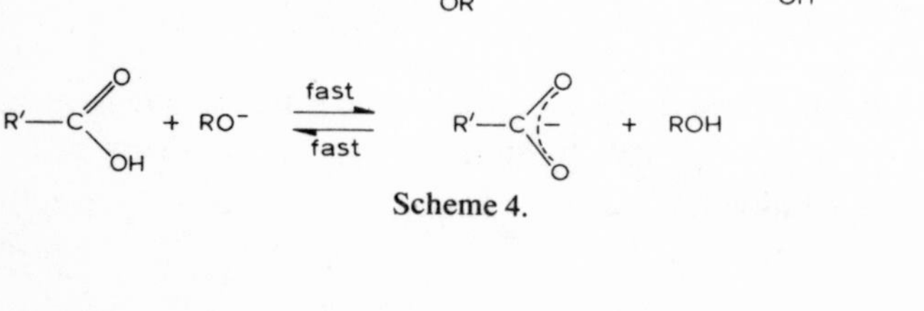

$$R'COOR + OH^- \underset{\text{fast}}{\overset{\text{slow}}{\rightleftharpoons}} R'C(O^-)(OH)(OR) \underset{\text{slow}}{\overset{\text{fast}}{\rightleftharpoons}} R'COOH + RO^-$$

$$R'COOH + RO^- \underset{\text{fast}}{\overset{\text{fast}}{\rightleftharpoons}} R'CO_2^- + ROH$$

Scheme 4.

The mechanism is formally reversible, but the equilibrium constant for the final proton transfer is so high that ester formation is not normally possible under alkaline conditions. Much of the evidence for acyl–oxygen cleavage is of long standing and will not be described in detail here. The alkaline hydrolysis of the esters of optically active alcohols is well known to give the alcohol with retention of configuration, indicating that the alkyl–oxygen bond remains intact. This reaction is so reliable that it is a stage in the resolution of alcohols by way of their phthalate half-esters[60].

In recent years, work with ^{18}O as a tracer has further strengthened the evidence for acyl–oxygen cleavage in alkaline hydrolysis[85,87–90,100,104,206–209], but the most important result of these investigations has been support for the existence of the tetrahedral addition compound. The evidence is that there is an exchange reaction between the oxygen atom of the carbonyl group of the ester and that of the solvent water, which occurs simultaneously, and often at a comparable rate, with the hydrolysis of the ester in alkali. A similar exchange reaction occurs together with hydrolysis under acidic conditions (see p. 106), but although the most reasonable interpretation is in terms of a tetrahedral addition compound in this case also (see p. 106) it is only for the alkaline reaction that the evidence specifically requires such a species to be present. In acid hydrolysis the reactive species is the conjugate acid, $R'{-}C(\overset{\cdots}{OH})(\overset{\cdots}{OR})^{+}$, and either H_2O or ROH might be lost by direct displacement. In alkali the only reasonable way in which the carbonyl oxygen can become a leaving group is by the formation of the tetrahedral addition compound, *viz*.

$$R'{-}C(={}^{18}O)OR + HO^- \rightleftharpoons R'{-}C({}^{18}O^-)(OH)OR \longrightarrow R'{-}C({}^{18}O)(O)^- + ROH$$

$$R'{-}C({}^{18}O^-)(OH)OR \rightleftharpoons R'{-}C({}^{18}OH)(O^-)OR$$

$$R'{-}C(=O)OR + H^{18}O^- \rightleftharpoons R'{-}C({}^{18}OH)(O^-)OR \longrightarrow R'{-}C({}^{18}O)(O)^- + ROH$$

Although the observation of the exchange reaction proves beyond reasonable doubt that the tetrahedral addition compound is formed during alkaline hydrolysis, it does not prove that it is actually an intermediate in the hydrolysis reaction. In other words, it remains to be shown that it is not a "blind-alley" intermediate.

Most of the available evidence is consistent with the tetrahedral addition compound being an intermediate in the hydrolysis reaction. In particular the ratio of the rates of hydrolysis and oxygen exchange, k_{hyd}/k_{exch}, is usually small and independent of the rate of hydrolysis. Also the ratio is generally higher for compounds with better leaving groups, as expected. Some relevant data, which include a few apparent anomalies, are given in Table 29. The circumstantial evidence is strong, though a formal proof that the tetrahedral addition compound is actually an intermediate in the hydrolysis reaction is still lacking, and it will be assumed in the following discussion that a tetrahedral intermediate is normally involved in nucleophilic and general base-catalyzed hydrolysis. It should be borne in mind, however, that a single transition state with appropriate properties could explain the observed results*.

The alkaline hydrolysis of esters has been studied in more detail, or at any rate more frequently, than any other chemical reaction, and results continue to appear in a steady stream. Since the gross mechanism is not in real doubt, this discussion will be limited mainly to the relationship between structure and reactivity in the reaction. The relationship can in most cases only be discussed for the overall reaction, which is, of course, of most immediate practical importance. Although it is of great interest to discover what effects structural changes may have on the separate steps of the reaction, it is only in the few cases where the oxygen exchange and hydrolysis reactions have been measured concurrently that this is possible.

The kinetically important steps of the hydrolysis reaction can be summarized by the simplified equation

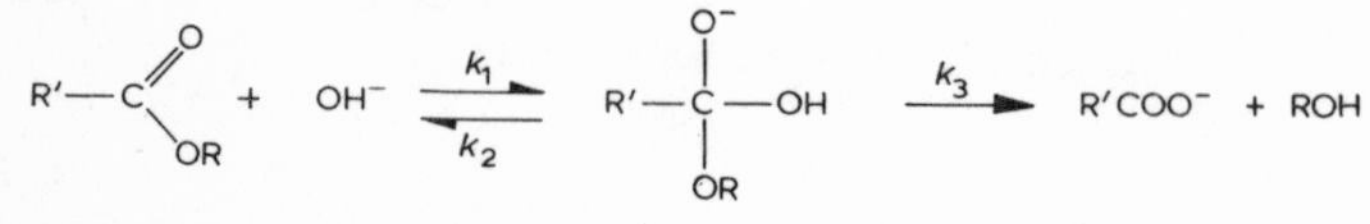

$$R'C(=O)OR + OH^- \underset{k_2}{\overset{k_1}{\rightleftharpoons}} R'C(O^-)(OH)(OR) \xrightarrow{k_3} R'COO^- + ROH$$

* To illustrate this point in general terms, it is not inconceivable that a transition state (26), of the type usually considered for the formation of a tetrahedral intermediate, might collapse with

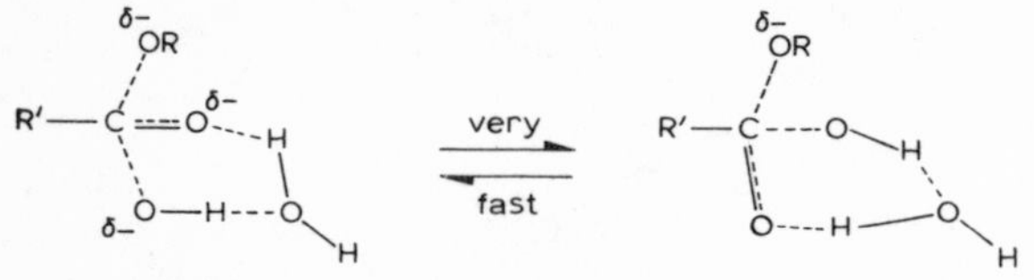

(26) $\underset{\text{fast}}{\overset{\text{very}}{\rightleftharpoons}}$ (27)

the loss of RO^- before the π-electrons of the C=O bond become fully localized on the oxygen atom. This process would appear to be most feasible where RO^- is a very good leaving group, as it rarely is in an ester. By only a small further stretch of the imagination it is possible to envisage proton transfers occurring at a rate much greater than that at which the other bonds are being formed and broken, so that the oxygen atoms of the carbonyl and entering groups become equivalent, as in (26) $\rightleftharpoons$ (27). The problem is not discussed here in more detail because of the availability of recent reviews of the evidence by Johnson[101] and by Samuel and Silver[210] who deal specifically with evidence from ^{18}O-exchange experiments.

TABLE 29

KINETIC DATA FOR THE HYDROXIDE-CATALYZED ^{18}O EXCHANGE REACTION OF ESTERS WITH ENRICHED WATER

Ester	*Conditions*	k_{exch} ($l \cdot mole^{-1} \cdot sec^{-1}$)	k_{hyd}/k_{exch}	*Ref.*
$PhCOOCH_3$	25°C, water, $\mu = 0.003$	2.85×10^{-3}	27.7	253
$PhCOOCH_3$	25°C, 33% dioxan–water, $\mu = 0.01$	3.0×10^{-4}	89	253
$PhCOOCH_3$	25°C, 33% dioxan–water, $\mu = 0.50$		[a,b]	253
$PhCOOC_2H_5$	25°C, water, $\mu = 0.003$	2.85×10^{-3}	12.6	253
$PhCOOC_2H_5$	25.1°C, water	6.2×10^{-3}	4.8	104
$PhCOOC_2H_5$	25°C, 33% dioxan–water	7.4×10^{-4}	11.3	208
$PhCOOC_2H_5$	25.1°C, 33% dioxan–water	8.2×10^{-4}	10.6	104
$PhCOOC_2H_5$	9°C, 33% dioxan–water	1.8×10^{-4}	14.7	208
$PhCOOC_2H_5$	40.2°C, 33% dioxan–water	2.55×10^{-3}	10.1	208
$PhCOOCH(CH_3)_2$	25.1°C, water	2.3×10^{-3}	2.7	104
$PhCOOCH(CH_3)_2$	62.5°C, 33% dioxan–water	4.9×10^{-3}	3.7	104
$PhCOOC(CH_3)_3$	62.5°C, 33% dioxan–water	1.7×10^{-4}	7.6	104
PhCOOPh	0°C, 50% dioxan–water		[a]	212
$PhCOOC_6H_4Cl$-*p*	24.8°C, 66% dioxan–water		[a]	212
p-$NO_2C_6H_4COOCH_3$	25°C, 33% dioxan–water		[a,b]	253
p-$NH_2C_6H_4COOCH_3$	25°C, 33% dioxan–water		[a,b]	253
(2,4,6-trimethylphenyl)–$COOCH_3$	60% dioxan–water		6.8	207

[a] Too high to measure, probably > 100.

[b] These data appear to supersede those of Bender and Thomas[211], who found much smaller ratios.

The step written as k_3 is complex, but the final proton transfers are assuredly very rapid, so that only the breakdown of the tetrahedral intermediate is kinetically important. Using the steady-state assumption for the tetrahedral intermediate, the rate coefficient for the overall reaction can readily be shown to be

$$k_{\text{hyd}} = \frac{k_1 k_3}{k_2 + k_3}$$

For good leaving groups ($k_3 \gg k_2$) k_{obs} becomes equal to k_1, and the addition step alone is rate-determining. But for the hydrolysis of alkyl esters the full expression must be used since k_3 and k_2 will be of the same order of magnitude referring as they do to the loss of an alkoxide and a hydroxide ion, respectively, from the tetrahedral intermediate. The observed rate coefficient for oxygen exchange is just half that for the breakdown of the tetrahedral intermediate to starting materials[208], *viz.*

$$k_{\text{exch}} = \frac{k_1 k_2}{2(k_2 + k_3)}$$

and from measurements of k_{hyd} and $k_{\text{hyd}}/k_{\text{exch}}$ it is thus possible to calculate both k_3/k_2 and k_1[208]. For ethyl benzoate in 33% dioxan–water Bender *et al.*[208] obtained the results in Table 30.

TABLE 30

RATE COEFFICIENTS FOR THE ALKALINE HYDROLYSIS OF ETHYL BENZOATE IN 33% DIOXAN–WATER

(From ref. 208)

T(°C)	k_{hyd} ($l \cdot mole^{-1} \cdot sec^{-1}$)	k_1	k_2/k_3
9.0	2.65×10^{-3}	3.02×10^{-3}	7.35
25.0	8.4×10^{-3}	9.83×10^{-3}	5.65
40.2	2.58×10^{-2}	3.10×10^{-2}	5.05

Both k_1 and the ratio k_2/k_3 gave linear Arrhenius plots. Activation parameters were calculated from these data as follows (at 25°C):

For k_{hyd}: $\Delta H^{\ddagger} = 12.3$ kcal·mole^{-1}, $\Delta S^{\ddagger} = -26.8$ eu
For k_1: $\Delta H^{\ddagger} = 12.6$ kcal·mole^{-1}, $\Delta S^{\ddagger} = -25.5$ eu

Clearly with ratios of $k_{\text{hyd}}/k_{\text{exch}}$ in the region observed for ethyl benzoate under these conditions, the rates and activation parameters are largely determined by the addition step. And in this case at least the small rate decrease due to the breakdown step is entirely an entropy effect. In several instances, notably the hydrolyses of methyl and substituted benzyl benzoates, phthalide and γ-butyrolactone, $k_{\text{hyd}} \gg k_{\text{exch}}$, since no concurrent oxygen exchange could be detected on alkaline hydrolysis[212,306].

These results do fall into a recognisable pattern. Methanol, and probably benzyl alcohols also, are slightly more acidic than ethanol, and the derived alkoxide ions are therefore better leaving groups than ethoxide. *iso*-Propoxide and *t*-butoxide, on the other hand, are poorer leaving groups than ethoxide, and less easily eliminated from the tetrahedral intermediate. Consequently, other things being equal, k_3 will be lower for esters of the more weakly acidic alcohols. Since k_2 should be relatively insensitive to the basicity of the leaving group the expected result is a decrease in $k_{\text{hyd}}/k_{\text{exch}}$ as is observed. (It is evident that the true picture is more complex than this, since hydroxide is slightly less basic than ethoxide ion, yet it is eliminated much less readily[253]).

From the available data it is thus reasonable to conclude that only in exceptional cases will the breakdown of the tetrahedral intermediate be a very significant factor in determining the rate of alkaline hydrolysis of an ester. Consequently the following discussion of the relationship between structure and reactivity can be simplified without serious error by the assumption that the rate-determining step is the addition of hydroxide ion to the carbonyl group of the ester.

(i) Structure and reactivity in alkaline hydrolysis

Most of the generalizations that can be made about the relationship between structure and reactivity in the alkaline hydrolysis of esters are familiar. They will therefore only be summarized here, and illustrated by a selection from the vast amount of available data (Tables 31 and 32).

For carboxylic esters the minimum in the pH–rate profile, or the mid-point in the pH-independent region, normally falls on the acid side of neutrality, indicating that acid catalysis is less efficient than catalysis by hydroxide ion. A further point of difference from the acid-catalyzed reaction is that the rate of alkaline hydrolysis is very sensitive to the polar effects of substituents in either the acyl or the leaving group of the ester: although, as has been suggested above (p. 154), it may yet be less sensitive than neutral hydrolysis. Because of this strong response to polar influences, steric effects are less apparent than in acid-catalyzed hydrolysis, although in fact they are quantitatively not very different in the two cases, as is shown by the considerable success of Taft's assumption[152] that they are identical.

(ii) Polar effects

The effects of polar substituents on the alkaline hydrolysis of esters are well-established. Since the rate of the reaction is determined largely by the rate of addition of hydroxide ion to the carbonyl group of the ester, any substituent which withdraws electrons from the carbonyl group will increase the reactivity of the ester. The most accessible quantitative measure of the effect is the Hammett or Taft reaction constant, and a large number of measurements are available. Taft *et al.*[240] found $\rho = 2.53$ for the base-catalyzed methanolysis of *meta*- and *para*-substituted (*l*)-menthyl benzoates, closely similar to the known value of $\rho = 2.37$ for the alkaline hydrolysis of substituted ethyl benzoates. Jones and Sloane's value[239], obtained with five esters, of $\rho = 2.41$ for the methoxyl exchange reaction of substituted methyl benzoates in methanol, is almost identical.

Slightly smaller ρ values are obtained with more polar solvents, where the medium is able to stabilize the transition state more effectively, and thus the response to the electronic effects of substituents is less critical. For the alkaline hydrolysis of the phenyl esters of substituted benzoates in 33% aqueous acetonitrile Kirsch *et al.*[235] find $\rho = 1.98$. The value is not significantly different for more or less reactive leaving groups: under the same conditions

References pp. 202–207

TABLE 31

KINETIC DATA FOR THE HYDROXIDE-ION-CATALYZED HYDROLYSIS OF CARBOXYLIC ESTERS (MOSTLY AT 25°C)

Ester	*Conditions*	*Temp.* (°C)	k_{OH} ($l \cdot mole^{-1} \cdot sec^{-1}$)	$\Delta H^{\ddagger}$ ($kcal \cdot mole^{-1}$)	$\Delta S^{\ddagger}$ (*eu*)	*Ref.*
A Esters of aliphatic esters						
A.1 Effect of varying leaving group						
$HCOOCH_3$	Water	25	38.4	9.81	−18.4	213
	Water	25	36.2	8.99	−21.3	214
$HCOOC_2H_5$	Water	25	2.82	13.86	−9.83	213
CH_3COOCH_3	Water	25	0.152	10.0	−30.1	213, 215
CH_3COOCH_3	70% Acetone	24.7	0.108	8.2	−35.3	216
$CH_3COOC_2H_5$	70% Acetone	24.7	4.66×10^{-2}	9.5	−32.6	216
$CH_3COO(CH_2)_2CH_3$	70% Acetone	24.7	2.70×10^{-2}	10.5	−30.3	216
$CH_3COO(CH_2)_3CH_3$	70% Acetone	24.7	2.30×10^{-2}	10.7	−30.8	216
$CH_3COOCH(CH_3)_2$	70% Acetone	24.7	7.06×10^{-3}	11.2	−30.3	216
$CH_3COOCH(CH_3)C_2H_5$	70% Acetone	24.7	3.27×10^{-3}	12.2	−29.4	216
$CH_3COOCH(CH_2)_5$	70% Acetone	24.7	4.56×10^{-3}	12.2	−28.5	216
$CH_3COOCH_2CH(CH_3)_2$	70% Acetone	24.7	1.82×10^{-2}	10.7	−30.8	216
$CH_3COOC(CH_3)_3$	70% Acetone	24.7	2.65×10^{-4}	12.9	−31.7	216
$CH_3COOCH_2C_6H_5$	60% Acetone	25	6.96×10^{-2}	11.7	−24.4	217
$CH_3COOCH_2C_6H_4CH_3$-*p*	60% Acetone	25	4.82×10^{-2}	11.9	−24.5	217
$CH_3COOCH_2C_6H_4NO_2$-*p*	60% Acetone	25	0.269	10.9	−24.5	217
$CH_3COO(CH_2)_2CH_3$	Water	20	6.5×10^{-2}	10.6	−27.7	218
$CH_3COOCH(CH_2)_4$	Water	20	2.6×10^{-2}	11.9	−25.2	218
$CH_3COOCH(CH_2)_2$	Water	20	0.178	12.1	−20.8	218
$CH_3COOCH{=}CH_2$	Water	20	3.28	12.8	−12.7	218
$CH_3COOC(CH_3){=}CH_2$	Water	20	0.301	11.6	−21.6	218
CH_3COO-(1-cyclopentenyl)	Water	20	0.855	10.5	−24.8	218
CH_3COO-(1-cyclohexenyl)	Water	20	0.178	10.5	−26.5	218

$CH_3COOC_6H_5$	60% Acetone	25	0.537	11.94	−19.5	217
$CH_3COOC_6H_4CH_3$-*p*	60% Acetone	25	0.319	12.30	−19.4	217
$CH_3COOC_6H_4NO_2$-*p*	60% Acetone	25	8.05	10.31	−19.4	217
$CH_3COOC_6H_4COO^-$-*p*	60% Acetone	25	0.929	11.42	−20.3	217
$CH_3COOC_6H_4NH_2$-*p*	60% Acetone	25	0.282	12.34	−19.5	217
$CH_3COOC_6H_5$	60% Acetone	1	$9{\cdot}03 \times 10^{-2}$	11.3	−22	219
$CH_3COOC_6H_4COOC_2H_5$-*p*	60% Acetone	1	0.606	9.9	−23	219
$CH_3COOC_6H_4COCH_3$-*p*	60% Acetone	1	0.752	10.4	−21	219
$CH_3COOC_6H_4CHO$-*p*	60% Acetone	1	1.21	6.6	−34	219
$CH_3COOC_6H_4CN$	60% Acetone	1	1.58	8.1	−28	219
$CH_3COOC_6H_4NO_2$-*p*	60% Acetone	1	2.2	7.4	−30	219
$CH_3COOC_6H_4NO_2$-*m*	60% Acetone	1	1.31	9.6	−22	219
$CH_3COOC_6H_4COCH_3$-*m*	60% Acetone	1	0.402	9.0	−27	219
$CH_3COOC_6H_4Cl$-*m*	60% Acetone	1	0.368	7.7	−32	219
$CH_3COOC_6H_4OCH_3$-*m*	60% Acetone	1	9.51×10^{-2}	11.6	−20	219
$CH_3COOC_6H_4CH_3$-*m*	60% Acetone	1	9.03×10^{-2}	11.3	−22	219
$CH_3COOC_6H_3(CH_3)_2$-3,5	60% Acetone	1	4.37×10^{-2}	13.1	−16	219
$C_2H_5COOCH_3$	70% Acetone	24.7	6.41×10^{-2}	8.8	−34.4	216
$C_2H_5COOC_2H_5$	70% Acetone	24.7	2.21×10^{-2}	10.1	−33.1	216
$C_2H_5COO(CH_2)_3CH_3$	70% Acetone	24.7	9.89×10^{-3}	11.3	−29.9	216
$C_2H_5COOCH(CH_3)_2$	70% Acetone	24.7	2.98×10^{-3}	11.9	−30.3	216
A.2 Effect of varying acyl group						
$HCOOC_2H_5$	Water	25	2.82	13.86	−9.83	213
$CH_3COOC_2H_5$	Water	25	0.111	10.7	−26.7	308
$C_2H_5COOC_2H_5$	Water	25	8.7×10^{-2}	10.13	−29.5	308
n-$C_3H_7COOC_2H_5$	Water	25	3.83×10^{-2}	9.63	−32.4	308
CH_3COOEt	70% Acetone	24.8	4.65×10^{-2}	9.2	−33.5	154
C_2H_5COOEt	70% Acetone	24.8	2.20×10^{-2}	10.0	−32.6	154
n-C_3H_7COOEt	70% Acetone	24.8	8.81×10^{-3}	11.1	−30.8	154
n-C_4H_9COOEt	70% Acetone	24.8	6.59×10^{-3}	11.8	−29.0	154
n-$C_6H_{13}COOEt$	70% Acetone	24.8	6.08×10^{-3}	11.8	−29.0	154
$(CH_3)_2CHCOOEt$	70% Acetone	24.8	5.50×10^{-3}	10.6	−33.1	154
$(CH_3)_2CHCH_2COOEt$	70% Acetone	24.8	2.18×10^{-3}	12.7	−28.1	154
$(CH_3)_3C{\cdot}COOEt$	70% Acetone	24.8	2.23×10^{-4}	12.4	−33.5	154
$(C_2H_5)_2CHCOOEt$	70% Acetone	24.8	8.3×10^{-5}	12.7	−31.3	154

TABLE 31 (CONTINUED)

KINETIC DATA FOR THE HYDROXIDE-ION-CATALYZED HYDROLYSIS OF CARBOXYLIC ESTERS (MOSTLY AT 25°C)

Ester	*Conditions*	*Temp.* (°C)	k_{OH} ($l \cdot mole^{-1} \cdot sec^{-1}$)	$\Delta H^{\ddagger}$ ($kcal \cdot mole^{-1}$)	$\Delta S^{\ddagger}$ (*eu*)	*Ref.*
CH_3COOMe	80% Methanol	24.95	4.00×10^{-3}			220
CH_3COOMe	80% Methanol	60	6.92×10^{-2}	15.4	−18.0	220
$C_6H_5CH_2COOMe$	80% Methanol	60	9.33×10^{-2}	15.2	−17.7	220
$(C_6H_5)_2CHCOOMe$	80% Methanol	60	1.44×10^{-2}	15.3	−21.4	220
$(C_6H_5)_3CCOOMe$	80% Methanol	60	5.93×10^{-5}			220
$ClCH_2COOMe$	Water, $\mu = 0.2$	25	61.6	9.08	−19.9	204
$BrCH_2COOMe$	Water, $\mu = 0.2$	25	91.1	7.65	−23.9	204
Cl_2CH_2COOMe	Water, $\mu = 0.2$	25	1500	6.15	−23.4	204
$ClCH_2COOEt$	Water, $\mu = 0.2$	25	33.2	8.31	−23.7	204
$BrCH_2COOEt$	Water, $\mu = 0.2$	25	49.7	6.49	−29.1	204
ICH_2COOEt	Water, $\mu = 0.2$	25	16.2	6.18	−32.2	204
$Cl_2CHCOOEt$	Water, $\mu = 0.2$	25	677	5.86	−25.9	204
$Br_2CHCOOEt$	Water, $\mu = 0.025$	25	202	6.6	−31	204
CF_3COOEt	Water, $\mu = 0.025$	15	*ca.* 15×10^5			204
CCl_3COOEt	Water, $\mu = 0.025$	25	2570	4.80	−38.5	204
$EtOOCCH_2COO^-$	Water	25	1.45×10^{-2}			221
$EtOOCCH_2SO_3^-$	Water	25	0.12			222
$EtOOCCH_2S^-$	Water	25	6.4×10^{-4}			223
$EtOOCCH_2SCH_3$	Water	25	0.92			223
$EtOOCCH_2SOCH_3$	Water	25	4.20			223
$EtOOCCH_2SO_2CH_3$	Water	25	12.8			223
$EtOOCCH_2S^+(CH_3)_2$	Water	25	205			223
$EtOOCCH_2N^+(C_2H_5)_3$	Water	25	32.0	12.1	−11.2	224
$EtOOCCH_2N^+(C_2H_5)_3$	Water, $\mu = 0.1$	25	20.8			225
$MeOOCCH_2N^+(CH_3)_3$	Water	25	98	10.4	−21.9	226
$MeOOCCH_2N^+(CH_3)_3$	Water, $\mu = 0.1$	25	60.8			226
$MeOOCCH_2NH_3^+$	Water	25	58			225
$EtOOCCH_2NH_3^+$	Water	25	60.5	8.3	−25	227

$EtOOCCH_2NH_3^+$	Water, $\mu = 0.1$	25	22.9			227
A.2 (a) Esters of α-amino-acids						
$EtOOCCH_2NH_2$	Water, $\mu = 0.1$	25	0.723	10.0	−25	227
$MeOOCCH_2NH_2$	Water, $\mu = 0.1$	25	1.28	9.5	−28	278
$MeOOCCH(NH_2)CH_3$	Water, $\mu = 0.1$	25	1.11			228
$MeOOCCH(NH_2)CH(CH_3)_2$	Water, $\mu = 0.1$	25	7.5×10^{-2}	7.8	−39.5	228
$MeOOCCH(NH_2)CH_2CH(CH_3)_2$	Water, $\mu = 0.1$	25	0.45			225
$MeOOCCH(NH_2)CH(CH_3)C_2H_5$	Water, $\mu = 0.1$	25	6.7×10^{-2}			228
$MeOOCCH(NH_2)CH_2C_6H_5$	Water, $\mu = 0.1$	25	0.55	9.2	−31	228
$MeOOCCH(NH_2)CH_2OH$	Water, $\mu = 0.1$	25	1.0			225
$MeOOCCH(NH_2)CH_2S^-$	Water, $\mu = 0.1$	25	7.36×10^{-2}			225
$MeOOCCH(NH_2)CH_2SCH_3$	Water, $\mu = 0.1$	25	1.07	7.4	−34	228
$MeOOCCH(NH_2)(CH_2)_2SCH_3$	Water, $\mu = 0.1$	25	0.77			228
$MeOOCCH(NH_2)CH_2C_6H_4OH$-*p*	Water, $\mu = 0.1$	25	0.27			228
$MeOOCCH(NH_2)CH_2$–(imidazole ring: N, NH)	Water, $\mu = 0.1$	25	0.62			228
$MeOOCCH(NH_2)CH_2$–(indole ring: N–H)	Water, $\mu = 0.1$	25	0.29			228
$MeOOCCH(NH_2)(CH_2)_2COOMe$	Water, $\mu = 0.1$	25	2.04			225
A.2 (b) Esters of unsaturated acids						
CH_3COOEt	85% Ethanol	25	6.98×10^{-3}	14.1	−21.2	229
$CH_2{=}CHCOOEt$	85% Ethanol	25	4.67×10^{-3}	14.8	−19.8	229
CH_3CH_2COOEt	85% Ethanol	25	3.65×10^{-3}	13.9	−23.0	229
$CH_2{=}CHCH_2COOEt$	85% Ethanol	25	3.63×10^{-3}	13.9	−23.0	229
$CH_2{=}CH(CH_2)_2COOEt$	85% Ethanol	25	3.62×10^{-3}	13.9	−23.0	229
trans-$CH_3CH{=}CHCOOEt$	85% Ethanol	25	6.25×10^{-4}	15.9	−19.8	229
$(CH_3)_2C{=}CHCOOEt$	85% Ethanol	25	9.59×10^{-5}	17.4	−19.4	229
trans-$C_6H_{13}CH{=}CHCOOEt$	85% Ethanol	25	6.60×10^{-4}	15.3	−21.7	229

TABLE 31 (CONTINUED)

KINETIC DATA FOR THE HYDROXIDE-ION-CATALYZED HYDROLYSIS OF CARBOXYLIC ESTERS (MOSTLY AT 25°C)

Ester	*Conditions*	*Temp. (°C)*	k_{OH} *(l·mole⁻¹·sec⁻¹)*	$\Delta H^{\ddagger}$ *(kcal·mole⁻¹)*	$\Delta S^{\ddagger}$ *(eu)*	*Ref.*
CH_2=CHCH=CHCOOEt	85% Ethanol	25	1.42×10^{-3}	16.2	−17.1	229
CH_3CH=CHCH=CHCOOEt	85% Ethanol	25	4.73×10^{-4}	17.0	−16.7	229
CH_2=C(CH_3)COOEt	85% Ethanol	25	9.33×10^{-4}	15.4	−20.7	229
cis-CH_3CH=C(CH_3)COOEt	85% Ethanol	25	1.76×10^{-4}	16.5	−20.7	229
cis-C_6H_5CH=CHCOOEt	87.8% Ethanol	30	1.90×10^{-3}			232
trans-C_6H_5CH=CHCOOEt	87.8% Ethanol	30	1.07×10^{-3}			232
trans-C_6H_5CH=CHCOOEt	85.4% Ethanol	24·8	1.39×10^{-3}	16.7	−14.8	230
B Aromatic esters						
B.1 Effect of varying leaving group						
B.1(a) Esters of aliphatic alcohols						
$PhCOOCH_3$	70% Dioxan	35	1.99×10^{-2}	12.6	−25.3	233
$PhCOOC_2H_5$	70% Dioxan	35	7.14×10^{-3}	12.8	−26.0	233
Methyl 1-naphthoate	70% Dioxan	35	7.22×10^{-3}	14.5	−20.1	233
Ethyl 1-naphthoate	70% Dioxan	35	2.68×10^{-3}	12.7	−29.0	233
Methyl 2-naphthoate	70% Dioxan	35	1.94×10^{-2}	12.8	−24.8	233
Ethyl 2-naphthoate	70% Dioxan	35	7.59×10^{-3}	14.0	−23.0	233
Methyl 1-anthroate	70% Dioxan	35	5.33×10^{-3}	15.6	−18.5	233
Ethyl 1-anthroate	70% Dioxan	35	2.76×10^{-3}	10.1	−37.8	233
Methyl 2-anthroate	70% Dioxan	35	2.68×10^{-2}	10.6	−31.2	233
Ethyl 2-anthroate	70% Dioxan	35	8.36×10^{-3}	13.6	−23.9	233
$C_6H_5COOCH_3$	60% Acetone	25	9.01×10^{-3}	13.6	−22.2	217
$C_6H_5COOC_2H_5$	60% Acetone	25	2.89×10^{-3}	14.8	−20.8	217
p-$NO_2C_6H_4COOCH_3$	60% Acetone	25	0.626	11.7	−20.1	217
p-$NO_2C_6H_4COOC_2H_5$	60% Acetone	25	0.246	11.8	−21.4	217
$C_6H_5COOCH_2C_6H_4OCH_3$-*p*	66% Dioxan	24.8	3.83×10^{-3}			212
$C_6H_5COOCH_2C_6H_5$	66% Dioxan	24.8	5.29×10^{-3}			212
$C_6H_5COOC_6H_4Cl$-*p*	66% Dioxan	24.8	9.44×10^{-3}			212
$C_6H_5COOC_6H_4NO_2$-*p*	66% Dioxan	24.8	3.05×10^{-3}			212
Phthalide	Water	24.8	0.255			212

$C_6H_5COOCH_3$	60% Dioxan	35	2.90×10^{-2}			164
$C_6H_5COOC_2H_5$	60% Dioxan	35	9.22×10^{-3}			164
$C_6H_5COO(CH_2)_2CH_3$	60% Dioxan	35	6.32×10^{-3}			164
$C_6H_5COO(CH_2)_3CH_3$	60% Dioxan	35	4.82×10^{-3}			164
$C_6H_5COOCH(CH_3)_2$	60% Dioxan	35	1.53×10^{-3}			164
$C_6H_5COOCH_2CH(CH_3)_2$	60% Dioxan	35	4.00×10^{-3}			164
$C_6H_5COOCH_2C_6H_5$	60% Dioxan	35	1.60×10^{-2}			164
$C_6H_5COO(CH_2)_2C_6H_5$	60% Dioxan	35	8.40×10^{-3}			164
B.1(b) Phenol esters						
$PhCOOC_6H_5$	60% Acetone, $\mu = 0.005$	25	5.07×10^{-2}	14.7	−15	234
$PhCOOC_6H_4CH_3$-*m*	60% Acetone, $\mu = 0.005$	25	2.80×10^{-2}	16.5	−10	234
$PhCOOC_6H_4Cl$-*m*	60% Acetone, $\mu = 0.005$	25	0.149	13.3	−18	234
$PhCOOC_6H_4COCH_3$-*m*	60% Acetone, $\mu = 0.005$	25	0.179	12.9	−19	234
$PhCOOC_6H_4CN$-*m*	60% Acetone, $\mu = 0.005$	25	0.520	12.5	−18	234
$PhCOOC_6H_4NO_2$-*m*	60% Acetone, $\mu = 0.005$	25	0.700	12.2	−18	234
$PhCOOC_6H_4COOEt$-*p*	60% Acetone, $\mu = 0.005$	25	0.261	11.8	−21	234
$PhCOOC_6H_4COOEt$-*p*	60% Acetone, $\mu = 0.01$	25	0.247			234
$PhCOOC_6H_4COOEt$-*p*	60% Acetone, $\mu = 0.02$	25	0.218			234
$PhCOOC_6H_4NO_2$-*p*	60% Acetone, $\mu = 0.005$	25	1.61	12.2	−16	234
$PcCOOC_6H_4CN$-*p*	60% Acetone, $\mu = 0.005$	25	0.850	11.1	−21	234
$PhCOOC_6H_4COCH_3$-*p*	60% Acetone, $\mu = 0.005$	25	0.377	11.1	−23	234
$PhCOOC_6H_4CH_3$-*p*	33% MeCN, $\mu = 0.3$	25	3.32×10^{-2}			235
$PhCOOC_6H_5$	33% MeCN, $\mu = 0.3$	25	5.57×10^{-2}			235
$PhCOOC_6H_4Cl$-*p*	33% MeCN, $\mu = 0.3$	25	0.124			235
$PhCOOC_6H_4NO_2$-*p*	33% MeCN, $\mu = 0.3$	25	0.805			235
$PhCOOC_6H_4NO_2$-*m*	33% MeCN, $\mu = 0.3$	25	0.459			235
p-$Me_2NC_6H_4COOC_6H_4CH_3$-*p*	33% MeCN, $\mu = 0.3$	25	7.95×10^{-4}			235
p-$Me_2NC_6H_4COOC_6H_5$	33% MeCN, $\mu = 0.3$	25	1.21×10^{-3}			235
p-$Me_2NC_6H_4COOC_6H_4NO_2$-*p*	33% MeCN, $\mu = 0.3$	25	1.49×10^{-2}			235
p-$Me_2NC_6H_4COOC_6H_4NO_2$-*m*	33% MeCN, $\mu = 0.3$	25	8.46×10^{-3}			235
p-$CH_3C_6H_4COOC_6H_4CH_3$-*p*	33% MeCN, $\mu = 0.3$	25	1.06×10^{-2}			235
p-$CH_3C_6H_4COOC_6H_5$	33% MeCN, $\mu = 0.3$	25	2.35×10^{-2}			235
p-$CH_3C_6H_4COOC_6H_4NO_2$-*p*	33% MeCN, $\mu = 0.3$	25	0.349			235
p-$CH_3C_6H_4COOC_6H_4NO_2$-*m*	33% MeCN, $\mu = 0.3$	25	0.187			235
p-$CH_3C_6H_4COOC_6H_4Cl$-*p*	33% MeCN, $\mu = 0.3$	25	5.11×10^{-2}			235

TABLE 31 (CONTINUED)

KINETIC DATA FOR THE HYDROXIDE-ION-CATALYZED HYDROLYSIS OF CARBOXYLIC ESTERS (MOSTLY AT 25°C)

Ester	*Conditions*	*Temp.* (°C)	k_{OH} ($l \cdot mole^{-1} \cdot sec^{-1}$)	$\Delta H^{\ddagger}$ ($kcal \cdot mole^{-1}$)	$\Delta S^{\ddagger}$ (*eu*)	*Ref.*
p-$ClC_6H_4COOC_6H_4CH_3$-*p*	33% MeCN, $\mu = 0.3$	25	9.70×10^{-2}			235
p-$ClC_6H_4COOC_6H_5$	33% MeCN, $\mu = 0.3$	25	0.175			235
p-$ClC_6H_4COOC_6H_4Cl$-*p*	33% MeCN, $\mu = 0.3$	25	0.306			235
p-$ClC_6H_4COOC_6H_4NO_2$-*p*	33% MeCN, $\mu = 0.3$	25	1.83			235
p-$ClC_6H_4COOC_6H_4NO_2$-*m*	33% MeCN, $\mu = 0.3$	25	0.135			235
p-$NO_2C_6H_4COOC_6H_4CH_3$-*p*	33% MeCN, $\mu = 0.3$	25	1.49			235
p-$NO_2C_6H_4COOC_6H_5$	33% MeCN, $\mu = 0.3$	25	1.79			235
p-$NO_2C_6H_4COOC_6H_4Cl$-*p*	33% MeCN, $\mu = 0.3$	25	4.55			235
p-$NO_2C_6H_4COOC_6H_4NO_2$-*p*	33% MeCN, $\mu = 0.3$	25	26.6			235
p-$NO_2C_6H_4COOC_6H_4NO_2$-*m*	33% MeCN, $\mu = 0.3$	25	16.4			235
B.2 Effect of varying acyl group						
p-$NH_2C_6H_4COOCH_3$	33% Dioxan	24.7	1.14×10^{-3}			211
p-$CH_3C_6H_4COOCH_3$	33% Dioxan	24.7	1.11×10^{-2}			211
$C_6H_5COOCH_3$	33% Dioxan	24.7	2.32×10^{-2}			211
p-$ClC_6H_4COOCH_3$	33% Dioxan	24.7	6.8×10^{-2}			211
p-$NO_2C_6H_4COOCH_3$	33% Dioxan	24.7	0.70			211
$C_6H_5COOCH_3$	60% Acetone	25	9.01×10^{-3}	13.57	−22.2	217
p-$CH_3C_6H_4COOCH_3$	60% Acetone	25	2.19×10^{-3}	14.83	−20.8	217
p-$NO_2C_6H_4COOCH_3$	60% Acetone	25	0.626	11.70	−20.1	217
C_6H_5COOEt	60% Acetone	25	2.89×10^{-3}	13.97	−23.1	217
p-$CH_3C_6H_4COOEt$	60% Acetone	25	1.14×10^{-3}	14.57	−23.0	217
p-$NO_2C_6H_4COOEt$	60% Acetone	25	0.246	11.81	−21.4	217
p-$NH_2C_6H_4COOEt$	60% Acetone	25	8.64×10^{-5}	16.11	−22.8	217
m-$CH_3C_6H_4COOEt$	60% Acetone	25	1.69×10^{-3}	14.28	−23.1	217
m-$NO_2C_6H_4COOEt$	60% Acetone	25	0.137	12.21	−21.4	217
m-$NH_2C_6H_4COOEt$	60% Acetone	25	1.66×10^{-3}	14.39	−22.8	217
$C_6H_5COOCH_3$	80% Methanol	34.8	8.33×10^{-2}	17.8	−13.6	151

m-$NO_2C_6H_4COOCH_3$	80% Methanol	34.5	3.24	16.5	−10.6	151
m-$ClC_6H_4COOCH_3$	80% Methanol	25	0.553	16.9	−13.2	151
m-$CH_3C_6H_4COOCH_3$	80% Methanol	25	4.39×10^{-2}	17.8	−14.4	151
C_6H_5COOEt	85% Ethanol	25	5.50×10^{-4}	17.1	−16.0	236
p-$NH_2C_6H_4COOEt$	85% Ethanol	25	1.27×10^{-5}	19.4	−16.0	236
p-$CH_3OC_6H_4COOEt$	85% Ethanol	25	1.15×10^{-4}	18.05	−15.9	236
p-$CH_3C_6H_4COOEt$	85% Ethanol	25	2.51×10^{-4}	17.6	−16.0	236
p-ClC_6H_4COOEt	85% Ethanol	25	2.37×10^{-3}	16.2	−16.2	236
p-IC_6H_4COOEt	85% Ethanol	25	2.78×10^{-3}	16.1	−16.2	236
p-BrC_6H_4COOEt	85% Ethanol	25	2.89×10^{-3}	16.2	−15.9	236
p-$NO_2C_6H_4COOEt$	85% Ethanol	25	5.67×10^{-2}	13.9	−17.1	236
m-$CH_3OC_6H_4COOEt$	85% Ethanol	25	8.05×10^{-4}	16.97		237
m-$CH_3C_6H_4COOEt$	85% Ethanol	25	4.30×10^{-4}	17.34		237
m-ClC_6H_4COOEt	85% Ethanol	25	4.63×10^{-3}	15.94		237
m-BrC_6H_4COOEt	85% Ethanol	25	4.92×10^{-3}	15.90		237
m-$NO_2C_6H_4COOEt$	85% Ethanol	25	4.15×10^{-2}	14.62		237
3,4-$(CH_3)_2C_6H_3COOEt$	85% Ethanol	25	1.84×10^{-4}	17.82		237
3,5-$(CH_3)_2C_6H_3COOEt$	85% Ethanol	25	3.02×10^{-4}	17.55		237
3,4-$(CH_3O)_2C_6H_3COOEt$	85% Ethanol	25	2.14×10^{-4}	17.76		237
3,5-$(CH_3O)_2C_6H_3COOEt$	85% Ethanol	25	1.04×10^{-3}	16.82		237
3,4-$Cl_2C_6H_3COOEt$	85% Ethanol	25	1.46×10^{-2}	15.26		237
3,5-$Cl_2C_6H_3COOEt$	85% Ethanol	25	3.97×10^{-2}	14.66		237
3,5-$Br_2C_6H_3COOEt$	85% Ethanol	25	3.96×10^{-2}	14.67		237
3-NO_2-4-$CH_3OC_6H_3COOEt$	85% Ethanol	25	6.36×10^{-2}	15.75		237
3,4,5-$(CH_3O)_3C_6H_2COOEt$	85% Ethanol	25	9.27×10^{-4}	16.89		237
$C_6H_5COOC_6H_5$	33% MeCN, $\mu = 0.3$	25	5.57×10^{-2}			235
p-$Me_2NC_6H_4COOC_6H_5$	33% MeCN, $\mu = 0.3$	25	1.21×10^{-3}			235
p-$CH_3C_6H_4COOC_6H_5$	33% MeCN, $\mu = 0.3$	25	2.35×10^{-2}			235
p-$ClC_6H_4COOC_6H_5$	33% MeCN, $\mu = 0.3$	25	0.175			235
p-$NO_2C_6H_4COOC_6H_5$	33% MeCN, $\mu = 0.3$	25	1.79			235
Effects of ortho substituents						
$C_6H_5COOCH_3$	80% Methanol	34.8	8.33×10^{-2}	17.8	−13.6	151
o-$CH_3C_6H_4COOCH_3$	80% Methanol	35.0	1.22×10^{-2}	18.8	−14.0	151
o-$C_2H_5C_6H_4COOCH_3$	80% Methanol	45.0	1.61×10^{-2}	20.2	−11.5	151
o-$(CH_3)_2CHC_6H_4COOCH_3$	80% Methanol	100.2	1.32	18.8	−16.4	151

TABLE 31 (CONTINUED)

KINETIC DATA FOR THE HYDROXIDE-ION-CATALYZED HYDROLYSIS OF CARBOXYLIC ESTERS (MOSTLY AT 25°C)

Ester	*Conditions*	*Temp.* (°C)	k_{OH} ($l \cdot mole^{-1} \cdot sec^{-1}$)	$\Delta H^{\ddagger}$ ($kcal \cdot mole^{-1}$)	$\Delta S^{\ddagger}$ (*eu*)	*Ref.*
o-$(CH_3)_3CC_6H_4COOCH_3$	80% Methanol	119.8	1.90×10^{-2}	24.1	−13.6	151
o-$FC_6H_4COOCH_3$	80% Methanol	35.0	0.387	17.0	−13.4	151
o-$ClC_6H_4COOCH_3$	80% Methanol	35.0	0.169	16.0	−18.4	151
o-$BrC_6H_4COOCH_3$	80% Methanol	34.9	0.120	16.5	−17.4	151
o-$IC_6H_4COOCH_3$	80% Methanol	34.8	8.32×10^{-2}	16.5	−18.1	151
$C_6H_5COOCH_3$	60% Dioxan	35.0	1.49	8.0	−23.9	151
o-$CH_3C_6H_4COOCH_3$	60% Dioxan	40.2	0.285	7.7	−25.5	151
o-$C_2H_5C_6H_4COOCH_3$	60% Dioxan	40.0	0.113	7.9	−24.5	151
o-$(CH_3)_2CHC_6H_4COOCH_3$	60% Dioxan	40.9	8.25×10^{-2}	7.2	−27.4	151
o-$(CH_3)_2CHC_6H_4COOCH_3$	60% Dioxan	114.6	4.7×10^{-2}	9.4	−16.9	151
2,6-$(CH_3)_2C_6H_3COOCH_3$	60% Dioxan	125	1.73×10^{-3}	21.0	−18.4	238
4-NH_2-2,6-$(CH_3)_2C_6H_2COOCH_3$	60% Dioxan	125	8.84×10^{-4}	19.9	−22.2	238
2,4,6-$(CH_3)_3C_6H_2COOCH_3$	60% Dioxan	125	1.35×10^{-3}	20.4	−20.9	238
2,4,6-$(CH_3)_3C_6H_2COOCH_3$	60% Dioxan	126	2.2×10^{-3}			207
4-Br-2,6-$(CH_3)_2C_6H_2COOCH_3$	60% Dioxan	125	6.82×10^{-3}	18.3	−22.5	238
4-NO_2-2,6-$(CH_3)_2C_6H_2COOCH_3$	60% Dioxan	125	3.52×10^{-2}	16.5	−23.8	238

TABLE 32

KINETIC DATA FOR METHOXIDE-CATALYZED TRANSESTERIFICATION IN METHANOL

Ester	k_{obs} *(l·mole^{-1}·sec^{-1})*	$\Delta H^{\ddagger}$ *(kcal·mole^{-1})*	$\Delta S^{\ddagger}$ *(eu)*
A. $ArCOOCH_3 + {}^*CH_3O^- \rightleftharpoons ArCOO^*CH_3 + CH_3O^-$ *(ref. 239) at 30.1°C*			
p-$CH_3OC_6H_4COOCH_3$	1.06×10^{-2}	14.3	−20.3
p-$CH_3C_6H_4COOCH_3$	1.86×10^{-2}	14.1	−20.1
$C_6H_5COOCH_3$	4.81×10^{-2}	12.7	−22.8
p-$BrC_6H_4COOCH_3$	0.212	11.0	−25.4
p-$NO_2C_6H_4COOCH_3$	4.16	8.1	−29.4
B. Methanolysis of substituted 1-menthyl benzoates at 30°C (ref. 240)			
C_6H_5COOR	5.19×10^{-5}	16.89	−22.5
o-$CH_3OC_6H_4COOR$	2.48×10^{-5}	17.36	−24.2
o-$CH_3C_6H_4COOR$	3.83×10^{-6}	17.54	−24.8
o-ClC_6H_4COOR	3.86×10^{-5}	16.42	−24.6
o-$NO_2C_6H_4COOR$	1.45×10^{-4}	16.23	−22.4
m-$CH_3OC_6H_4COOR$	6.85×10^{-5}	16.74	−22.5
m-$CH_3C_6H_4COOR$	3.88×10^{-5}	17.12	−22.3
m-ClC_6H_4COOR	5.81×10^{-4}	15.76	−21.4
m-BrC_6H_4COOR	5.75×10^{-4}	15.76	−21.4
m-$NO_2C_6H_4COOR$	4.18×10^{-3}	14.68	−21.0
p-$CH_3OC_6H_4COOR$	1.12×10^{-5}	18.04	−21.8
p-$CH_3C_6H_4COOR$	2.18×10^{-5}	17.41	−22.5
p-ClC_6H_4COOR	2.36×10^{-4}	16.14	−21.8
p-BrC_6H_4COOR	2.83×10^{-4}	16.08	−21.8
p-$NO_2C_6H_4COOR$	5.89×10^{-3}	14.19	−21.8

$\rho = 2.03$ for the *p*-methylphenyl esters[235], 2.04 for the *p*-chlorophenyl and *n*-nitrophenyl esters[235], and 2.01 for the *p*-nitrophenyl esters[235,242] all at 25°C.

A significantly smaller ρ-value was found by Newman *et al.*[238] for the hydrolysis of methyl 4-substituted 2,6-dimethylbenzoates in 60% dioxan at 125°C. Their value of 1.64 corresponds to a figure of 2.20 at 25°C[241] (the usual inverse relationship between ρ and temperature can be presumed to hold: for example, ρ for the hydrolysis of *m*-substituted methyl benzoates in 80% aqueous methanol at 100°C is 1.94[151], equivalent to 2.43 at 25°C). This low ρ was originally taken to mean that the two *ortho* methyl groups inhibit attack on the carbonyl group so effectively that attack is diverted to the ester methyl group[238], but hydrolysis in ^{18}O-enriched water shows that this is not so[207], since the hydrolysis of methyl 2,4,6-trimethylbenzoate involves acyl–oxygen cleavage. The low sensitivity to substituents must therefore have another explanation; one possibility is that the transmission of polar effects from the aromatic ring to the carboxyl group is hindered. Conjugation is fully effective only when the two are coplanar, and this is not possible when the carboxyl group is flanked by two *ortho*-methyl substituents.

References pp. 202–207

TABLE 33

HAMMETT REACTION CONSTANTS FOR THE ALKALINE HYDROLYSIS OF SUBSTITUTED AROMATIC AND ARYLALIPHATIC ESTERS

Ester	*Conditions*	*Temp. (°C)*	*Reaction constant, ρ*	*Ref.*
$ArCOOCH_3$	60% Acetone	25	2.37	189, 243
$ArCOOC_2H_5$	85% Ethanol	25	2.51	241, 243
$ArCOOC_2H_5$	85% Ethanol	25	2.61	189, 243
$ArCOOC_2H_5$	88% Ethanol	30	2.55	189, 243
$ArCOOC_2H_5$	60% Ethanol	25	2.47	189, 243
$ArC{\equiv}C{\cdot}COOC_2H_5$	88% Ethanol	20	2.05	231, 241
trans-$ArCH{=}CHCOOC_2H_5$	85% Ethanol	25	1.24	237, 243
trans-$ArCH{=}CHCOOC_2H_5$	88% Ethanol	30	1.23	247, 243
cis-$ArCH{=}CHCOOC_2H_5$	88% Ethanol	30	1.09	232, 241
$ArC_6H_4COOC_2H_5$	88% Ethanol	25	0.60	241
$ArCH_2COOC_2H_5$	85% Ethanol	25	1.03	149, 241
$ArCH_2COOC_2H_5$	60% Acetone	25	0.96	241, 248
$ArCH_2CH_2COOC_2H_5$	88% Ethanol	30	0.64	241, 249
cis-ArCH——CHCOOC$_2$H$_5$ (bridged by CH_2)	88% Ethanol	30	1.10	241, 249
trans-ArCH——CHCOOC$_2$H$_5$ (bridged by CH_2)	88% Ethanol	30	0.85	241, 249

When the substituent is more remote from the carboxyl group than in substituted benzoate esters, smaller values of ρ are to be expected. Thus, $\rho = 1.21$ for the alkaline hydrolysis of *ortho*-substituted ethyl cinnamates, ArCH=CHCOOEt, in 87.8% aqueous ethanol at 30°C[148,149].

When the substituent is directly on the double bond, as in the substituted acrylate esters $XCH{=}CHCOOCH_3$, studied by Bowden[166], its effect is increased again. *Cis*-substituents in the 3 position, as might be expected, affect reactivity in much the same way as do *ortho*-substituents in benzoate esters. The *trans*-substituents, on the other hand, certainly cannot interact sterically, and the data are correlated by the Hammett equation, with the very high ρ value of 3.74 at 18.8°C[166]. This figure is almost twice as large as ρ for *meta*- and *para*-substituted methyl benzoates under the same conditions (2.05 at 18.8°C[166]), and represents the highest measured sensitivity to the polar effects of substituents in the alkaline hydrolysis of esters, although it is probable that the transmission of polar effects is more efficient still for *cis*-substituents in acrylic esters[166], and perhaps also for *ortho*-substituents in benzoic acid derivatives[166,244].

Taft reaction constants, ρ^*, for the hydrolysis of aliphatic esters, are available for only a few systems. Using independently determined values of E_s and σ^*, Bowden *et al.*[220], found that the rates of alkaline hydrolysis of simple methyl arylaliphatic carboxylates in 80% methanol are correlated very well by the modified Taft equation[152]

$$\log (k/k_0) = \sigma^*\rho^* + \delta E_s \tag{6}$$

with $\rho^* = 1.64$ and $\delta = 0.87$. Hancock *et al.*[163] obtained a similar value of $\rho^* = 1.75$ by fitting their data for the hydrolysis of nine aliphatic methyl esters (in 40% dioxan at 35°C) to the more complex equation

$$\log (k/k_0) = \sigma^*\rho^* + \delta E_s^c + h(n-3) \tag{7}$$

where E_s^c is a corrected steric substituent constant, h the "reaction constant" for hyperconjugation, and n the number of α hydrogen atoms.

For more remote substitution ρ^* is expected to be smaller. Thus Hay and Porter[228] found $\rho^* = 0.6$ for the alkaline hydrolysis of the methyl esters of α-amino acids, $RCH(NH_2)COOCH_3$ in water at 25°C. However, higher sensitivities are possible if the geometry of the system is such that the direct transmission of polar effects becomes more important. For example, Roberts and Moreland[245] found that the effects of substituents in the 4 position of ethyl bicyclo-[2.2.2]octane-1-carboxylates (28) on the hydrolysis in 88% ethanol at 30°C are comparable to those observed for *meta*- and *para*-substituted benzoates ($\rho = 2.24$), and a similar, though smaller, effect, is observed for methyl *trans*-4-substituted cyclohexanecarboxylates (29).

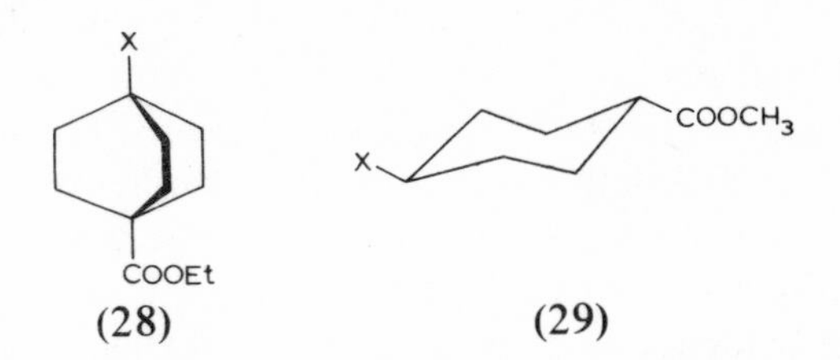

(28) (29)

The effects of substituents in the leaving group are considerably smaller than those described above for substitution in the acyl group. For their five series of substituted phenyl benzoates Kirsch *et al.*[235] found ρ close to 1.24 (in 33% acetonitrile at 25°C). This is in the region expected if substitution in the leaving group affects reactivity almost exclusively by its effect on the rate of addition of hydroxide anion to the carbonyl group, since the interposition of the oxygen atom decreases polar effects by a factor calculated by Bowden as 0.61[241]. For the hydrolysis of substituted phenyl acetates[219] and benzoates[246] in 60% acetone at 20° and 25°C, respectively, Humffray and Ryan[219] found that the data are correlated better by the Yukawa–Tsuno equation

$$\log (k/k_0) = \rho[\sigma + r(\sigma^- - \sigma)]$$

than by the simple Hammett relation, although the best values of r were small

(0.20 and 0.3, respectively). Using this treatment, the data for both *meta*- and *para*-substituted compounds fall on the same straight line, and accurate values were found of $\rho = 1.69$ for the acetates, and $\rho = 1.76$ for the benzoates. For the acetates of nine aliphatic alcohols Hancock *et al.*[163] found $\rho^* = 1.80$ when the data were fitted to the Taft equation, falling to $\rho^* = 1.40$ when their own modified eqn. (7) was used. A similar value of $\rho^* = 1.52$ was found[164] for the hydrolysis of alkyl benzoates in 60% dioxan at 35°C. A good comparison of the relative effects of substitution in the acyl group and leaving group is available from this work[164]; the rates of alkaline hydrolysis of 35 alkyl benzoates, substituted in the ring, or the alkyl group, or both, are accurately correlated by the equation

$$\log k = 0.174 + 2.22\sigma + 1.53\sigma^* + 0.668E_s^{\,c} \qquad (8)$$

This equation contains perhaps the most accurate comparison of the relative effects of substitution on the acyl and leaving groups in alkaline hydrolysis, and is consistent with a rate-determining step involving little or no bond-breaking. The ratio of coefficients of σ^* and σ, $1.53/2.22 = 0.69$, from eqn. (8) is not significantly larger than the transmission coefficient of 0.61 calculated by Bowden[241]. If a significant amount of bond-breaking were involved the reaction constants for varying the leaving group should be larger. Furthermore they should be correlated, in the case of substituted phenyl esters, better by σ^- than by σ. In fact, as described above, the correlation is improved by allowing for a small dependence on σ^-, since the data for substituted phenyl acetates and benzoates fit the Yukawa–Tsuno equation better than the simple Hammett relation (p. 179), but only a relatively small correction is necessary.

Perhaps the most important conclusion to be drawn from this discussion of polar effects is that there is no evidence of extensive bond-breaking in the transition state, as there surely should be if a direct displacement mechanism were involved. But the evidence is entirely consistent with a mechanism involving the largely rate-determining formation of a tetrahedral addition intermediate.

(iii) Steric effects

Steric effects on the alkaline hydrolysis of esters were taken by Taft[152] to be much the same as for acid-catalyzed hydrolysis, since the transition states differ by the presence or absence of just two protons, *viz.*

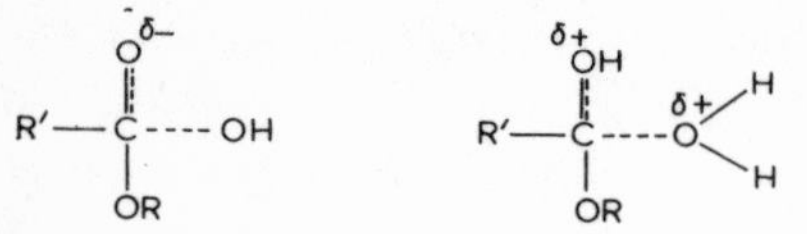

R'—C(=O$^{\delta-}$)(OR)----OH R'—C(=OH$^{\delta+}$)(OR)----O$^{\delta+}$H$_2$

This argument remains approximately valid even if the breakdown of the tetrahedral intermediate is partially rate-determining in one case and not in the other. And the considerable success of the Taft–Ingold separation of polar and steric effects in aliphatic systems is good evidence that the assumption is fundamentally sound. The treatment, and some of the results, have been discussed for acid hydrolysis (p. 132). The most recent work in this area has been with *ortho*-substituted benzoate esters[151,250,254] and has culminated in Charton's conclusion[250] that steric effects are unimportant in this system. In contrast to the acid-catalyzed reaction, reactivity in alkaline ester hydrolysis depends largely on σ_I rather than σ_R. It seems likely that this represents principally a field effect, rather than the inductive effect, as it may also do in the similar case of the ionization of 2-substituted benzoic acids[252].

The same considerations presumably apply to the reactivity of *cis*-substituted acrylate esters (31a) studied by Bowden[166].

H, H / C=C / X, COOCH₃ (31a) X, H / C=C / H, COOCH₃ (31b)

In this work the rates of alkaline hydrolysis in 70% dioxan–water and those of acid-catalyzed ester formation in methanol were compared for the *cis* (31a) and *trans* (31b) substituted compounds. This was expected to isolate the steric effects of the *cis* substituents, since inductive and resonance effects should be similar in the *cis* and the *trans* compounds. The results are summarized in Table 34, and show the trend already observed for *ortho*-substituted

TABLE 34

STERIC EFFECTS OF *cis*-SUBSTITUENTS IN $XCH{=}CHCOOCH_3$

(Ref. 166)

	$\log (k_{cis}/k_{trans})$	
Substituent X	*Ester formation (methanol, 35°C)*	*Ester hydrolysis (70% dioxan, 18.8°C)*
H	0.0	0.0
CH_3	−0.37	−0.25
Cl	−0.18	−0.61
Br	−0.20	−0.96
I	−0.40	−1.03

benzoates[151], of larger apparent steric effects in alkaline hydrolysis than in the acid-catalyzed reaction. The reactivity of the *cis*-substituted acrylates accurately parallels that of the corresponding *ortho*-substituted benzoates, and so these figures should be explicable in terms of the field effects of substituents[254].

References pp. 202–207

The greater sensitivity of the alkaline reaction to changes in substituent is readily explained in this way—the comparison of *cis* and *trans* compounds isolates not only the steric effect but also the direct field effect—but the correlation is poor with σ_I as well as with r_v values. This set of results is too small to allow firm conclusions, and more careful work is required to identify or rule out steric effects in this system.

(iv) Medium effects

The alkaline hydrolysis of esters normally involves the rate-determining attack of an anion upon a neutral molecule, and is not therefore expected to be very sensitive to changes in ionic strength. Some data taken from Table 31 support this prediction: the second-order rate coefficient for the hydroxide-catalyzed hydrolysis of *p*-carbethoxyphenyl benzoate in 60% acetone–water, for example, falls from 0.261 to 0.247 to 0.218 at 25°C when the ionic strength is increased successively from 5×10^{-3} to 10^{-2} to 2×10^{-2} M[234], and similar results are found for phenyl benzoate itself[206,234] and for other substituted phenyl benzoates[234]. In each case a negative effect on the rate is observed when the ionic strength is increased, and there are indications that the effect is greater for the less reactive esters; *p*-nitrophenyl benzoate, for example, is hydrolyzed at almost the same rate at ionic strength 5×10^{-3} as it is at 10^{-2} M[161].

There are not sufficient data available for a useful discussion of the effects of added salts on neutral esters. Some of the most interesting work done in recent years has involved the hydrolysis of charged compounds. Hoppé and Prue[255], for example, studied the effects of metal cations on the alkaline hydrolysis of the ethyl half-esters of a series of dicarboxylic acids, $EtOOC(CH_2)_nCOO^-$, and found a relatively large negative salt effect for alkali metal chlorides when the negatively-charged centre was close to the reacting ester group. The effect diminished regularly with increasing values of *n*. Several authors have studied the effects of added salts on the alkaline hydrolysis of positively charged esters[224,226,237,256,257]. Here, too, a negative salt effect is observed, which is larger than those found for neutral and anionic esters (Table 31).

Solvent effects on the alkaline hydrolysis of esters have been studied systematically by a number of authors. The addition of a less polar solvent to an aqueous medium generally has a marked negative effect on the rate of hydrolysis[258–260], as may be discerned from an examination of the data given in Table 31 for reactions in different solvents. Some data collected specifically for the investigation of solvent effects are given in Table 35.

The decrease in rate with decreasing polarity of the medium is a function of the dielectric constant, and plots of $\log k$ against $1/D$ are approximately linear over a wide range of mixtures of water with acetone, dioxan, and

TABLE 35

RATE COEFFICIENTS (l·mole^{-1}·sec^{-1}) FOR THE ALKALINE HYDROLYSIS OF ESTERS IN VARIOUS SOLVENTS AT 25°C

A. Hydrolysis of ethyl haloacetates in aqueous ethanol (ref. 260)

Ester	*Water*	*49.75% Ethanol*	*91.77% Ethanol*
FCH_2COOEt	12.3±0.4	6.79±2	1.16±0.1
$ClCH_2COOEt$	25.8±2	11.3±4	1.95±0.2
$Cl_2CHCOOEt$		240±23	19.3±4
$Cl_3CCOOEt$		370±20	32.7±2

B. Ethyl benzoates in aqueous ethanol (ref. 258)

Ester	*30% Ethanol*	*50% Ethanol*	*70% Ethanol*
C_6H_5COOEt	7.25×10^{-3}	1.60×10^{-3}	1.02×10^{-3}
p-$NO_2C_6H_4COOEt$	0.230	0.124	0.110

C. Ethyl benzoates in aqueous dioxan (ref. 258)

Ester	*Dioxan (%)*	k_2	$\Delta H^{\ddagger}$ *(kcal·mole^{-1})*	$\Delta S^{\ddagger}$ *(eu)*
C_6H_5COOEt	20	12.4×10^{-3}	11.95	−27.1
	30	8.84×10^{-3}	12.13	−27.2
	40	6.78×10^{-3}	12.33	−27.1
	50	5.08×10^{-3}	12.38	−27.5
p-$NH_2C_6H_4COOEt$	30	4.57×10^{-4}	13.43	−28.7
	40	2.68×10^{-4}	14.34	−26.7
	50	1.70×10^{-4}	15.10	−25.1
p-$NO_2C_6H_4COOEt$	30	0.385	11.23	−22.9
	40	0.340	11.31	−22.7
	50	0.319	11.20	−23.0

various alcohols[258,260,261]. Qualitatively the effect is the opposite of that expected for a reaction in which charge is dispersed in the transition state. In simple cases, such as many S_N2 reactions between anionic nucleophiles and neutral compounds, a small increase in rate is found, as expected (ref. 1, pp. 347–9) as the polarity of the solvent is reduced. Addition to the carbonyl group, however, is a very different reaction. The breakdown of the tetrahedral intermediate is normally a rapid reaction, and it follows from the Hammond postulate[262] that bond formation is well advanced in the transition state. The charge is consequently much less completely dispersed than it is in the more symmetrical transition state of an S_N2 reaction. Further, the two negative centres in the transition state for addition to a carbonyl group are much closer in space (32) and their solvation shells cannot be independent, as they must be in the linear transition state of an S_N2 reaction (33), *viz.*

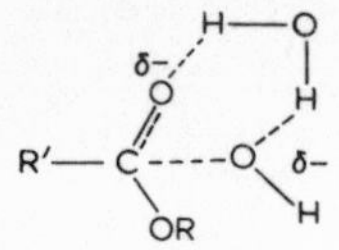

(32) 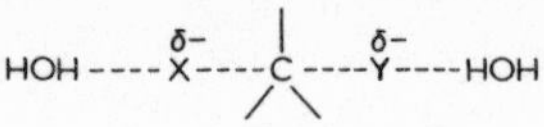 (33)

The situation is thus sufficiently complex that it is not surprising that the behaviour with different solvents is different from that predicted by the simple theory. The differences are in fact small, though in opposite directions in the two cases. Laidler and Landskroener[126] have had a degree of success in fitting data for the alkaline hydrolysis of esters to their theory of solvent effects (which predicts a linear dependence of log k on the reciprocal of the dielectric constant of the medium, as do other theories concerned with ion-dipolar molecular interactions[263]). These authors assumed a transition state in which a molecule of water is involved bridging the attacking hydroxide ion and the leaving group[126]. This particular mechanism now seems unlikely, since there is much evidence that bond-breaking has not proceeded to a significant extent in the transition state. But clearly any mechanism which includes a molecule of water, however it is involved, could explain a decrease in rate as the water concentration is decreased. Some authors have attempted to measure a kinetic "order" in water by plotting log k_{OH} against log $[H_2O]$ for various solvent mixtures, and a slope of 1.0 is found, for example, for the alkaline hydrolysis of ethyl acetate in aqueous acetone and dioxan[264]. However, these plots, although often linear, do not generally give slopes close to 1.0, but values varying widely between 0 and 3[258,265–7].

A notable exception to the solvent effects described above is the behaviour of esters on alkaline hydrolysis in aqueous dimethyl sulphoxide. In this, and other dipolar aprotic solvents, the reactivity of the hydroxide ion is considerably greater than in water[268], and this effect is much larger than the more usual small negative effects of most organic solvents. The rate of hydrolysis of an ester is generally reduced by the addition of low concentrations of dimethyl sulphoxide to the aqueous solvent, but at sufficiently high concentrations of the organic solvent the rate of alkaline hydrolysis increases sharply with decreasing water content[259,269]. Ethyl acetate is an exception, in that the rate is increased even by small amounts of dimethyl sulphoxide[270] (Fig. 15), but its behaviour is otherwise similar to that of other esters.

This behaviour is that expected if the positive effect of the dipolar aprotic solvent on the activity of the hydroxide ion is superimposed on the normal negative effect of an organic solvent, and should be observed also for other dipolar aprotic solvents. In fact this type of behaviour is found for the alkaline hydrolysis of ethyl p-nitrobenzoate in mixtures of water and acetone, which is less polar than dimethyl sulphoxide. Other esters show intermediate behaviour in this solvent[259] (Fig. 16). It should be stressed that this represents special

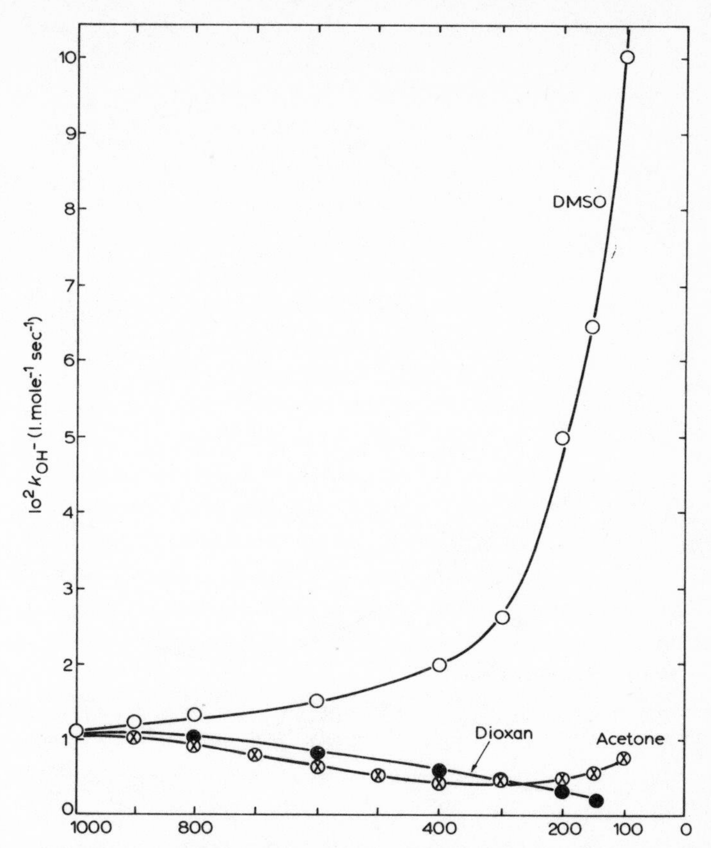

Fig. 15. Effects of added solvents on the rate coefficients for alkaline hydrolysis of ethyl acetate at 25°C. Data from refs. 259 and 270.

behaviour. In the presence of less strongly polar, or protic organic solvents, such as dioxan and the lower aliphatic alcohols, a steady decrease in the rate of alkaline hydrolysis is observed as the proportion of the organic solvent is increased. These various types of behaviour are illustrated in Figs. 15 and 16.

Roberts[269] has studied the hydrolysis of series of ethyl esters[271,272] and alkyl benzoates[273] in a limited range of water–dimethyl sulphoxide mixtures in some detail, and finds that the data for the hydrolysis of aliphatic ethyl esters (at 35°C in 85% dimethyl sulphoxide–water) fit the modified Taft equation (eqn. 5, p. 132). The values of $\rho^* = 1.88$ and $\delta = 0.88$ can be compared with $\rho^* = 2.39$ and $\delta = 1.04$ for 85% EtOH–water at the same temperature. The polar reaction constant is reduced in the dipolar aprotic solvent, consistent with a reduced degree of bond formation in the transition state, expected if the activity of the hydroxide ion is increased. However, Roberts considers that the sensitivity to steric effects, as measured by δ, would be reduced more substantially if bond formation were less advanced. It is difficult to accept this argument, since we

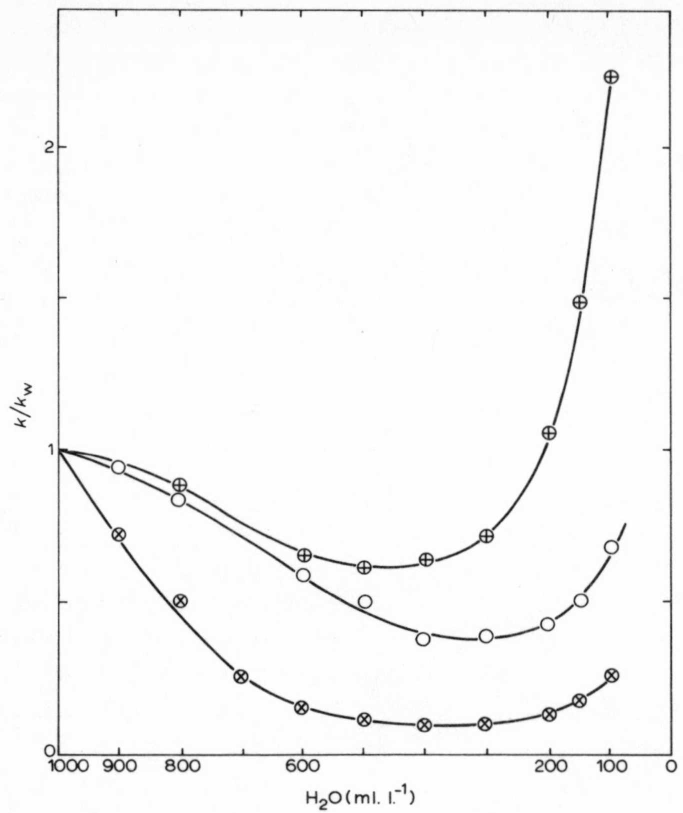

Fig. 16. Effects of added acetone on the alkaline hydrolysis of substituted ethyl benzoates. Data for ethyl *p*-nitrobenzoate (⊕) and benzoate (⊗) at 25°C (from ref. 259). The central curve (○) is for ethyl acetate.

know little about the sensitivities of either ρ^* or δ to the degree of bond formation, or the consequent conclusion that the rate enhancement on the addition of dimethyl sulphoxide is due not to the increased activity of hydroxide ion but to more efficient solvation of the transition state[272].

(*b*) *Catalysis by other oxyanions*

Oxyanions of sufficient basicity will catalyze the hydrolysis of all but the least reactive esters; but since the latter include the esters of the common aliphatic alcohols, early attempts to detect the reaction were negative or inconclusive. Dawson and Lowson[274], claimed to have detected catalysis by acetate ion of the hydrolysis of ethyl acetate as early as 1927, but the extent of catalysis observed was too small to rule out the possibility that salt or solvent effects were, in fact, responsible.

Curiously, the clue to the successful observation of general base and nucleophilic catalysis of ester hydrolysis appeared in the literature only a few months later. This was Kilpatrick's classical work on anhydride hydrolysis. In 1928 he showed[275] that formate ion strongly catalyzes the hydrolysis of acetic anhydride. Catalysis by acetate ion is also observed, but is much weaker, while propionate and butyrate inhibit hydrolysis. Similar results are observed with propionic anhydride[276] and acetic propionic anhydride[277], except that in these cases acetate is a strong catalyst, comparable with formate. Similar observations have been made more recently, with other anhydrides, for instance by Emery and Gold[278]. These results are explained by the so-called anhydride-exchange reaction, which involves a nucleophilic displacement on the original anhydride by the carboxylate anion involved, *viz.*

$$\mathrm{R'CO{\cdot}O{\cdot}COR' + R''COO^- \rightleftharpoons R'CO{\cdot}O{\cdot}COR'' + R'COO^-}$$
$$\mathrm{R'CO{\cdot}O{\cdot}COR'' + R''COO^- \rightleftharpoons R''CO{\cdot}O{\cdot}COR'' + R'COO^-}$$

If the anhydride derived from R″COOH is more reactive than the original anhydride, catalysis of hydrolysis is observed. If it is less reactive, hydrolysis is inhibited. The first case amounts to nucleophilic catalysis of hydrolysis.

Catalysis of the hydrolysis of acetic anhydride by acetate ion cannot be explained in this way, since nucleophilic displacement simply generates another molecule of acetic anhydride. The mechanism of this reaction is presumed to be general base catalysis, usually written as

$$\mathrm{CH_3COO^{\ominus}\ H{-}O(H)\ \ CH_3C(=O){-}O{-}C(=O)CH_3 \longrightarrow 2\,CH_3COOH + CH_3COO^{\ominus}}$$

The dividing line between esters and anhydrides is an arbitrary one. 2,4-Dinitrophenol is a stronger acid than acetic acid, so its acyl derivatives, usually regarded as aryl esters, may be considered alternatively as mixed anhydrides. As such they would be expected to undergo nucleophilic attack by oxyanions, to generate a molecule of a carboxylic anhydride, *viz.*

$$\mathrm{R''COO^- + R'CO{\cdot}OAr \rightleftharpoons R''{\cdot}CO{\cdot}O{\cdot}CO{\cdot}R' + ArO^-}$$

If the anhydride formed is more reactive than the ester, as is generally the case, nucleophilic catalysis of hydrolysis should result. There is ample evidence supporting this mechanism.

(i) Nucleophilic catalysis

As expected, carboxylate and other oxyanions catalyze the hydrolysis of esters with good leaving groups. And evidence specifically implicating the nucleophilic mechanism is available from work using labelled substrates. Bender and Neveu[279] hydrolyzed 2,4-dinitrophenyl benzoate in the presence of ^{18}O-labelled acetate ion. The acetyl group is well known to be more reactive towards nucleophilic attack than the benzoyl group, so the intermediate acetic benzoic anhydride should be hydrolyzed predominantly by attack at the acetyl carbonyl group. Thus ^{18}O should appear in the benzoate ion produced on hydrolysis, *viz.*

$$CH_3C(^{18}O)(^{18}O)^- + PhCO{-}O{-}C_6H_3(NO_2)_2 \rightleftharpoons H_3C{-}C(={}^{18}O){-}{}^{18}O{-}C(=O){-}Ph + ArO^{\ominus}$$

$$\xrightarrow{H_2\ddot{O}} CH_3{-}C(^{18}O)(O)^- + Ph{-}C(^{18}O)(O)^- + 2H^+$$

When hydrolysis was carried out in 50% dioxan–water 1 *M* in acetic acid–acetate buffer of pH 5.5, it was found that some 75% of the ^{18}O derived from one of the oxygen atoms of the labelled acetate ion appeared in the benzoic acid produced.

A similar result was obtained by Schowen and Behn[280], for the methanolysis of *p*-nitrophenyl acetate in the presence of tritium-labelled acetate ion. In this case the intermediate anhydride is effectively symmetrical, and methanol will attack equally at the labelled and unlabelled acetyl group. As expected for nucleophilic catalysis, 50% (within experimental error) of the methyl acetate produced was labelled $T_3CCOOCH_3$.

A second approach demonstrating the existence of the intermediate anhydride is exemplified by the aniline trapping technique of Gold *et al.*[281]. Low concentrations of added amines, particularly aniline, react much more rapidly with the anhydride than does the solvent. Thus, if the hydrolysis of, for example, 2,4-dinitrophenyl acetate catalyzed by acetate ion, is carried out in the presence of low concentrations of aniline, acetanilide is formed[281], *viz.*

$$CH_3COO^- + CH_3CO\cdot OAr \rightleftharpoons CH_3\cdot CO\cdot O\cdot COCH_3 + ArO^-$$

$$\xrightarrow{PhNH_2} CH_3CONHPh + CH_3COOH$$

When small corrections are made to allow for the direct reaction of aniline with the ester, it can be shown that a quantitative yield of acetanilide is produced, so that nucleophilic catalysis accounts quantitatively for the observed reaction.

When this technique is applied to esters of less acidic phenols a quantitative yield of acetanilide is often not obtained. Thus Butler and Gold[282] estimated in their earliest experiments using this method that in the acetate-catalyzed hydrolysis of *p*-nitrophenyl acetate the nucleophilic mechanism accounted for less than half the observed catalysis. More recent results[281], using an improved technique, have shown clearly that the proportion of nucleophilic catalysis depends critically on the leaving group. The results are given in full in Table 36.

TABLE 36

NUCLEOPHILIC *versus* GENERAL BASE CATALYSIS BY ACETATE ION OF THE HYDROLYSIS OF ARYL ACETATES IN WATER AT pH 5.0 AND IONIC STRENGTH 0.5 *M*

(Data from ref. 281)

Acetate of	pK_a *of phenol*	$k_0{}^a$ (sec^{-1})	k_{AcO^-} ($l \cdot mole^{-1} \cdot sec^{-1}$)	*Fraction of nucleophilic catalysis*
2,6-Dinitrophenol	3.71	1.42×10^{-5}	1.25×10^{-3}	0.9
2,4-Dinitrophenol	4.11	1.18×10^{-5}	5.46×10^{-4}	0.8
2,3-Dinitrophenol	4.96	5.62×10^{-6}	2.56×10^{-4}	1.0
3,4-Dinitrophenol	5.42	4.60×10^{-6}	1.02×10^{-4}	0.8
4-Nitrophenol	7.15	8.46×10^{-7}	5.64×10^{-6}	0.7
2-Nitrophenol	7.21	7.64×10^{-7}	4.13×10^{-6}	?
3-Nitrophenol	8.40	3.02×10^{-7}	3.36×10^{-6}	0
4-Chlorphenol	9.38	6.3×10^{-8}	8.97×10^{-7}	0–0.2
Phenol	10.0	6.6×10^{-8}	3.84×10^{-7}	0
4-Methylphenol	10.26	3.9×10^{-8}	2.63×10^{-7}	0

[a] Rate coefficient for spontaneous hydrolysis.

These results show that nucleophilic catalysis predominates when the leaving group is less basic than acetate, or of comparable basicity. Where the aryl-oxide is more than about 3 p*K* units more basic, however, the nucleophilic mechanism becomes insignificant. And for *p*-nitrophenyl acetate, which appears to lie close to the borderline, nucleophilic and general base catalysis proceed at comparable rates.

These results can be rationalised most simply, though not uniquely, in terms of a mechanism for nucleophilic catalysis involving a tetrahedral intermediate, *viz.*

$$CH_3COO^- + CH_3{-}\overset{O}{\overset{\|}{C}}{-}OAr \underset{k_2}{\overset{k_1}{\rightleftharpoons}} CH_3COO{-}\overset{O^-}{\underset{CH_3}{C}}{-}OAr \xrightarrow{k_3} CH_3CO{\cdot}O{\cdot}COCH_3 + ArO^-$$

If the substituted phenolate ion is less basic than acetate ion it will be lost preferentially from the tetrahedral intermediate, which will be rapidly converted to acetic anhydride. On the other hand, if the substituted phenolate ion is a much poorer leaving group than acetate, the tetrahedral intermediate will revert almost exclusively to products; nucleophilic catalysis will then be small, and in particular smaller relative to any alternative mechanism of catalysis.

References pp. 202–207

These principles appear to hold for nucleophilic catalysis of hydrolysis by other species also. The evidence for catalysis by amino-compounds is discussed below. Catalysis by a wide variety of oxyanions (and other anions) has been measured by several authors, particularly of the hydrolysis of *p*-nitrophenyl acetate. This is a convenient substrate kinetically, since the release of *p*-nitrophenoxide is easily followed spectrophotometrically at 400 nm, but perhaps not ideal mechanistically, since, as described above, at least some of its reactions involve a mixture of mechanisms. A selection of data, obtained under the same conditions in one laboratory[283], is given in Tables 37 and 38. Some of these data are plotted logarithmically (in Fig. 17) against the

TABLE 37

SECOND-ORDER RATE COEFFICIENTS FOR THE REACTIONS OF OXYANIONS WITH PHENYL ACETATE (PA), *p*-NITROPHENYL ACETATE (pNPA) AND 2,4-DINITROPHENYL ACETATE (DNPA), IN WATER AT 25°C AND IONIC STRENGTH 1.0 (Ref. 283)

		k_2 ($l \cdot mole^{-1} \cdot min^{-1}$)		
Oxyanion	*pK_a*	*PA*	*pNPA*	*DNPA*
HO^-	15.75	76	570	3220
CH_3O^-	15.5	5200	2.9×10^4	1.92×10^5
$HC{\equiv}CCH_2O^-$	13.55	1410	1.08×10^4	6.8×10^4
$CF_3CH_2O^-$	12.37	410	3850	2.4×10^4
HOO^-	11.6	3.2×10^4	2.2×10^5	9.75×10^5
CH_3OO^-	11.5	6300	6×10^4	3.6×10^5
p-$CH_3C_6H_4O^-$	10.07	2.4	113	1240
$C_6H_5O^-$	9.96	1.7	58	730
CH_3CONHO^-	9.37	190	7000	6.2×10^4
p-$ClC_6H_4O^-$	9.28		41	570
p-$NO_2C_6H_4O^-$	7.14	2.4×10^{-3}		
$(CH_3)_3CCOO^-$	4.86			0.107
CH_3COO^-	4.61	2.1×10^{-5}	3.8×10^{-4}	0.034
$CH_3OCH_2COO^-$	3.43			6.8×10^{-3}
H_2O	−1.75	2×10^{-8}	4.7×10^{-7}	1.2×10^{-5}
Other anions				
CN^-	9.3	1.27	10.8	33
N_3^-	4.0	0.014	1.82	57
F^-	3.1	1.5×10^{-5}	1.37×10^{-3}	0.19
NO_2^-	3.4		1.37×10^{-3}	0.53

pK_a of the conjugate acid of the oxyanion nucleophile. This is a Bronsted plot, but differs very obviously from Bronsted plots for the general base catalysis in that there are more deviations from the best line than points falling on it. This is as expected for nucleophilic catalysis. Bond formation to carbon is a quite different process to attack on a hydrogen atom, and is much more sensitive to

TABLE 38

FURTHER DATA FOR THE REACTIONS OF NUCLEOPHILES WITH *p*-NITROPHENYL ACETATE AT 25°C

Nucleophile	pK_a	k_2 ($l \cdot mole^{-1} \cdot min^{-1}$)	*Ref.*
$CH_3OCH_2CH_2O^-$	14.8	5.3×10^3	284
$ClCH_2CH_2O^-$	14.3	5.6×10^3	284
$F_2CHCF_2CH_2O^-$	12.74	3.64×10^3	284
$CCl_3CH_2O^-$	12.24	1.85×10^3	284
$(CH_3)_2C{=}NO^-$	12.4	3.6×10^3	285
CO_3^{2-}	10.4	1.0	285
$CCl_3CH(OH)O^-$	9.66	51	286
SO_3^{2-}	7.1	46	285
ClO^-	7.1	1.6×10^3	285
HPO_4^{2-}	6.9	7.4×10^{-3}	285
$HAsO_4^{2-}$	6.8	4.1×10^{-2}	285
N-hydroxyphthalimide	6.1	28.9	285
$(CH_3)_3N^+{-}O^-$	4.6	8.8×10^{-4}	285
Pyridine-N-oxide	0.8	2.7×10^{-4}	285

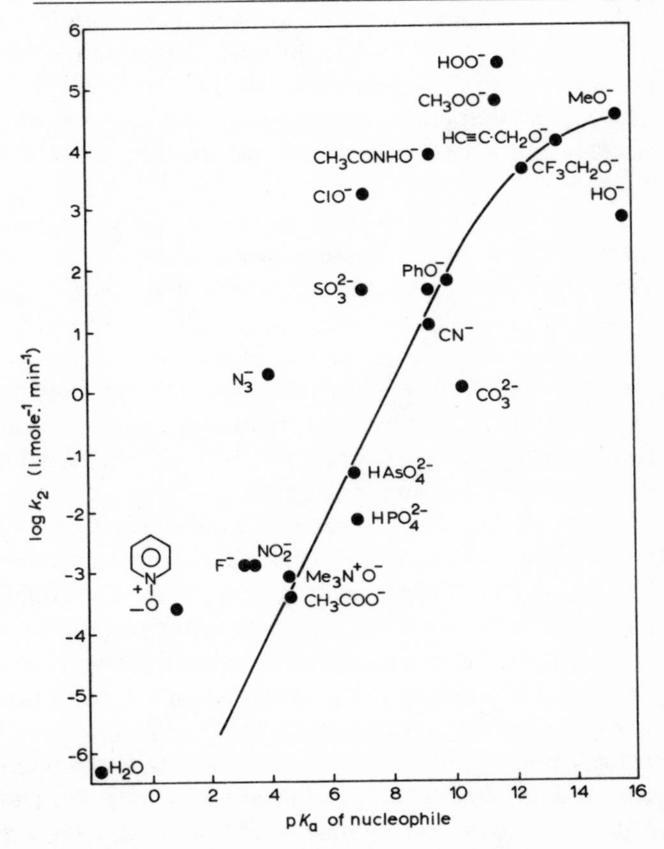

Fig. 17. Bronsted plot for the reactions of nucleophiles with *p*-nitrophenyl acetate in water at 25°C. These reactions involve simple nucleophilic catalysis of hydrolysis. (Data from ref. 283).

steric effects, and other differences in the nucleophilic centre. But, as in classical general base catalysis, there is a clear dependence of nucleophilicity on basicity, and it is generally possible to account for major deviations from normal behaviour.

The most common deviation is the exceptionally high reactivity of nucleophiles, such as hydroperoxide, hypochlorite and hydroxamate ions, with atoms bearing lone-pair electrons next to the nucleophilic centre. This phenomenon, known as the alpha-effect[287], is found for aminolysis reactions of esters also[285], and is commonly observed for attack at electrophilic centres where reactivity depends fairly strongly on the basicity of the nucleophile. Negative deviations may be evidence of steric hindrance, or in a few cases, in particular that of hydroxide ion, may reflect special solvation effects on the pK_a or the nucleophilicity (or both) of the nucleophile.

The central, linear part of this plot (Fig. 17) was measured first, and a Bronsted exponent, given by the slope of this line, in the region of 0.8, found for attack by a wide range of nucleophiles[285,288]. But it soon became apparent that this linear relationship did not extend to strongly basic species such as the alkoxides. Jencks and Gilchrist[284] studied the reactions of *p*-nitrophenyl acetate with the anions derived from a number of alcohols with pK_a's lying in the region 12–14, and found that reactivity depended only very weakly on basicity in this region. This phenomenon accounts for the curvature in the plot of Fig. 17 at high pK_a's.

The demonstration that the mechanism of catalysis may change to general base catalysis for weakly basic nucleophiles further complicates the interpretation of such plots. The most useful generalizations that can be extracted from the data for aryl acetates are illustrated in Fig. 18. This is a plot of data for the reactions of oxyanions with three esters, phenyl, 4-nitrophenyl, and 2,4-dinitrophenyl acetates under the same conditions. Only nucleophiles showing normal reactivity are included: points for hydroxide ion and α-effect nucleophiles have been excluded. The data are those of Jencks and Gilchrist[283], who published a slightly different version of this plot.

The data for each ester can be correlated by a line of small curvature, which could be drawn straight, in the central region, but more marked curvature is observed in each case for reactions with strongly basic nucleophiles, as described above for *p*-nitrophenyl acetate. This behaviour is interpreted by Jencks and Gilchrist in terms of a change in character of the transition state from one resembling products, in the case of nucleophiles much less basic than the leaving group, to one resembling starting materials, for nucleophiles which are much more strongly basic than the leaving group. In the latter case the formation of the tetrahedral intermediate, if one is involved, must be rate-determining, and bond-formation to the nucleophile would be at a relatively early stage, thus accounting for the low limiting sensitivity to the basicity of the nucleophile.

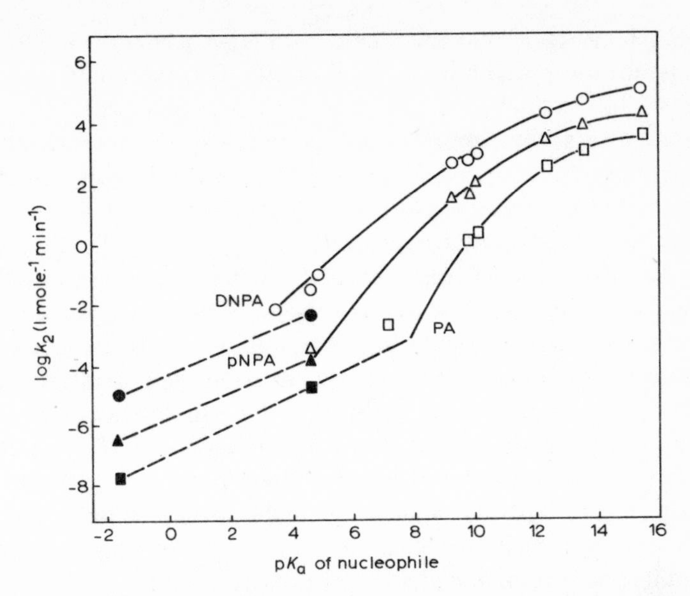

Fig. 18. Bronsted plot for the reactions of nucleophiles with phenyl acetate (PA, □) *p*-nitrophenyl acetate (PNPA, △) and 2,4-dinitrophenylacetate (DNPA, ○) at 25°C. The open symbols (and full lines) represent data for the total reaction, which is in most cases nucleophilic catalysis. The closed symbols (and broken lines) represent general base catalysis of hydrolysis.

For weakly basic nucleophiles the mechanism changes to general base catalysis, as discussed above. Acetate ion, for example, acts almost exclusively as a general base to catalyze the hydrolysis of phenyl acetate, and the two mechanisms occur together for *p*-nitrophenyl acetate[281]. Assuming that the Bronsted plot for general base catalysis is a good straight line, and passes through the point for water, the data of Jencks and Gilchrist[283] and Gold *et al.*[281] can be combined to yield this line for each of the three esters of Fig. 18. These are the broken lines in the figure: they have similar slopes, lower in each case than the slope of the line for nucleophilic catalysis, so that the complete curve is sigmoid, with a change of slope some 2 pK units below the pK_a of the leaving group. The difference in slope depends apparently on the pK_a of the leaving group, being smallest for the most reactive ester, so that a sharp change in slope is not a necessary condition for a change of mechanism. This point is made also by Gold *et al.*[281], who show that the change in slope of the logarithmic plot of their data for the acetate catalyzed hydrolysis of aryl acetates against the pK_a of the conjugate acid of the leaving group is very small, although there is no doubt that a change of mechanism from nucleophilic to general base catalysis occurs.

(ii) General base catalysis

The simplest generalization to emerge from the work described above on nucleophilic catalysis, is that the mechanism is only important when the leaving group is not too much more strongly basic than the nucleophile. Accordingly, general base catalysis is likely to be involved in reactions where the hydrolysis of esters with poor leaving groups, for example, those of simple aliphatic alcohols, is catalyzed by nucleophiles of low basicity. Such catalysis is observed only for very reactive esters, that is, those with strongly electron-withdrawing substituents in the acyl group.

In 1957 Brouwer *et al.*[289], reported that imidazole catalyzes the hydrolysis of dimethyl oxalate. In the light of our present knowledge it would appear that this is a system where general base catalysis would be expected, although recently Vuori and Koskikallio[290] were unable to detect catalysis of dimethyl oxalate hydrolysis by acetate or phosphate. The first clear demonstration of general base catalysis came with the work of Jencks and Carriuolo[291] on the hydrolysis of several ethyl haloacetates. They found that a range of general bases catalyzes the hydrolysis of ethyl dichloroacetate and difluoroacetate (Table 39), and catalysis is also observed with ethyl chloro and trichloroacetate, ethyl oxamide, and protonated glycine ethyl ester.

The experimental evidence that these reactions do, in fact, represent general base catalysis, rather than nucleophilic catalysis of hydrolysis, is very strong. The solvent deuterium isotope effect of 2–3 (Table 39) is consistent with the breaking of a bond to hydrogen in the rate-determining step of the reaction:

TABLE 39

SECOND-ORDER RATE COEFFICIENTS FOR THE GENERAL BASE-CATALYZED HYDROLYSIS OF ESTERS AT 25°C AND IONIC STRENGTH 1.0[291]

General base	k_2 $(l \cdot mole^{-1} \cdot min^{-1})$	k_H/k_D
Ethyl difluoroacetate		
H_2O	6.1×10^{-5}	2.1
$PhNH_2$	9.2×10^{-3}	
CH_3COO^-	3.7×10^{-2}	2.7
Imidazole	6.2×10^{-1}	2.8
Hydroxide	2.7×10^{5}	
Ethyl dichloroacetate		
Water	5.3×10^{-6}	5
$HCOO^-$	1.9×10^{-3}	
$PhNH_2$	1.0×10^{-3}	
CH_3COO^-	3.0×10^{-3}	2.2
Pyridine	1.2×10^{-2}	
Succinate	1.4×10^{-2}	
4-Picoline	1.7×10^{-2}	
Phosphate	4.4×10^{-2}	
Imidazole	8.2×10^{-2}	3.0
Hydroxide	5.3×10^{4}	

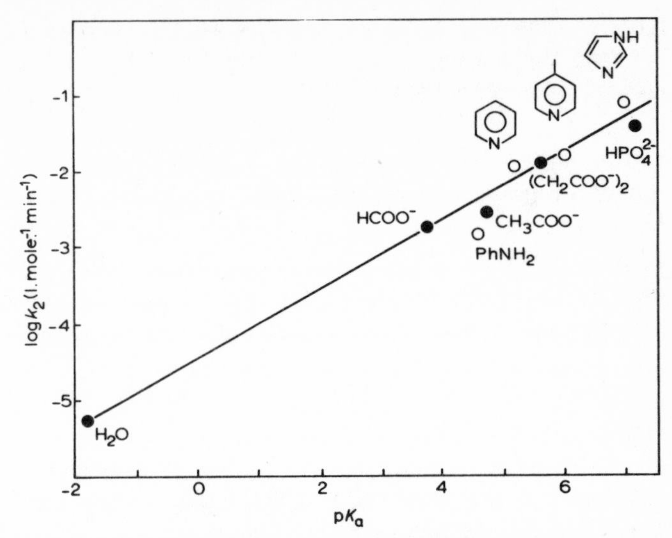

Fig. 19. Bronsted plot of the rate coefficients for the general base-catalyzed hydrolysis of ethyl dichloroacetate at 25°C. (Data from ref. 291).

significant deuterium isotope effects are not normally observed for reactions involving nucleophilic catalysis[292]. The regular Bronsted plot (Fig. 19) conspicuously lacks the irregularities associated with such plots for nucleophilic reactions (*e.g.*, Fig. 17), and the Bronsted exponent, β, of 0.47[291] is much lower than expected for reactions in which the nucleophile displaces a leaving group of much greater basicity[283]. And, quite conclusively for the aniline reaction, the product is not the anilide, which would be stable under the reaction conditions, but the carboxylate anion[291].

Yet another criterion which has been used to distinguish nucleophilic from general base catalysis is the value of the entropy of activation for the catalyzed reaction. This is expected to be less favourable for the necessarily termolecular general base-catalyzed reaction, and what little data is available is consistent with this expectation. General base catalysis by acetate ion of phenyl[279] and *p*-tolyl[293] acetate hydrolysis has an entropy of activation close to −32 eu in each case, whereas the values for similar nucleophilic reactions are considerably smaller. Gold *et al.*[293] find a value of −10 eu for the acetate-catalyzed hydrolysis of 3,4-dinitrophenyl acetate, and find that the figure for the corresponding reaction of *p*-nitrophenyl acetate (−28.7 eu[279]) is made up of contributions of −39 eu for the general base-catalyzed reaction and −16 eu for nucleophilic catalysis of hydrolysis. The slightly higher figure of −18 eu is found for nucleophilic catalysis of the hydrolysis of acetyl 3,5-dinitrosalicylate, also by acetate ion[294]. Thus the generalization that emerges from this limited set of data is that general base-catalyzed reactions are characterised by entropies of activation of less than −30 eu, while the figure for nucleophilic catalysis, though also negative, is greater than −20 eu. These figures do not necessarily apply for reactions other than the acetate-catalyzed hydrolysis of aryl acetates, but are likely to be of more general application.

(c) Catalysis by amines

The reaction of an amine with an ester generally produces an amide, and catalysis of hydrolysis occurs only in two special situations. Tertiary amines cannot give amides, and may act as nucleophiles or general bases to catalyze ester hydrolysis, and general base catalysis of hydrolysis may be observed with any amine, as long as competing reactions (in particular aminolysis) are relatively slow.

(i) Nucleophilic catalysis by tertiary amines

Nucleophilic catalysis by tertiary amines of the hydrolysis of esters with good leaving groups was demonstrated in 1957, independently by Bender and Turnquest[295] and Bruice and Schmir[296]. The proof of mechanism is simpler than for the corresponding oxyanion reactions, because the intermediates, and particularly N-acyl imidazoles, are relatively stable compounds. Thus Bender and Turnquest[295] observed the formation and disappearance of N-acetylimidazole spectroscopically during the imidazole-catalyzed hydrolysis of *p*-nitrophenyl acetate. They also[297] demonstrated catalysis of the hydrolysis of this and other aryl acetates, and of ethyl thiolacetate, by pyridine and substituted pyridines, N-methyl imidazole, and trimethylamine. N-acetyl derivatives of these tertiary amines are highly reactive species, and have not been detected as intermediates in ester hydrolysis, although the N-acetylpyridinium ion has been observed spectroscopically during the pyridine-catalyzed hydrolysis of acetic anhydride[298]. The back reaction of acetate ion with the N-acetylpyridinium ion is kinetically significant in this reaction, and similar back reactions (k_2 in the scheme below) are observed between the same (presumed) intermediate and the displaced aryloxide ions in pyridine-catalyzed ester hydrolysis[283], *viz.*

$$R_3N + CH_3 \cdot CO \cdot OAr \underset{k_2}{\overset{k_1}{\rightleftharpoons}} R_3N^+\text{–}COCH_3 + ArO^-$$

$$R_3N^+\text{–}COCH_3 \xrightarrow{k_3,\ H_2O} R_3N + CH_3COO^-$$

Jencks and Gilchrist[283] estimate that the *p*-nitrophenolate ion is some 10^6 times more reactive than water towards the N-acetyl-4-methylpyridinium ion, and the back reaction is also significant in the pyridine-catalyzed hydrolysis of phenyl acetate[283]. This result illustrates the important point that nucleophilic catalysis by tertiary amines is a more favourable reaction than the corresponding reaction of oxyanions. Acetate ion, which is only slightly less basic than pyridine, acts exclusively as a general base to catalyze the hydrolysis of phenyl acetate, and a criterion which has been used to help distinguish nucleophilic from general base catalysis is that imidazole and phosphate, which have similar pK_a's, are of similar reactivity in general base-catalyzed hydrolysis, whereas imidazole is more reactive, often by several orders of magnitude, in nucleophilic catalysis[291,292].

Structure–reactivity relationships in these reactions have been studied by several authors. Bruice and Schmir[299] found that a series of substituted imidazoles react with *p*-nitrophenyl acetate at rates which depend on their basicity, and are correlated by the Bronsted equation with a β-value of 0.8: although the simple relationship is complicated by the increasing importance of reaction with the anion of more acidic imidazoles. The reaction of imidazole itself with a series of substituted phenyl acetates is correlated by the Hammett equation with $\rho = 1.90$[296].

Data for the reactions of several cyclic tertiary amines with phenyl, 4-nitrophenyl and 2,4-dinitrophenyl acetates, at 25°C and ionic strength 1.0, appear in Table 40, and as a Bronsted plot in Fig. 20. The usual irregularities of such plots for nucleophilic attack are evident. Linear relationships between log k and pK_a are generally found for groups of compounds of closely similar structure, as for the substituted pyridines in Fig. 20. The data for the two tricyclic amines fall on separate curves, and the points for imidazole clearly fall on neither of the first two sets of lines. The separate lines for the reactions of particular classes of nucleophile are approximately parallel, as is usually found.

TABLE 40

REACTIONS OF TERTIARY AMINES WITH PHENYLACETATE (PA), 4-NITROPHENYLACETATE (pNPA) AND 2,4-DINITROPHENYLACETATE (DNPA) AT 25°C AND IONIC STRENGTH 1.0[283]

		k_2 ($l \cdot mole^{-1} \cdot min^{-1}$)		
Amine	*pK_a*	*PA*	*pNPA*	*DNPA*
Nicotinamide	3.55	1.2×10^{-5}	2.4×10^{-3}	3.2
Pyridine	5.52	4.4×10^{-4}	0.17	148
4-Methylpyridine	6.33	3.2×10^{-3}	1.6	780
Imidazole	7.21	0.52	35	350
Triethylenediamine	9.20	9.0×10^{-3}	3.06	4,300
Quinuclidinol	10.07	7.6×10^{-2}	15	7,900

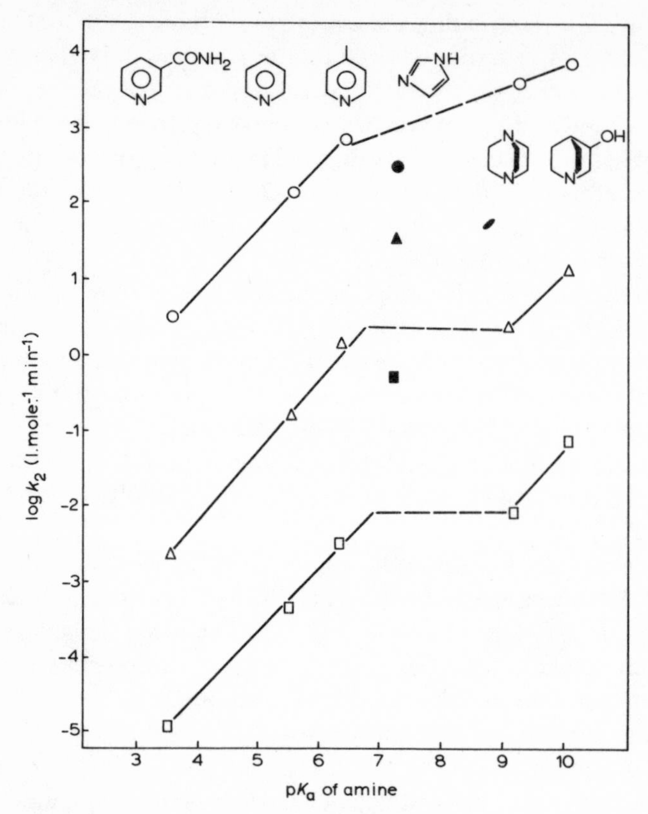

Fig. 20. Bronsted plot of the rate coefficients for the tertiary amine-catalyzed hydrolysis of phenyl (○), *p*-nitrophenyl (△) and 2,4-dinitrophenyl (□) acetates at 25°C. (Data from ref. 283).

(ii) General base catalysis

General base catalysis of ester hydrolysis is less favoured for amine nucleophiles, relative to nucleophilic catalysis, than is the corresponding reaction with oxyanions, presumably because the C–N bond is stronger relative to C–O, than is the N–H bond relative to O–H. General base catalysis by simple amines is important only for esters with very poor leaving groups, and detectable only for those compounds with reactive acyl groups. Thus aniline, pyridine and imidazole catalyze the hydrolysis of ethyl dichloroacetate and similar compounds[291], and the data for this ester are correlated by a Bronsted plot which includes data for catalysis by oxyanions also (Fig. 19, p. 195). Since the anilide, which is stable under the reaction conditions, is not a product, the mechanism is clearly general base catalysis.

The transition from nucleophilic to general base catalysis of ester hydrolysis has been demonstrated by Kirsch and Jencks[300] for attack by imidazole on a series of acetate esters. Their data are given in Table 41, and include a figure for the catalysis by imidazole of the hydrolysis of ethyl acetate. This must

TABLE 41

IMIDAZOLE CATALYSIS OF ACETATE ESTER HYDROLYSIS AT 25°C AND IONIC STRENGTH 1.0

(Ref. 300)

Ester	k_{OH} $(l \cdot mole^{-1} \cdot sec^{-1})$	k_{Im} $(l \cdot mole^{-1} \cdot sec^{-1})$
(Acetic anhydride)	970	147
2,4-Dinitrophenyl acetate	53.5	5.85
4-Nitrophenyl acetate	9.5	0.58
3-Nitrophenyl acetate	6.86	0.23
4-Chlorophenyl acetate	2.20	2.68×10^{-2}
Phenyl acetate	1.26	8.75×10^{-3}
4-Methylphenyl acetate	0.98	3.56×10^{-3}
4-Methoxyphenyl acetate	1.05	3.25×10^{-3}
$CH_3CON(CH_3)OCOCH_3$	14.2	7.12×10^{-2}
Trifluoroethyl acetate	1.87	3.96×10^{-6}
Acetoxime acetate	0.97	1.15×10^{-5}[a]
Ethyl acetate	0.113	1.25×10^{-7}
Ethyl chloroacetate	36.6	3.7×10^{-5}
Ethyl dichloroacetate	880	1.37×10^{-3}
Ethyl difloroacetate	4.5×10^{3}	1.03×10^{-2}
Ethyl oxamide	420	2.5×10^{-4}
Ethyl glycine cation	20	6.3×10^{-6}

[a] Rate coefficient measured for the reaction with N-methyl imidazole. Imidazole reacts as the anion.

represent general base catalysis, and, as expected, is similar to the rate coefficient measured by Holland and Miller[301] for catalysis by phosphate.

In Fig. 21 these data are plotted logarithmically against the rate coefficient for alkaline hydrolysis of each ester. This procedure is designed to eliminate irregularities due to steric effects. And since alkaline hydrolysis is believed to go by the same mechanism, involving rate-determining nucleophilic attack by hydroxide on the carbonyl group for all the esters studied, it is a convenient measure of the effects of structural change on reactivity towards nucleophilic attack.

The lower line of Fig. 21 is drawn through the points for ethyl esters with activated acyl groups, measured by Jencks and Carriuolo[291]: these reactions are known to involve general base catalysis. The upper curve is drawn through the points for acetates with different leaving groups. For esters with good leaving groups the curve approximates to a straight line, of slope not greatly different than that of the lower line: these points represent esters known to be hydrolyzed by nucleophilic catalysis. But for esters of less acidic phenols the line takes on a sharp downward curvature, and the points for trifluoroethyl and ethyl acetates, for example, deviate by several orders of magnitude from the upper straight line. In fact they fall on or close to the line for general base-catalyzed reactions, and it seems clear that the curvature represents the transi-

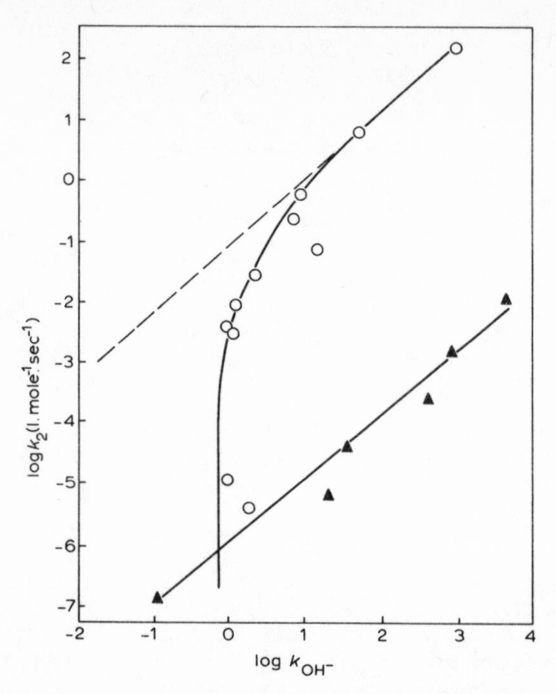

Fig. 21. Logarithmic plot of the second-order rate coefficients (k_2) for catalysis by imidazole of the hydrolysis of various esters, against the rate coefficients for alkaline hydrolysis. The most reactive compound is acetic anhydride: the other open circles represent results for acetate esters of phenols, except for the two least reactive compounds, trifluorethyl acetate, and the acetate of acetone-oxime. The closed triangles represent data for ethyl esters with activated acyl groups, with the exception of the least reactive compound, which is ethyl acetate.

tion from nucleophilic to general base catalysis of hydrolysis by imidazole.

(*d*) *Catalysis by other nucleophiles*

From the discussion so far it is apparent that nucleophilic catalysis of ester hydrolysis is to be expected from any nucleophile that is sufficiently basic to displace the leaving group, but not so basic that the acylated derivative formed is stable. Thus Jencks *et al.*[283,285] have shown that the hydrolysis of *p*-nitrophenyl acetate is catalyzed by such anions as cyanide, azide, fluoride, nitrite, sulphite and various thiolates. Most of these reactions probably involve nucleophilic catalysis. Cyanide, for example, does not catalyze the hydrolysis of ethyl benzoates[302], but does displace the better leaving groups of phenyl acetate[283] and ethyl thiolacetate[302], *e.g.*

$$CH_3COSEt + CN^- \rightleftharpoons EtS^- + CH_3COCN \xrightarrow{H_2O} CH_3COO^- + CN^- + 2\,H^+$$

The very small solvent deuterium isotope effect in this case is consistent with nucleophilic catalysis, which would anyway be expected because of the high basicity of cyanide.

The reactions of thiolate anions with esters are of particular interest because thiol groups are essential to the action of certain hydrolytic enzymes, such as papain and ficin, and there is evidence that the thiol group is acylated by certain substrates (ref. 3, p. 104). Such reactions have been studied by several authors[303,304]. Whitaker[303] was able to detect a thiolester intermediate in the reaction of *p*-nitrophenyl acetate with cysteine at low pH, and Ogilvie *et al.*[304] measured the rates of attack on this ester of a series of thiolate anions. The linear Bronsted plot had a slope $\beta = 0.38$, significantly lower than the values found for amine and oxyanion nucleophiles. A generalization that emerges from this work is that just as amines are more strongly nucleophilic towards the ester group than are oxyanions, so thiolate anions are more strongly nucleophilic than amines. This generalization holds, however, only for thiolate anions of sufficiently low basicity, since the low Bronsted β means that the reactivities of amines and thiolate anions rapidly converge at higher pH's (near pH 9).

General base catalysis of ester hydrolysis by nucleophiles other than amines and oxyanions has not been characterized, though it would be expected to be closely similar to the reaction involving these bases. Cyanide ion, for example, would be expected to catalyze the hydrolysis of ethyl esters with activated acyl groups, such as ethyl dichloroacetate, by this mechanism.

4. Intramolecular catalysis

Nucleophilic and general species catalysis are most efficient when the catalytic group is held in close, properly orientated proximity to the group under attack. These mechanisms are thus much more important, relative to alkaline and acid catalysis, in intramolecular reactions, and in some cases mechanism not observed in intermolecular reactions become important in intramolecular catalysis. The generalizations described in the previous section are still applicable, however, to intramolecular catalysis, and no new principles are involved that are well-understood at present.

Intramolecular catalysis has become a large subject in its own right in recent years, and is discussed adequately elsewhere. As so often in the study of the mechanisms of chemical reactions, esters have been used more often than any other type of compound in the study of intramolecular catalysis. So the reader is referred to the comprehensive survey of intramolecular catalysis by Bruice and Benkovic (ref. 3, pp. 119–211) and the more recent review by Kirby and Fersht[305], which brings the topic up to date (December 1969).

REFERENCES

1 C. K. Ingold, *Structure and Mechanism in Organic Chemistry*, Bell, London, 1953.
2 M. L. Bender, *Chem. Rev.*, **60** (1960) 153.
3 T. C. Bruice and S. J. Benkovic, *Bioorganic Mechanisms*, Benjamin, New York, 1966, Vol. 1.
4 A. Hantzsch, *Z. Phys. Chem.*, **61** (1908) 257; **65** (1909) 41.
5 H. P. Treffers and L. P. Hammett, *J. Am. Chem. Soc.*. **59** (1937) 1708.
6 J. A. Leisten, *J. Chem. Soc.*, (1956) 1572.
7 R. Stewart and K. Yates, *J. Am. Chem. Soc.*, **82** (1960) 4059.
8 G. A. Olah and A. M. White, *J. Am. Chem. Soc.*, **89** (1967) 3591.
9 T. Birchall and R. J. Gillespie, *Can. J. Chem.*, **43** (1965) 1045.
10 G. Fraenkel, *J. Chem. Phys.*, **34** (1961) 1466.
11 H. Hogeveen and A. F. Bickel, *Rec. Trav. Chim.*, **86** (1967) 687; *Chem. Commun.*, (1966) 898.
12 H. Hogeveen, *Rec. Trav. Chim.*, **86** (1967) 809.
13 M. Brookhart, G. C. Levy and S. Winstein, *J. Am. Chem. Soc.*, **84** (1962) 1735.
14 G. A. Olah, D. H. O'Brien and A. M. White, *J. Am. Chem. Soc.*, **89** (1967) 5694.
15 C. B. Anderson, E. C. Friedrich and S. Winstein, *Tetrahedron Letters*, (1963) 2037.
16 D. G. Ramsay and R. W. Taft, *J. Am. Chem. Soc.*, **88** (1966) 3058.
17 M. S. Newman, *J. Am. Chem. Soc.*, **63** (1941) 24.
18 M. S. Newman, H. G. Kuivila and A. B. Garrett, *J. Am. Chem. Soc.*, **67** (1945) 704.
19 L. P. Kuhn, *J. Am. Chem. Soc.*, **71** (1949) 1575.
20 N. C. Deno, C. U. Pittman and M. J. Wisotsky, *J. Am. Chem. Soc.*, **77** (1955) 4370.
21 G. A. Olah, W. S. Tolgyesi, S. J. Kuhn, M. E. Moffatt, I. J. Bastien and E. B. Baker, *J. Am. Chem. Soc.*, **85** (1963) 1328; **84** (1962) 2733.
22 G. A. Olah and M. W. Meyer, in *Friedel-Craft's and Related Reactions*, G. A. Olah (Ed.), Interscience, New York, 1963, Vol. 1, p. 665.
23 G. Oulevy and B. P. Susz, *Helv. Chim. Acta*, **48** (1965) 1965.
24 F. Seel, *Z. Anorg. Allgem. Chem.*, **250** (1943) 331.
25 F. P. Boer, *J. Am. Chem. Soc.*, **88** (1966) 1572.
26 R. Weisz and J.-M. Le Carpentier, *Compt. Rend.*, **265** (1967) 797.
27 R. Weisz and J.-M. Le Carpentier, *Chem. Commun.*, (1968) 596.
28 H. Hogeveen, *Rec. Trav. Chim.*, **86** (1967) 896.
29 H. Hogeveen, *Rec. Trav. Chim.*, **86** (1967) 289.
30 L. L. Schaleger and F. A. Long, *Advan. Phys. Org. Chem.*, **1** (1963) 1.
31 M. S. Dewar, *Hyperconjugation*, Ronald Press, New York, 1962.
32 S. Ehrenson, *J. Am. Chem. Soc.*, **86** (1964) 847.
33 D. N. Kershaw and J. A. Leisten, *Proc. Chem. Soc.*, (1960) 84.
34 R. Stewart and M. R. Granger, *Can. J. Chem.*, **39** (1961) 2508.
35 R. Stewart and K. Yates, *J. Am. Chem. Soc.*, **82** (1960) 4059.
36 E. M. Arnett, *Progr. Phys. Org. Chem.*, **1** (1963) 223.
37 K. Yates and R. A. McClelland, *J. Am. Chem. Soc.*, **89** (1967) 2686.
38 C. A. Lane, M. F. Cheung and G. F. Dorsey, *J. Am. Chem. Soc.*, **86** (1964) 2521; **90** (1968) 6492.
39 A. R. Goldfarb, A. Mele and N. Gutstein, *J. Am. Chem. Soc.*, **77** (1955) 6195.
40 J. T. Edward and K. Wang, *Can. J. Chem.*, **40** (1962) 966.
41 W. M. Schubert, J. Donohue and J. D. Gardner. *J. Am. Chem. Soc.*, **76** (1954) 9.
42 J. Hine and R. P. Bayer, *J. Am. Chem. Soc.*, **84** (1962) 1989.
43 M. J. Jorgenson and D. R. Hartter, *J. Am. Chem. Soc.*, **85** (1963) 878.
44 R. P. Bell, A. L. Dowding and J. A. Noble, *J. Chem. Soc.*, (1965) 3106.
45 D. Jaques, *J. Chem. Soc.*, (1965) 3874.
46 P. Salomaa, *Suomen Kemistilehti*, **32B** (1959) 145.
47 M. L. Bender, H. Ladenheim and M. C. Chen, *J. Am. Chem. Soc.*, **83** (1961) 123.
48 C. T. Chmiel and F. A. Long, *J. Am. Chem. Soc.*, **78** (1956) 3326.
49 C. A. Bunton, D. H. James and J. B. Senior, *J. Chem. Soc.*, (1960) 3364.

50 J. F. BUNNETT, *J. Am. Chem. Soc.*, **83** (1961) 4968.
51 J. F. BUNNETT AND F. P. OLSEN, *Can. J. Chem.*, **44** (1966) 1917.
52 A. R. OLSON AND J. R. MILLER, *J. Am. Chem. Soc.*, **60** (1938) 2687.
53 A. R. OLSON AND J. L. HYDE, *J. Am. Chem. Soc.*, **63** (1941) 2459.
54 F. A. LONG AND M. PURCHASE, *J. Am. Chem. Soc.*, **72** (1950) 3267.
55 M. L. BENDER AND M. C. CHEN, *J. Am. Chem. Soc.*, **85** (1963) 37.
56 C. A. BUNTON, J. H. CRABTREE AND L. ROBINSON, *J. Am. Chem. Soc.*, **90** (1968) 1258.
57 J. G. GRAHAM AND E. D. HUGHES, quoted in ref. 1, p. 771.
58 J. F. BUNNETT, *J. Am. Chem. Soc.*, **83** (1961) 4956.
59 J. N. E. DAY AND C. K. INGOLD, *Trans. Faraday Soc.*, **37** (1941) 686.
60 A. G. DAVIES AND J. KENYON, *Quart. Rev.*, **9** (1955) 203.
61 S. G. COHEN AND A. SCHNEIDER, *J. Am. Chem. Soc.*, **63** (1941) 3382.
62 C. A. BUNTON, E. D. HUGHES, C. K. INGOLD AND D. F. LEIGH, *Nature*, **166** (1950) 680.
63 C. A. BUNTON AND J. L. WOOD, *J. Chem. Soc.*, (1955) 1522.
64 K. R. ADAM, I. LAUDER AND V. R. STIMSON, *Australian J. Chem.*, **20** (1962) 467.
65 T. YRJÄNÄ, *Suomen Kemistilehti*, **37B** (1964) 108.
66 B. CAPON, *Quart. Rev.*, **18** (1964) 45.
67 C. A. BUNTON AND A. KONASIEWICZ, *J. Chem. Soc.*, (1955) 1354.
68 L. P. HAMMETT, *Physical Organic Chemistry*, McGraw-Hill, New York, 1940, pp. 273–277.
69 G. S. HAMMOND, *J. Am. Chem. Soc.*, **77** (1955) 334.
70 A. R. OSBORN AND E. WHALLEY, *Can. J. Chem.*, **39** (1961) 1094.
71 V. R. STIMSON, *J. Chem. Soc.*, (1955) 2673.
72 V. R. STIMSON, *J. Chem. Soc.*, (1955) 2010.
73 V. R. STIMSON AND E. J. WATSON, *J. Chem. Soc.*, (1954) 2848.
74 C. A. BUNTON, A. E. COMYNS, J. GRAHAM AND J. R. QUAYLE, *J. Chem. Soc.*, (1955) 3817.
75 G. S. HAMMOND AND J. T. RUDESWILL, *J. Am. Chem. Soc.*, **72** (1950) 2769.
76 V. R. STIMSON, *J. Chem. Soc.*, (1955) 4020.
77 W. B. S. NEWLING AND C. N. HINSHELWOOD, *J. Chem. Soc.*, (1936) 1357.
78 P. SALOMAA, *Acta Chem. Scand.*, **14** (1960) 577.
79 A. MOFFATT AND H. HUNT, *J. Am. Chem. Soc.*, **80** (1958) 2985.
80 P. SALOMAA, *Acta Chem. Scand.*, **11** (1957) 125, 132, 141, 235, 239.
81 P. SALOMAA AND R. LINNANTIE, *Acta Chem. Scand.*, **14** (1960) 586.
82 P. SALOMAA, *Acta Chem. Scand.*, **11** (1957) 247.
83 H. T. LIANG AND P. D. BARTLETT, *J. Am. Chem. Soc.*, **80** (1958) 3585.
84 C. G. SWAIN, T. E. C. KNEE AND A. MACLACHLAN, *J. Am. Chem. Soc.*, **82** (1960) 6101.
85 C. A. BUNTON, J. N. E. DAY, R. H. FLOWER, P. SHEEL AND J. L. WOOD, *J. Chem. Soc.*, (1957) 963.
86 G. J. HARVEY AND V. R. STIMSON, *J. Chem. Soc.*, (1956) 3629.
87 C. A. BUNTON, G. ISRAEL, M. M. MHALA AND D. L. H. WILLIAMS, *J. Chem. Soc.*, (1956) 3718.
88 C. A. BUNTON AND T. HADWICK, *J. Chem. Soc.*, (1957) 3043.
89 C. A. BUNTON AND T. HADWICK, *J. Chem. Soc.*, (1961) 943.
90 C. A. BUNTON AND T. HADWICK, *J. Chem. Soc.*, (1958) 3248.
91 T. C. BRUICE AND B. HOLMQUIST, *J. Am. Chem. Soc.*, **89** (1967) 4028.
92 L. A. KIPRIANOVA AND A. F. REKASHEVA, *Dokl. Akad. Nauk. SSSR*, **144** (1962) 386.
93 J. A. LANDGREBE, *J. Org. Chem.*, **30** (1965) 2997.
94 T. YRJÄNÄ, *Suomen Kemistilehti*, **39B** (1966) 81.
95 B. HOLMBERG, *Chem. Ber.*, **45** (1912) 2997.
96 C. K. INGOLD AND E. H. INGOLD, *J. Chem. Soc.*, (1932) 758.
97 O. R. QUAYLE AND H. M. NORTON, *J. Am. Chem. Soc.*, **62** (1940) 1170.
98 S. C. DATTA, J. N. E. DAY AND C. K. INGOLD, *J. Chem. Soc.*, (1939) 838.
99 I. ROBERTS AND H. C. UREY, *J. Am. Chem. Soc.*, **61** (1939) 2584.
100 F. A. LONG AND L. FRIEDMAN, *J. Am. Chem. Soc.*, **72** (1950) 3692.
101 S. L. JOHNSON, *Advan. Phys. Org. Chem.*, **5** (1967) 273.
102 E. H. CORDES, *Progr. Phys. Org. Chem.*, **4** (1967) 1.
103 C. A. BUNTON AND R. H. DEWOLFE, *J. Org. Chem.*, **30** (1965) 1371.

104 M. L. Bender, *J. Am. Chem. Soc.*, **73** (1951) 1026.
105 W. Ostwald, *J. Prakt. Chem.*, **28** (1883) 449; **30** (1884) 93.
106 H. Goldschmidt, *Chem. Ber.*, **30** (1895) 3218.
107 H. B. Friedman and G. V. Elmore, *J. Am. Chem. Soc.*, **63** (1941) 864.
108 H. Sadek, F. M. Abdel-Halim and F. Y. Khalil, *Suomen Kemistilehti*, **37B** (1964) 47.
109 H. Sadek and F. Y. Khalil, *Suomen Kemistilehti*, **38B** (1965) 55.
110 J. B. Hyne, *J. Am. Chem. Soc.*, **82** (1960) 5129: E. S. Amis, *Solvent Effects on Reaction Rates and Mechanisms*, Academic Press, New York, 1966, p. 184.
111 L. Zucker and L. P. Hammett, *J. Am. Chem. Soc.*, **61** (1939) 2791.
112 F. A. Long and M. A. Paul, *Chem. Rev.*, **57** (1957) 935.
113 M. Duboux and A. De Sousa, *Helv. Chim. Acta*, **23** (1941) 1381.
114 F. A. Long, W. F. McDevit and F. B. Dunkle, *J. Phys. Chem.*, **55** (1951) 813.
115 F. A. Long, W. F. McDevit and F. B. Dunkle, *J. Phys. Chem.*, **55** (1951) 829.
116 M. A. Paul and F. A. Long, *Chem. Rev.*, **57** (1957) 1.
117 J. F. Bunnett, *J. Am. Chem. Soc.*, **83** (1961) 4973.
118 J. F. Bunnett, *J. Am. Chem. Soc.*, **83** (1961) 4978.
119 J. F. Bunnett and F. P. Olsen, *Can. J. Chem.*, **44** (1966) 1899.
120 E. M. Arnett and G. W. Mach, *J. Am. Chem. Soc.*, **88** (1966) 1177.
121 K. Yates, J. B. Stevens and A. R. Katritzky, *Can. J. Chem.*, **42** (1964) 1957.
122 R. L. Hinman and J. Lang, *J. Am. Chem. Soc.*, **86** (1964) 3796.
123 J. N. Bronsted, *Z. Physik. Chem.*, **102** (1922) 169.
124 K. Yates and J. B. Stevens, *Can. J. Chem.*, **43** (1965) 529.
125 P. Salomaa, L. L. Schaleger and F. A. Long, *J. Am. Chem. Soc.*, **86** (1964) 1.
126 K. J. Laidler and P. A. Landskroener, *Trans. Faraday Soc.*, **52** (1956) 200.
127 J. Butterworth, D. D. Eley and G. S. Stone, *Biochem. J.*, **53** (1953) 30.
128 E. W. Timm and C. S. Hinshelwood, *J. Chem. Soc.*, (1938) 862.
129 H. Kwart and M. B. Price, *J. Am. Chem. Soc.*, **82** (1960) 5123.
130 R. B. Martin, *J. Am. Chem. Soc.*, **84** (1962) 4130.
131 R. E. Strube, *Organic Syntheses, Collected*, **4** (1963) 417: A. L. McCloskey, G. S. Fonken, R. W. Kluiber and W. S. Johnson, *Organic Syntheses, Collected*, **4** (1963) 261.
132 G. W. Anderson and F. M. Callahan, *J. Am. Chem. Soc.*, **82** (1960) 3359; R. Roeske, *J. Org. Chem.*, **28** (1963) 1251.
133 H. Meerwein, G. Hinz, P. Hofmann, E. Kroning and E. Pfeif, *J. Prakt. Chem.*, **147** (1937) 257.
134 J. H. Brewster and C. J. Ciotti, *J. Am. Chem. Soc.*, **77** (1955) 6214.
135 R. A. More O'Ferrall, *Advan. Phys. Org. Chem.*, **5** (1967) 331.
136 M. L. Bender, R. R. Stone and R. S. Dewey, *J. Am. Chem. Soc.*, **78** (1956) 319.
137 I. Roberts and H. C. Urey, *J. Am. Chem. Soc.*, **61** (1939) 2580.
138 M. L. Bender and R. S. Dewey, *J. Am. Chem. Soc.*, **78** (1956) 317.
139 D. R. Llewellyn and C. O'Connor, *J. Chem. Soc.*, (1964) 545.
140 C. O'Connor and D. R. Llewellyn, *J. Chem. Soc.*, (1965) 2197.
141 C. O'Connor and D. R. Llewellyn, *J. Chem. Soc.*, (1965) 2669.
142 R. J. Hartman and A. M. Borders, *J. Am. Chem. Soc.*, **59** (1937) 2107.
143 R. J. Hartman and A. G. Gassmann, *J. Am. Chem. Soc.*, **62** (1940) 1559.
144 H. H. Jaffé, *Chem. Rev.*, **53** (1953) 191.
145 N. B. Chapman, M. C. Rodgers and J. Shorter, *J. Chem. Soc. B*, (1968) 157.
146 R. J. Hartman, H. M. Hoogsteen and J. A. Moede, *J. Am. Chem. Soc.*, **66** (1944) 1714.
147 N. B. Chapman, M. C. Rodgers and J. Shorter, *J. Chem. Soc. B*, (1968) 164.
148 M. Charton, *Can. J. Chem.*, **38** (1960) 2493.
149 K. Kindler, *Annalen*, **452** (1927) 90; **464** (1928) 278.
150 M. S. Newman and S. H. Merrill, *J. Am. Chem. Soc.*, **77** (1955) 5552.
151 N. B. Chapman, J. Shorter and J. H. P. Utley, *J. Chem. Soc.*, (1963) 1291.
152 R. W. Taft, in *Steric Effects in Organic Chemistry*, M. S. Newman (Ed.), Wiley, New York, 1956, p. 598.
153 H. A. Smith and R. B. Hurley, *J. Am. Chem. Soc.*, **72** (1950) 112.

154 G. Davies and D. P. Evans, *J. Chem. Soc.*, (1940) 339.
155 H. A. Smith and J. H. Steele, *J. Am. Chem. Soc.*, **63** (1941) 3466.
156 H. A. Smith and R. R. Myers, *J. Am. Chem. Soc.*, **64** (1942) 2362.
157 H. A. Smith and C. H. Reichardt, *J. Am. Chem. Soc.*, **63** (1941) 605.
158 H. A. Smith and J. Burn, *J. Am. Chem. Soc.*, **66** (1944) 1494.
159 H. A. Smith, *J. Am. Chem. Soc.*, **62** (1940) 1136.
160 H. A. Smith, *J. Am. Chem. Soc.*, **61** (1939) 1176.
161 G. Gorin, O. R. Pierce and E. T. McBee, *J. Am. Chem. Soc.*, **75** (1953) 5622.
162 M. S. Newman, *J. Am. Chem. Soc.*, **72** (1950) 4783.
163 C. K. Hancock, E. A. Myers and B. J. Yager, *J. Am. Chem. Soc.*, **83** (1961) 4211.
164 C. K. Hancock and C. P. Falls, *J. Am. Chem. Soc.*, **83** (1961) 4214.
165 C. K. Hancock, B. J. Yager, C. P. Falls and J. O. Schreck, *J. Am. Chem. Soc.*, **85** (1963) 1297.
166 K. Bowden, *Can. J. Chem.*, **44** (1966) 661.
167 M. Harfenist and M. Baltzly, *J. Am. Chem. Soc.*, **69** (1947) 362.
168 L. A. Wiles, *J. Chem. Soc.*, (1953) 996.
169 L. P. Hammett, ref. 68, p. 358.
170 A. Moffatt and H. Hunt, *J. Am. Chem. Soc.*, **79** (1957) 54.
171 M. L. Bender and Y. L. Chow, *J. Am. Chem. Soc.*, **81** (1959) 3929.
172 J. G. Traynham, *J. Am. Chem. Soc.*, **74** (1952) 4277.
173 M. Charton, *J. Org. Chem.*, **29** (1964) 1222.
174 L. R. Fedor and T. C. Bruice, *J. Am. Chem. Soc.*, **87** (1965) 4138.
175 M. L. Bender and H. d'A. Heck, *J. Am. Chem. Soc.*, **89** (1967) 1211.
176 E. Taschner and E. Libereck, *Roczniki Chem.*, **30** (1956) 323.
177 F. Elsinger, J. Schreiber and A. Eschenmoser, *Helv. Chim. Acta*, **43** (1960) 113.
178 L. P. Hammett and H. L. P. Fluger, *J. Am. Chem. Soc.*, **55** (1933) 4079.
179 S. L. Johnson, *J. Am. Chem. Soc.*, **86** (1964) 3819.
180 J. F. Bunnett, M. M. Robison and F. C. Pennington. *J. Am. Chem. Soc.*, **72** (1950) 2378.
181 W. A. Cowdrey, E. D. Hughes, C. K. Ingold, S. Masterman and A. D. Scott, *J. Chem. Soc.*, (1937) 1264.
182 P. D. Bartlett and P. N. Rylander, *J. Am. Chem. Soc.*, **73** (1951) 4273.
183 J. Hine, *Physical Organic Chemistry*, McGraw-Hill, New York, 1962, p. 161.
184 R. H. Mills, M. W. Farrar and O. J. Weinkauff, *Chem. Ind. (London)*, (1962) 2144.
185 R. C. Fuson, J. Corse and E. C. Horning, *J. Am. Chem. Soc.*, **61** (1939) 1290.
186 W. P. Jencks and J. Carriuolo, *J. Am. Chem. Soc.*, **83** (1961) 1743.
187 A. Skrabal and A. Zahorka, *Monatsh. Chem.*, **53–54** (1929) 562.
188 A. Skrabal, *Z. Elektrochem.*, **33** (1927) 322.
189 H. H. Jaffé, *Chem. Rev.*, **53** (1953) 191.
190 T. C. Bruice, T. H. Fife, J. J. Bruno and J. E. Brandon, *Biochemistry*, **1** (1962) 7.
191 V. Gold, D. G. Oakenfull and T. Riley, *J. Chem. Soc. B*, (1968) 515.
192 E. K. Euranto, *Suomen Kemistilehti*, **38A** (1965) 25.
193 B. M. Anderson, E. H. Cordes and W. P. Jencks, *J. Biol. Chem.*, **236** (1961) 455.
194 E. J. Salmi and T. Suonpää, *Chem. Ber.*, **73** (1940) 1126.
195 H. Palomaa, E. J. Salmi and R. Korte, *Chem. Ber.*, **72** (1939) 790.
196 E. K. Euranto and N. J. Cleve, *Acta Chem. Scand.*, **17** (1963) 1584.
197 N. J. Cleve and E. K. Euranto, *Suomen Kemistilehti*, **37B** (1964) 126.
198 E. K. Euranto, *Suomen Kemistilehti*, **35B** (1962) 18.
199 A. Moffatt and H. Hunt, *J. Am. Chem. Soc.*, **81** (1959) 2082.
200 K. Koehler, R. Skora and E. H. Cordes, *J. Am. Chem. Soc.*, **88** (1966) 3577.
201 V. Gold, N. Fuller, S. G. Perry and V. J. Shiner, *Chem. Ind. (London)*, (1960) 1130.
202 J. Koskikallio, D. Pouli and E. Whalley, *Can. J. Chem.*, **37** (1959) 360.
203 D. R. Llewellyn and C. O.'Connor, *J. Chem. Soc.*, (1964) 4400.
204 E. K. Euranto and A.-L. Moisio, *Suomen Kemistilehti*, **37B** (1964) 92.
205 H. Zimmerman and J. Rudolph, *Angew. Chem. Intern. Ed. Engl.*, **4** (1965) 40.
206 C. A. Bunton and D. A. Spatcher, *J. Chem. Soc.*, (1956) 1079.

207 M. L. Bender and R. S. Dewey, *J. Am. Chem. Soc.*, **78** (1956) 317.
208 M. L. Bender, R. D. Ginger and J. P. Unik, *J. Am. Chem. Soc.*, **80** (1958) 1044.
209 M. L. Bender and K. C. Kemp, *J. Am. Chem. Soc.*, **79** (1957) 111.
210 D. Samuel and B. L. Silver, *Advan. Phys. Org. Chem.*, **3** (1965) 157.
211 M. L. Bender and R. J. Thomas, *J. Am. Chem. Soc.*, **83** (1961) 4189.
212 M. L. Bender, H. Matsui, R. J. Thomas and S. W. Tobey, *J. Am. Chem. Soc.*, **83** (1961) 4193.
213 H. M. Humphreys and L. P. Hammett, *J. Am. Chem. Soc.*, **78** (1956) 521.
214 R. Leimu, R. Korte, E. Laaksonen and U. Lehmsukoski, *Suomen Kemistilehti*, **19B** (1946) 93.
215 R. A. Fairclough and C. N. Hinshelwood, *J. Chem. Soc.*, (1937) 538.
216 R. W. A. Jones and J. D. R. Thomas, *J. Chem. Soc. B*, (1966) 661.
217 E. Tommila and C. N. Hinshelwood, *J. Chem. Soc.*, (1938) 1801.
218 C. H. DePuy and L. R. Mahoney, *J. Am. Chem. Soc.*, **86** (1964) 2653.
219 A. A. Humffray and J. J. Ryan, *J. Chem. Soc. B*, (1966) 842.
220 K. Bowden, N. B. Chapman and J. Shorter, *J. Chem. Soc.*, (1964) 3370.
221 W. J. Svirbely and I. C. Mador, *J. Am. Chem. Soc.*, **72** (1950) 5699.
222 R. P. Bell and G. A. Wright, *Trans. Faraday Soc.*, **59** (1961) 1377.
223 R. P. Bell and B. A. W. Collier, *Trans. Faraday Soc.*, **61** (1965) 1445.
224 R. P. Bell and J. F. Lindars, *J. Chem. Soc.*, (1954) 4601.
225 R. W. Hay, L. J. Porter and P. J. Morris, *Australian J. Chem.*, **19** (1966) 1197.
226 M. Robson Wright, *J. Chem. Soc. B*, (1968) 548.
227 M. Robson Wright, *J. Chem. Soc. B*, (1967) 1265.
228 R. W. Hay and L. J. Porter, *J. Chem. Soc. B*, (1967) 1261.
229 J. D. R. Thomas and H. B. Watson, *J. Chem. Soc.*, (1956) 3958.
230 B. Jones and J. G. Watkinson, *J. Chem. Soc.*, (1958) 4064.
231 J. D. Roberts and R. A. Carboni, *J. Am. Chem. Soc.*, **77** (1955) 5554.
232 J. J. Bloomfield and R. Fuchs, *J. Org. Chem.*, **26** (1961) 2991.
233 P. H. Gore, E. C. Vignes and A. Feistein, *Chem. Ind. (London)*, (1958) 1514.
234 A. A. Humffray and J. J. Ryan, *J. Chem. Soc. B*, (1967) 468.
235 J. F. Kirsch, W. Clewell and A. Simon, *J. Org. Chem.*, **33** (1968) 127.
236 C. K. Ingold and W. S. Nathan, *J. Chem. Soc.*, (1936) 222.
237 B. Jones and J. Robinson, *J. Chem. Soc.*, (1955) 3845.
238 H. L. Goering, T. Rubin and M. S. Newman, *J. Am. Chem. Soc.*, **76** (1954) 787.
239 L. B. Jones and T. M. Sloane, *Tetrahedron Letters*, (1966) 831.
240 R. W. Taft, M. S. Newman and F. H. Verhoek, *J. Am. Chem. Soc.*, **72** (1950) 4511.
241 K. Bowden, *Can. J. Chem.*, **41** (1963) 2781.
242 M. Caplow and W. P. Jencks, *Biochemistry*, **1** (1962) 883.
243 H. van Bekkum, P. E. Verkade and B. M. Wepster, *Rec. Trav. Chim.*, **78** (1959) 815.
244 K. Bowden, *Can. J. Chem.*, **43** (1965) 3354.
245 J. D. Roberts and W. T. Moreland, *J. Am. Chem. Soc.*, **75** (1953) 2167.
246 D. J. Ringshaw and H. J. Smith, *J. Chem. Soc.*, (1964) 1559.
247 H. H. Jaffé, *Chem. Rev.*, **53** (1953) 191.
248 R. O. C. Norman, G. K. Radda, D. A. Brimacombe, P. D. Ralph and E. M. Smith, *J. Chem. Soc.*, (1961) 3247.
249 J. J. Bloomfield and R. Fuchs, *J. Org. Chem.*, **28** (1963) 910.
250 M. Charton, *J. Am. Chem. Soc.*, **91** (1969) 615.
251 M. Charton, *J. Am. Chem. Soc.*, **91** (1969) 619.
252 M. Charton and B. I. Charton, *J. Org. Chem.*, **33** (1968) 3872.
253 S. A. Shain and J. F. Kirsch, *J. Am. Chem. Soc.*, **90** (1968) 5848.
254 M. Charton, *J. Am. Chem. Soc.*, **91** (1969) 624.
255 J. I. Hoppé and J. E. Prue, *J. Chem. Soc.*, (1957) 1775.
256 G. Aksnes and J. E. Prue, *J. Chem. Soc.*, (1959) 103.
257 R. P. Bell and M. Robson, *Trans. Faraday Soc.*, **60** (1964) 893.
258 E. Tommila and M.-L. Savolainen, *Suomen Kemistilehti*, **40B** (1967) 212.
259 E. Tommila, *Suomen Kemistilehti*, **37B** (1964) 117.

260 G. J. NOLAN AND E. S. AMIS, *J. Phys. Chem.*, **65** (1961) 1556.
261 E. TOMMILA, A. KOIVISTO, J. PHYRRA, K. ANTELL AND S. HEIMO, *Ann. Acad. Sci. Fennicae AII* (1952) No. 47.
262 G. S. HAMMOND, *J. Am. Chem. Soc.*, **77** (1955) 334.
263 E. S. AMIS, *Solvent Effects on Reaction Rates and Mechanisms*, Academic Press, New York, 1966, Chap. 2.
264 E. TOMMILA, M. TIILIKAINEN AND A. VOIPIO, *Ann. Acad. Sci. Fennicae, AII* (1953) No. 65.
265 E. TOMMILA, E. PAAKKALA, U. K. VIRTANEN, A. EVVA AND S. VANILA, *Ann. Acad. Sci. Fennicae, AII* (1959) No. 65.
266 E. TOMMILA, *Suomen Kemistilehti*, **25B** (1952) 37.
267 E. TOMMILA, L. TAKANEN AND K. SALONEN, *Suomen Kemistilehti*, **31B** (1958) 37.
268 A. J. PARKER, *Advan. Phys. Org. Chem.*, **5** (1967) 173.
269 D. D. ROBERTS, *J. Org. Chem.*, **29** (1964) 2039.
270 E. TOMMILA AND M. L. MURTO, *Acta Chem. Scand.*, **17** (1963) 1947.
271 D. D. ROBERTS, *J. Org. Chem.*, **30** (1965) 3516.
272 D. D. ROBERTS, *J. Org. Chem.*, **29** (1964) 2714.
273 D. D. ROBERTS, *J. Org. Chem.*, **31** (1966) 4037.
274 H. M. DAWSON AND W. LOWSON, *J. Chem. Soc.*, (1927) 2444.
275 M. KILPATRICK, *J. Am. Chem. Soc.*, **50** (1928) 2891.
276 M. KILPATRICK, *J. Am. Chem. Soc.*, **52** (1930) 1410.
277 M. KILPATRICK AND M. L. KILPATRICK, *J. Am. Chem. Soc.*, **52** (1930) 1418.
278 A. R. EMERY AND V. GOLD, *J. Chem. Soc.*, (1950) 1447.
279 M. L. BENDER AND M. C. NEVEU, *J. Am. Chem. Soc.*, **80** (1958) 5388.
280 R. L. SCHOWEN AND C. G. BEHN, *J. Am. Chem. Soc.*, **90** (1968) 5839.
281 V. GOLD, D. G. OAKENFULL AND T. RILEY, *J. Chem. Soc. B*, (1968) 515.
282 A. R. BUTLER AND V. GOLD, *J. Chem. Soc.*, (1962) 1334.
283 W. P. JENCKS AND M. GILCHRIST, *J. Am. Chem. Soc.*, **90** (1968) 2622.
284 W. P. JENCKS AND M. GILCHRIST, *J. Am. Chem. Soc.*, **84** (1962) 2910.
285 W. P. JENCKS AND J. CARRIUOLO, *J. Am. Chem. Soc.*, **82** (1960) 1778.
286 O. GAWRON AND F. DRAUS, *J. Am. Chem. Soc.*, **80** (1958) 5392.
287 J. O. EDWARDS AND R. G. PEARSON, *J. Am. Chem. Soc.*, **84** (1962) 16.
288 T. C. BRUICE, T. H. FIFE, J. J. BRUNO AND N. E. BRANDON, *Biochemistry*, **1** (1962) 7.
289 D. M. BROUWER, M. J. VAN DER VLUGT AND E. HAVINGA, *Koninkl. Nederl. Akad. Wetenschap. Proc. Ser. B*, **60** (1957) 275.
290 J. VUORI AND J. KOSKIKALLIO, *Suomen Kemistilehti*, **42** (1969) 136.
291 W. P. JENCKS AND J. CARRIUOLO, *J. Am. Chem. Soc.*, **83** (1961) 1743.
292 S. L. JOHNSON, *Advan. Phys. Org. Chem.*, **5** (1967) 237.
293 D. G. OAKENFULL, T. RILEY AND V. GOLD, *Chem. Commun.*, (1966) 385.
294 A. R. FERSHT AND A. J. KIRBY, *J. Am. Chem. Soc.*, **90** (1968) 5818.
295 M. L. BENDER AND B. W. TURNQUEST, *J. Am. Chem. Soc.*, **79** (1957) 1652.
296 T. C. BRUICE AND G. L. SCHMIR, *J. Am. Chem. Soc.*, **79** (1957) 1663.
297 M. L. BENDER AND B. W. TURNQUEST, *J. Am. Chem. Soc.*, **79** (1957) 1656.
298 A. R. FERSHT AND W. P. JENCKS, *J. Am. Chem. Soc.*, **91** (1969) 2125.
299 T. C. BRUICE AND G. L. SCHMIR, *J. Am. Chem. Soc.*, **80** (1958) 148.
300 J. F. KIRSCH AND W. P. JENCKS, *J. Am. Chem. Soc.*, **86** (1964) 837.
301 J. M. HOLLAND AND J. G. MILLER, *J. Phys. Chem.*, **65** (1961) 463.
302 F. HIBBERT AND D. P. N. SATCHELL, *J. Chem. Soc. B*, (1968) 565.
303 J. R. WHITAKER, *J. Am. Chem. Soc.*, **84** (1962) 1800.
304 J. W. OGILVIE, J. T. TILDON AND B. S. STRAUCH, *Biochemistry*, **3** (1964) 754.
305 A. J. KIRBY AND A. R. FERSHT, in *Progress in Bioorganic Chemistry*, E. T. KAISER AND F. J. KEZDY (Eds.), Vol. 1, Wiley, 1971, New York, p. 1.
306 D. S. NOYCE AND R. M. POLLACK, *J. Am. Chem. Soc.*, **91** (1969) 119.
307 A. SYKES, J. C. TATLOW AND C. R. THOMAS, *J. Chem. Soc.*, (1956) 835.
308 A. M. WHITE AND G. A. OLAH, *J. Am. Chem. Soc.*, **91** (1969) 2943.
309 R. F. BORSCH, *J. Am. Chem. Soc.*, **90** (1968) 5303.

Chapter 3

The Hydrolysis of Carboxylic Acid Derivatives

R. J. E. TALBOT

1. Introduction

1.1 SCOPE OF THIS CHAPTER

This chapter deals with the kinetics and mechanisms of the hydrolysis of carboxylic acid derivatives of general formula RCOX. These include carboxylic acid halides, amides, and anhydrides with small sections on carboxylic acid cyanides etc. Many recent developments in this field have been made with acid derivatives in which R is not an aliphatic or aromatic group, for example, carbamic acid derivatives, and these are reported where relevant, as are reactions such as ethanolysis, aminolysis, etc. where they throw light on the mechanisms of hydrolysis.

Various aspects of the hydrolysis of carboxylic acid derivatives have been reviewed by Bender[1], Hudson[2], Johnson[3] and Bruice and Benkovic[4].

Interest in the field has existed since the late 19th century and to attempt a comprehensive review would be a gargantuan task. This chapter attempts to draw together some of the principles involved and to deal only in some detail with the more recent mechanistic studies.

1.2 ELECTRONIC STRUCTURE OF CARBONYL COMPOUNDS

The electronic structure of the carbonyl group is perturbed by the rest of the molecule. In classical terms, a carbonyl compound may be represented as a resonance hybrid of the canonical structure I to V, *viz.*

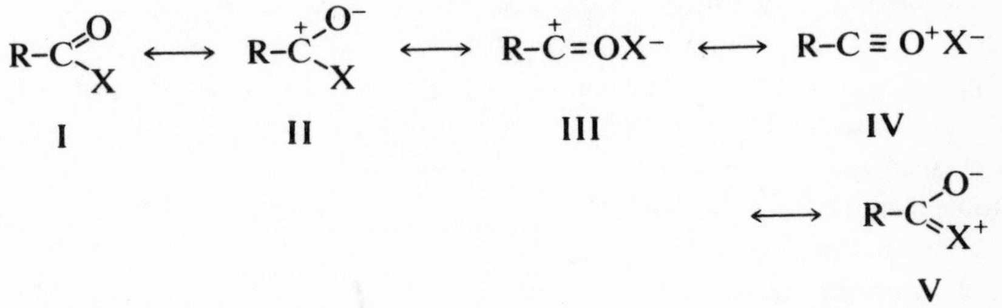

$$\underset{\text{I}}{R\text{-}C\begin{smallmatrix}\nearrow O\\ \searrow X\end{smallmatrix}} \longleftrightarrow \underset{\text{II}}{R\text{-}\overset{+}{C}\begin{smallmatrix}\nearrow O^-\\ \searrow X\end{smallmatrix}} \longleftrightarrow \underset{\text{III}}{R\text{-}\overset{+}{C}{=}OX^-} \longleftrightarrow \underset{\text{IV}}{R\text{-}C\equiv O^+X^-} \longleftrightarrow \underset{\text{V}}{R\text{-}C\begin{smallmatrix}\nearrow O^-\\ \searrow X^+\end{smallmatrix}}$$

References pp. 287–293

Structures III and IV assist ionisation of the C–X bond, whereas structure II facilitates nucleophilic addition and consequently a bimolecular displacement of X. The various derivatives of carboxylic acids form a series with varying degrees of resonance stabilisation decreasing in the following order[5]

$$-\mathrm{C}\overset{\delta-}{\underset{\mathrm{O}^-}{\overset{\mathrm{O}}{\lessgtr}}} > -\mathrm{C}\overset{\delta+}{\underset{\mathrm{O}^-}{\overset{\mathrm{NR_2}}{\lessgtr}}} > -\mathrm{C}\overset{\delta+}{\underset{\mathrm{O}^-}{\overset{\mathrm{OR}}{\lessgtr}}} > -\mathrm{C}\overset{\delta+}{\underset{\mathrm{O}^-}{\overset{\mathrm{Cl}}{\lessgtr}}} > -\mathrm{C}{\underset{\mathrm{O}}{\overset{\mathrm{R}}{\lessgtr}}} > -\mathrm{C}{\underset{\mathrm{O}}{\overset{\mathrm{H}}{\lessgtr}}}$$

Although this order is qualitatively reasonable, it is difficult to specify quantitatively the order given above. For example, the resonance energies calculated from heats of combustion for acetic acid, ethyl acetate, acetamide and acetic anhydride vary from[6] 13 to 16 kcal·mole^{-1} and from[7] 14 to 18 kcal·mole^{-1}. The errors involved in these sets of calculations are such that the differences in resonance energies between the various molecules cannot be said to be real or the absolute values of significance[8]. The lack of free rotation about the C–X bond is evident in esters and amides. Dipole moment measurements on the esters of saturated monohydric alcohols and saturated monocarboxylic acids show that the average values of 1.7–1.9 D are nearly invariant over temperature ranges of up to 190°C. Predicted moments for VI and VII

$$\underset{\mathrm{VI}}{\mathrm{R{-}C}{\underset{\mathrm{O{-}R}}{\overset{\mathrm{O}}{\lessgtr}}}} \qquad \underset{\mathrm{VII}}{\mathrm{R{-}C}{\underset{\mathrm{O}\diagup\mathrm{R}}{\overset{\mathrm{O}}{\lessgtr}}}}$$

are 1.53 and 3.53 D, and the observed moment is much too small to be an average of these values. VI is, therefore, thought to be the correct structure of simple esters, and is supported by the observation that γ-butyro-lactone has a dipole moment of 4.12 D[8,9].

The amide group is resonance-stabilised to a greater extent than the ester group, as shown by the extensive evidence of X-ray analysis of the peptide bond[10] and the carbon–nitrogen bond distance in amides[8].

The electronic distribution of the carbonyl group is influenced by inductive effects, dipolar field effects and strain. A number of linear free energy correlations have been demonstrated between the infrared frequency shifts of carbonyl compounds[11,12]. In particular, the relationship between the carbonyl stretching frequency of substituted acetophenones and the Hammett sigma constants for the substituent[13]; a relationship between the Taft polar substituent constants and both the carbonyl stretching frequency of methyl ketones[14] and the intensity of the carbonyl stretching frequency of substituted ethyl acetates. Dipolar field effects are observed in the infrared spectra of α-halogen substituents in keto steroids and cyclohexanones[15,16]. Effects of strain of the C–C(O)–C bond on the infrared spectra have been rationalised by Foote[17].

1.3 MECHANISMS OF THE REACTION AT THE CARBONYL GROUP

The reactions of carbonyl compounds, and in particular derivatives of carboxylic acids, may be compared with reactions of carbon–carbon double bonds or alkyl halides. Addition of a nucleophile to a carbonyl group is aided by the polarisation of the carbonyl group in the sense $\overset{\delta+}{C}=\overset{\delta-}{O}$. The bond broken is a carbon–oxygen pi bond, weaker than a carbon–carbon pi bond, and steric hindrance is considerably reduced in comparison with both alkenes and alkyl halides by the ability of the nucleophile to interact with the carbon atom at right angles to the axis of the pi bond. The mechanism of nucleophilic attack on a carbonyl group has been the subject of a great deal of research. Substitution reactions at a carbonyl carbon atom may theoretically proceed through one of three mechanisms, a direct displacement mechanism, an addition–elimination mechanism and an elimination–addition mechanism. The transition state in a bimolecular displacement may proceed through two alternative transition states depending upon the relative π-bond energy of the carbonyl bond and on the σ-bond energy of the bond formed. A planar transition state VIII which is formed if the π-bond is retained[18] is similar both electronically and sterically to an S_N2 transition state. This square-planar transition state has

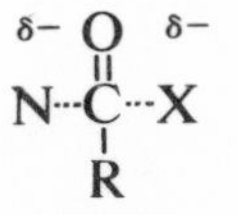

δ− O δ−
N···C···X
R

VIII

O⁻
R–C–N
X

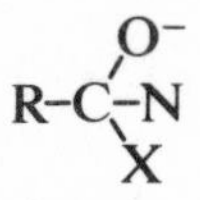

IX

been criticised by Bender as unlikely, citing the evidence of Schowen[19] that back-bonding into the π^* orbital can be significant. In the sp^3 hybridised structure IX, the N–C σ-bond energy largely compensates for the loss of π-bond energy as in species such as chloral hydrate[20] and trifluoroacetone hydrate[21]. There is little doubt that the initial attack is nucleophilic attack on the carbonyl group, the work of Lapworth and Manske[22] showed that in the formation of cyanohydrins the rate-controlling step is nucleophilic addition by the cyanide ion, *viz.*

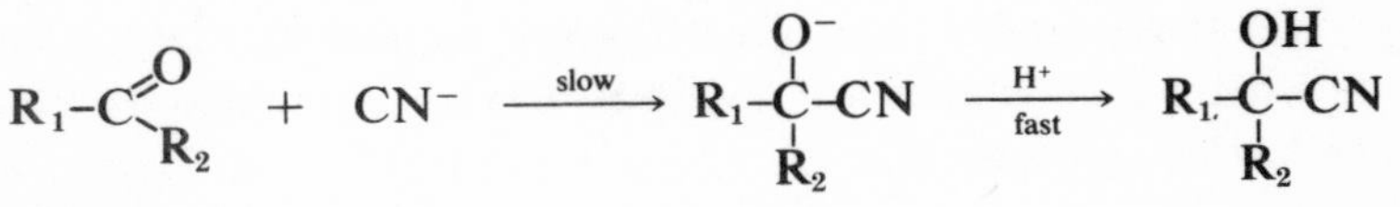

$$R_1\text{–}C(=O)R_2 + CN^- \xrightarrow{\text{slow}} R_1\text{–}C(O^-)(R_2)\text{–}CN \xrightarrow[\text{fast}]{H^+} R_1\text{–}C(OH)(R_2)\text{–}CN \qquad (1)$$

References pp. 287–293

However, detection of the tetrahedral intermediate in the addition of a nucleophile to an ester, acid halide, amide or anhydride must be adduced from kinetic evidence, in particular the evidence of oxygen exchange in such an intermediate. Such tracer work has established the presence of symmetrical addition compounds in the hydrolysis of esters[23], amides and acid chlorides[24]. Since the attempts to detect such intermediates have played a considerable part in the development of hydrolysis studies, it is worthwhile considering this point in some detail.

1.4 THE TETRAHEDRAL ADDITION INTERMEDIATE

As pointed out by Johnson[3] the presence of a tetrahedral addition intermediate may be inferred from four kinds of evidence.

(*i*) ^{18}O exchange into a reactant.

(*ii*) Breaks in the pH–rate profiles for the reaction which cannot be attributed to ionisation processes of the reactants.

(*iii*) Breaks in the buffer concentration *versus* rate curves for the reaction.

(*iv*) Breaks in reactivity rate relationship.

Direct evidence for the existence of a tetrahedral intermediate was found by Zaugg *et al.*[25] in the ammoniolysis of 3-(β-bromoethyl)-3-phenyl-2-benzofuranone (X) in acetonitrile. Approximately 10% of the trapping product XI was isolated, resulting from an intramolecular displacement of bromide ion in the tetrahedral intermediate.

CH_2CH_2Br C_6H_5 X $\xrightarrow{NH_3}$ NH_2 O^- CH_2CH_2Br C_6H_5 $\longrightarrow$ NH_2 C_6H_5 XI

$\downarrow$

O^- CH_2Br $CONH_2$ C_6H_5

The kinetic evidence for two-step reaction, hinges on an analysis of the observed kinetics in terms of the rates of formation and partitioning of the intermediate. Johnson[3] has analysed the hydrolysis of acid derivatives, RCOL as involving either "symmetrical" or "unsymmetrical" mechanisms. Symmetrical mechanisms involve a similar type of catalysis in partitioning the tetrahedral intermediate whereas unsymmetrical catalysis involves dissimilar partitioning of the tetrahedral intermediate, *viz.*

Symmetrical

$$\mathrm{R{-}\overset{\displaystyle O}{\overset{\|}{C}}{-}L + H_2O \underset{k_{-1}}{\overset{k_1}{\rightleftharpoons}} R{-}\underset{\displaystyle OH}{\overset{\displaystyle OH}{C}}{-}L \underset{k_{-2}}{\overset{k_2}{\rightleftharpoons}} RCO_2H + LH} \tag{2}$$

Unsymmetrical

$$\mathrm{R{-}\overset{\displaystyle O}{\overset{\|}{C}}{-}L \underset{k_{-1}}{\overset{k_1[OH^-]}{\rightleftharpoons}} R{-}\underset{\displaystyle OH}{\overset{\displaystyle O^-}{C}}{-}L \underset{k_{-2}[B][H_2O]}{\overset{k_2[BH]}{\rightleftharpoons}} RCO_2H + LH} \tag{3}$$

In a symmetrical mechanism, no break in buffer concentration–rate curve or a pH–rate profile will be observed, whereas for the unsymmetrical mechanism the forward rate is given by

$$\frac{k_1 k_2 K_W[B]}{(k_{-1}+k_2[BH])K_{BH}} \tag{4}$$

where K_W is the self ionisation constant of water and K_{BH} the acid dissociation constant of BH. At constant pH and low base concentration $k_{-1} \gg k_2[BH]$ and the catalytic coefficient is given by

$$\frac{k_1 k_2 K_W}{k_{-1} K_{BH}} \tag{5}$$

As the concentration of BH increases, the observed catalytic coefficient will decrease until, when $k_2[BH] \gg k_{-1}$, the catalytic coefficient equals $k_1[OH^-]$ and the rate-determining step is the addition of hydroxide ion to the substrate. Choice may be made between a number of unsymmetrical mechanisms depending upon the rate dependence upon hydrogen ion, hydroxide ion or water concentrations at high buffer concentrations or [B] or [BH] at low buffer concentrations. Johnson has tabulated the 18 kinetic possibilities and the 13 different types of kinetic behaviour of general acid–base-catalysed reaction, pointing out that this tabulation uses only one ionic form for the tetrahedral intermediate.

Simpler evidence for the presence of a tetrahedral intermediate is adduced from a study of the kinetics of alkaline hydrolysis of amides such an anilides[26–28], chloroacetamide[30], N,N-diacylamines[31], and urea[32]. The rate equations for these reactions contain both first- and second-order terms in hydroxide ion. A reasonable explanation is that the hydrolysis mechanism involves a tetrahedral intermediate, rather than that the second-order term is due to base catalysis of the addition of the hydroxide ion to the carbonyl group. Such a mechanism is

$$
\begin{array}{ccccc}
 & & R\text{-}\overset{O}{\overset{\|}{C}}\text{-}N & & \\
 & & k_1[OH^-]\downarrow\uparrow k_2 & & \\
R\text{-}C(O^-)(O^-)\text{-}N\lesssim^{H} & \overset{K_y}{\rightleftharpoons} & R\text{-}C(O^-)(OH)\text{-}N & \underset{K_x[OH^-]}{\rightleftharpoons} & R\text{-}C(O^-)(O^-)\text{-}N \\
\text{XIV} & & \text{XII} & & \text{XIII} \\
\downarrow k_3 & & & & \downarrow k_4 \\
RCO_2^- + HN\lesssim & & & & RCO_2^- + {}^-N\lesssim
\end{array}
\quad (6)
$$

The rate equation is written with the tautomer (XIV) of the initial addition intermediate XII as being the source of the first-order hydroxide term, giving the expression

$$
k_{obs} = \frac{k_1k_3K_y[OH^-]+k_1k_4K_x[OH^-]}{k_2+k_4K_x[OH^-]+k_3K_y} \quad (7)
$$

Bender and Thomas[26] found that, in the case of ring-substituted acetanilides, the overall hydrolytic term has a Hammett ρ value of approximately 0.1. In elegant work using ^{18}O exchange techniques, the partition ρ value for the ratio k_3/k_2, $\rho_{3/2}$, was found to be approximately -1.0 and ρ_1 is equal to $+1.0$. Although the sign and size of ρ_1 agree with what might be expected for the addition step, neither the sign nor the size of the value for $\rho_{3/2}$ agrees with the expulsion of an anilide ion from XII. Protonation of the anilide nitrogen before expulsion must, therefore, be necessary giving rise to a substantial positive value of $\rho_{3/2}$. In the alkaline hydrolysis of N-methyl-2,2,2-trifluoracetanilide[29,33] and 2,2,2-trifluoracetanilide[28] there is a change in the rate-determining step in the reaction from the addition of hydroxide ion to the carbonyl carbon ($k_1[OH^-]$) at high pH to the decomposition of intermediates XII and XIV, (*viz.* when $k_4K_x[OH^-] < (k_2+k_3K_y)$, at lower pH. This evidence must be taken to mean that the tetrahedral intermediate lies on the reaction path and is kinetically significant.

The work of Zerner and Bender[34] on the hydrolysis of *o*-carboxyphthalimide (XV below) shows the presence of a two-step mechanism. The hydrolysis of *o*-carboxyphthalimide is considerably faster than that of phthalimide at pH 1.5–4, but of similar rate in regions where hydroxide and hydrogen ion catalysis is important. This extra reactivity of *o*-carboxyphthalimide has a maximum at pH 2.9, a value which is not directly related to the acidities of the reagents (3.65 for XV and 15.7 for H_2O). A pH–rate maximum cannot be explained by concurrent acid and base catalysis since both are additive, and since two steps are involved in the conversion of XV to its hydrolysis product it follows that an intermediate must be formed. Two kinetically indistinguishable pathways are possible, both involving general acid–base catalysis. Path I involves general base-catalysed hydration to form the anionic intermediate XVI followed by some form of acid-catalysed decomposition of XVI. Path 2 is the converse of path 1, the formation of the tetrahedral intermediate is catalysed by the neighbouring carboxylic acid group, followed by a base-catalysed decomposition of the intermediate XVII, *viz.*

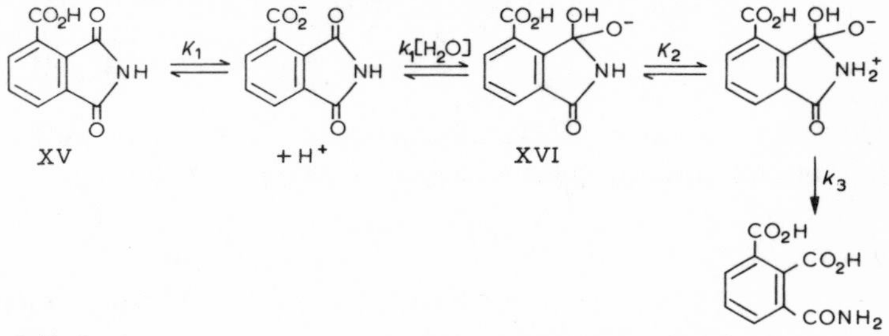

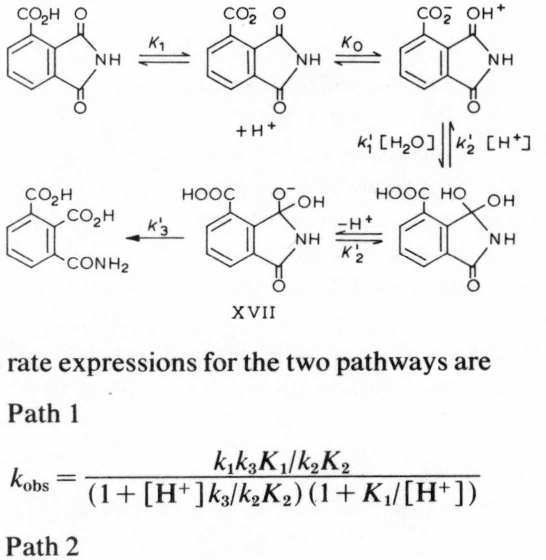

The rate expressions for the two pathways are

Path 1

$$k_{obs} = \frac{k_1 k_3 K_1/k_2 K_2}{(1+[H^+]k_3/k_2K_2)(1+K_1/[H^+])} \tag{8}$$

Path 2

$$k_{obs} = \frac{k_1' K_0}{(1+[H^+]k_2'/k_3'K_2')(1+K_1/[H^+])} \tag{9}$$

Both expressions predict a rate maximum (as observed), but the exact mechanism cannot be inferred from the evidence above. Other cases of a similar kind are the hydroxylaminolysis of formamide[35] which exhibits a pH–rate maximum at pH 6.2–6.5 and the reaction of amines with imido esters to give amidines[36]. Both reactions show kinetics consistent with the presence of a tetrahedral intermediate on the reaction path, but in both cases attempts to detect the presence of the intermediate spectrophotometrically were unsuccessful.

Of particular interest is the observation of Bender and Heck[37] who find from ^{18}O exchange that the value of k_e/k_h (the ratio of the rate coefficients for ^{18}O exchange and for hydrolysis) is in agreement with the value of k_2'/k_3' from the data of Fedor and Bruice[38] on the hydrolysis of ethyl trifluorothiolacetate. The mechanism for this reaction is

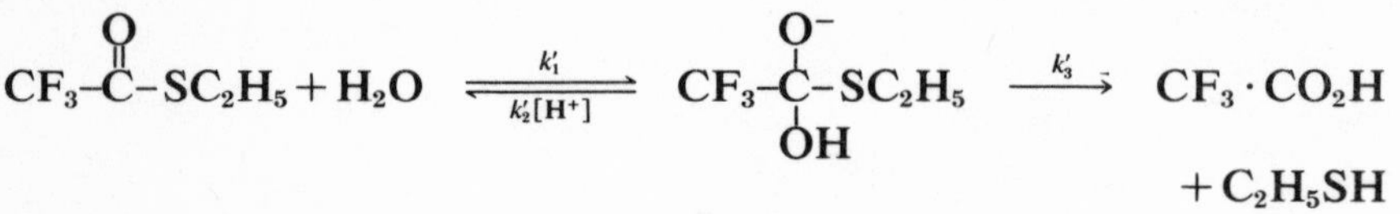

$$CF_3{-}\overset{O}{\overset{\|}{C}}{-}SC_2H_5 + H_2O \underset{k_2'[H^+]}{\overset{k_1'}{\rightleftharpoons}} CF_3{-}\underset{OH}{\underset{|}{\overset{O^-}{\overset{|}{C}}}}{-}SC_2H_5 \xrightarrow{k_3'} CF_3 \cdot CO_2H + C_2H_5SH \quad (10)$$

This is the first demonstration that ^{18}O exchange measures the same properties of the tetrahedral intermediate as given by kinetic measurements.

1.5 OXYGEN-18 EXCHANGE EVIDENCE FOR A TETRAHEDRAL ADDITION INTERMEDIATE

The exchange of ^{18}O into the intermediate in the hydrolysis of a carboxylic acid provides valuable evidence about the existence and lifetime of such an intermediate. Bender[23] has shown that in ester hydrolyses by the $B_{AC}2$ and $A_{AC}2$ mechanisms (see Chapter 2), the carbonyl O partly exchanged with the oxygen of the solvent water. The result can be explained by postulating that, in the course of an individual bimolecular displacement, the oxygen of the water and of the carbonyl group are associated simultaneously with the carbonyl carbon atom for a period of time long enough to permit either direct or indirect proton transfer to occur between the oxygen atoms. The amount of ^{18}O recovered from the unhydrolysed ester then reflects the partitioning of the tetrahedral intermediate assuming this to lie on the reaction path to products. This is shown below, *viz.*

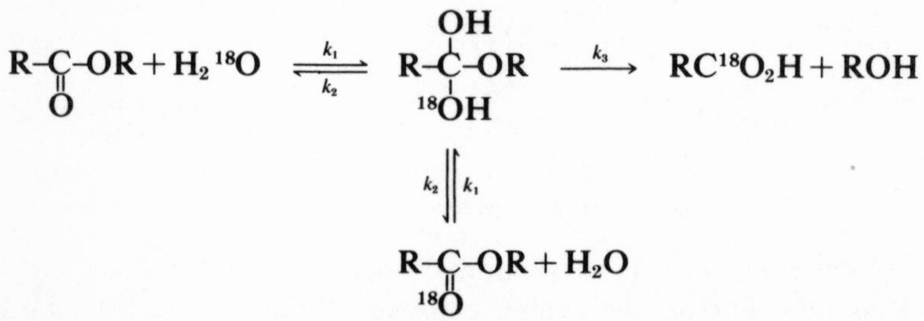

$$R{-}\underset{O}{\underset{\|}{C}}{-}OR + H_2{}^{18}O \underset{k_2}{\overset{k_1}{\rightleftharpoons}} R{-}\underset{^{18}OH}{\underset{|}{\overset{OH}{\overset{|}{C}}}}{-}OR \xrightarrow{k_3} RC^{18}O_2H + ROH$$

$$\Big\downarrow\!\uparrow \; k_2,\, k_1$$

$$R{-}\underset{^{18}O}{\underset{\|}{C}}{-}OR + H_2O \quad (11)$$

The ratio k_e/k_h, ^{18}O exchange to hydrolysis (p. 129), is obviously equal to $k_2/2k_3$. As k_3 increases, the ratio k_e/k_h would be in the order amides > esters >

anhydrides > acid halides. That is, the better the leaving group, the larger is the value of k_e/k_h. The data of Bender *et al.*[23,37,39] and Bunton *et al.*[24,40,41] (Table 1) clearly show this tendency.

TABLE 1

RATIO OF RATES OF ^{18}O EXCHANGE TO RATES OF HYDROLYSIS FOR CARBOXYLIC ACID DERIVATIVES

Compound	*Solvent*	*Temp.* (°C)	k_h/k_e		*Ref.*
$C_6H_5CONH_2$	H_2O	109	0.21	Hydroxide reaction	39
$C_6H_5CO_2C_2H_5$	33% Dioxan	25	10.6	Hydroxide reaction	23
$(CH_3CO)_2O$	60% Dioxan	25	>300	Water reaction	40
$(C_6H_5CO)_2O$	60% Dioxan	25	*ca.* 200	Water reaction	40
Camphoric anhydride	87.5% Dioxan	101	*ca.* 200	Water reaction	40
C_6H_5COCl	66% Dioxan	25	18	Water reaction	24
$CH_3\underset{\overset{\|}{O}}{C}N\overset{+}{N}H$	H_2O	25	>100	Water reaction	40
$CF_3CO_2C_2H_5$	25% acetonitrile	25	2.0	Water reaction	37
$(CH_3CO)_2O$	H_2O	0	No exchange	Acid reaction	41
$(C_6H_5CO)_2O$	H_2O	25	No exchange	Acid reaction	41
$C_6H_5CONH_2$	H_2O	109	>374	Acid reaction	39
$C_6H_5CO_2C_2H_5$	H_2O	109	5.2	Acid reaction	39

The rate equation for the hydrolysis of an acid derivative, as exemplified by the ester in reaction (11), is given by

$$k_{obs} = \frac{k_1 k_3}{k_2 + k_3} = \frac{k_1}{(k_2/k_3)+1} \tag{12}$$

Values of $2k_e/k_h$ which are smaller than unity show that the addition step is rate-determining. Examples are the alkaline, neutral and acid hydrolyses of esters and the acid hydrolyses of anhydrides and amides and the neutral hydrolyses of acid chlorides, anhydrides and the N-acetylimidazolium ion. When $2k_e/h_h$ is larger than unity the rate-determining step is the breakdown of the tetrahedral intermediate, and it is in reactions in which this condition holds that detection of the tetrahedral intermediate would be expected. Such reactions are the alkaline hydrolyses of amides but there are as yet no reports of the detection of an intermediate in these hydrolyses. The tetrahedral intermediate has been directly observed only under non-hydrolytic conditions[42]. Bender[42] found that when metal alkoxides are added to halo-acetic acid esters or diethyl oxalate in di-*n*-butyl ether, the resulting heterogeneous system shows a diminution in the intensity of the carbonyl stretching frequency in the infrared. The parent haloacids do not show this behaviour.

The general difficulty encountered in the detection of intermediates in the reactions of nucleophiles with carboxylic acid derivatives is a function of their ground-state stabilisation. Such stabilisation decreases in the order $O^- >$ $NR_2 > OR > SR > F > Cl > Br > I >$ alkyl as is shown by the restricted rotation about nitrogen in amides[43–45]. The tetrahedral intermediate should, therefore, be looked for in the reactions of acyl halides. Entelis and Nesterov[46] have reported the spectrophotometric detection of transient intermediates during the reaction of acid chlorides with amines in alkane solvents. In order to study the kinetics of fast reactions their experiments were carried out with very low concentrations of reactants (10^{-6}–10^{-5} mole·l^{-1}) using long-path-length cells (5 cm) and hence the absorption changes were very small (~0.06 units). It is interesting to note that they observed that the extinction coefficient for the intermediate is smaller than for the halide and for the product in the case of benzoyl chloride with both piperidine and piperazine but is greater than that of the halide and greater than the product in the reactions of pthaloyl chloride and terephthaloyl chloride in the reaction with piperazine, where in the latter cases the process observed is the attack on the second acyl group.

The work of Bender and Thomas[47] has shown that the lifetime of the tetrahedral intermediate in the alkaline hydrolysis of *p*-substituted methyl benzoates must be of the order of 10^{-9} sec or less, corresponding to a value of 6 kcal for the depth of the well in the reaction profile leading to products. The dielectric relaxation time for water is *ca.* 10^{-11} sec[48], and intermediates with lifetimes of 10^{-11} sec or smaller will therefore not be in solvation equilibrium with solution. Proton transfer producing ^{18}O exchange could take place to the tetrahedral intermediate in a time comparable with the lifetime of a transition state. While the structural details of the solvation in aqueous solution of such short-lived species remains a matter of considerable speculation[49], it is perhaps worth noting that Eigen[50] has proposed a one-encounter mechanism in the acid-catalysed dehydration of acetaldehyde hydrate, in which concerted nucleophilic attack and acid catalysis occur. Because of unsuspected order in the solvent, a reaction which is normally written as a two-step process may take place in one step. As the lifetime of the tetrahedral intermediate decreases these considerations will become more important.

Two further properties which must be intimately related with the problems of the tetrahedral intermediate are the solvent isotope effect and the entropy of activation for the hydrolysis of RCOX. Table 2 lists the values of entropies of activation for the neutral hydrolyses of esters, anhydrides, and acid chlorides. Values of −30 to −40 eu are common to the hydrolyses of esters and anhydrides, but for acid halides the value is down to ~ −10 eu. The only slight dependence of these values on the composition of the solvent reflects only the differences in structure of esters, acid chlorides, etc., and variation in rate with variation in the composition of the solvent has been used as a criteria for

TABLE 2

ENTROPIES OF ACTIVATION FOR THE NEUTRAL HYDROLYSES OF CARBOXYLIC ACID DERIVATIVES

Compound	$\Delta S^{\ddagger}$ *(eu)*	*Solvent*	*Ref.*
sec-Butyltrifluoroacetate	−49.1	29.3% water–70.7% acetone	53
Phenyl trifluoroacetate	−45.9	29.3% water–70.7% acetone	53
Acetic anhydride	−37.6[a]	Water	54
	−42.3[b]		
	−45.2[c]		
Acetic anhydride	−43	40% water–60% dioxan	40
Succinic anhydride	−28.7	Water	55
Succinic anhydride	−31	40% water–60% dioxan	40
Glutaric anhydride	−31.5	Water	
3,3-Dimethylglutaric anhydride	−30.3	Water	55
Acetyl chloride	−14	25% water–75% dioxan	56
Benzoyl chloride	−7.1	50% water–50% acetone	
Dimethylcarbamoyl chloride	+5.6	Water	57
Acetylimidazolium ion	−30.2	Water	58
1-(N,N-Dimethylcarbamoyl)-pyridinium ion	−27	Water	59
Ethyl chloroformate	−12.4	Water	57

[a] 0–19.4°C.
[b] 19.4–29.6°C.
[c] 29.6–39.9°C.

mechanism in the case of acyl halides. However, Arnett *et al.*[51,52] have shown that, in mixtures of ethanol or *t*-butanol in water, the partial molar volumes of the solute pass through distinct minima in the region of 0.9 mole fraction of water. This and other evidence is a strong indication that the structure of the solvent suffers considerable changes in such a situation and this is reflected by the heat of solvation (ΔH_s) of a third compound added to the solvent mixture. The determinations of ΔH_s by Arnett *et al.*[52] for the initial state solvation of a wide range of neutral compounds passes through a maximum corresponding to the point of change of the physical properties of the solvent mixture. Arnett's studies also indicate that this effect on ΔH_s is absent for the smaller ions, but that at least 95% of the variation of $\Delta H^{\ddagger}$ for *t*-butyl chloride solvolyses in the region of the solvent minimum is accounted for by change in ground-state solvation. Similar observations have been made by Hyne[60] on the hydrolyses of sulphonium salts. In this connection, it is worth noting that $\Delta C_p^{\ddagger'}$, the heat capacity of activation, for the solvolysis of *t*-butyl chloride in ethanol/water mixtures[61] (Table 3) clearly shows a minimum, but there is simply a decrease in $\Delta S^{\ddagger}$. As pointed out by Robertson[49], $\Delta C_p^{\ddagger}$ is a more sensitive probe into the effects of solvent change than $\Delta S^{\ddagger}$ and he points out that the breakdown of

TABLE 3

ENTROPY AND HEAT CAPACITY OF ACTIVATION FOR THE SOLVOLYSIS OF *t*-BUTYLCHLORIDE IN A SERIES OF ETHANOL/WATER MIXTURES[61]

Solvent	$\Delta C_p^{\ddagger}$ *(cal·mole⁻⁵·deg⁻¹)*	$\Delta S^{\ddagger}$ *(eu)*
Water	−83	14.4
27/73(v/v)EtOH/H_2O	−116	2.7
27/63(v/v)EtOH/H_2O	−49	−3.2
50/50(v/v)EtOH/H_2O	−34	−2.9

water structure involves a value[62] of $\Delta C_p^{\ddagger}$ of −33 cal·mole^{-1}·deg^{-1}, as determined from measurements of viscous flow and dielectric relaxation. This value is very similar to that determined for the hydrolysis of sulphonic esters and suggests the possibility that the disruption of a unit structure of water is involved in the activation process. Without $\Delta C_p^{\ddagger}$ data for the hydrolysis of acyl derivatives speculation as to the meaning of the high negative entropy of activation must be limited to a postulate that either a number of molecules are highly constrained in the transition state or that the transition state is highly polar and causes a bulk change in the electrical properties of the solvent. In this context, Hudson *et al.*[63,64] have shown that the rate of alcoholysis of acetyl chloride in dilute solutions of alcohols in non-polar solvents is proportional to the concentration of the alcohol in the associated state over a wide concentration range. This evidence suggests that the transition state is associated with a discrete number of alcohol molecules to give an assembly (XVIII) which collapses to give the reaction products, *viz.*

$$(R'OH)_n + RCOCl \rightleftharpoons RCOCl(R'OH)_n \rightarrow RCOOH + HCl(R'OH)_{n-1}$$

Similar hydrogen-bonded complexes have been detected kinetically in ester hydrolyses[65].

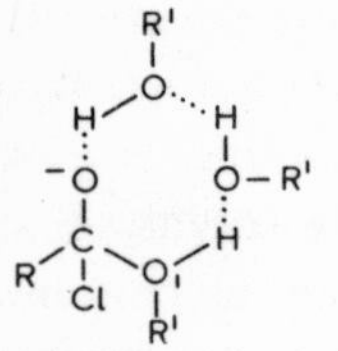

XVIII

Values for k_{H_2O}/k_{D_2O}, listed in Table 4, range from 1.0 (N,O-diacetylserinamide) to 5 ± 1 for the acetyl pyridinium ion and ethyl dichloracetate. Such high values must be the result of binding at least two molecules of water in the transition state.

TABLE 4

SOLVENT ISOTOPE EFFECTS FOR NEUTRAL WATER REACTIONS OF CARBOXYLIC ACID DERIVATIVES

Compound	*Solvent* % *Water*	% *Dioxan*	*Temp.* (°C)	k_{H_2O}/k_{D_2O}	*Ref.*
Acetic anhydride	100	0	25	2.9	66, 67
Acetic anhydride	40	60	25	2.6	40
Benzoic anhydride	100	0	25	3.9	68
Benzoic anhydride	40	60	25	2.75	69
Benzoic anhydride	40	60	35	2.9	70
Glutaric anhydride	40	60	25	2.95	69
Glutaric anhydride	100	0	25	2.84	69
Succinic anhydride	100	0	25	2.5	68
Succinic anhydride	100	0	25	2.8	40
Acetic benzoic anhydride	40	60	25	3.4	40
Acetic mesitoic anhydride	40	60	25	2.8	40
p-Methylbenzoyl chloride	9	91		1.67	71
p-Methylbenzoyl chloride	27	73		1.41	71
p-Methylbenzoyl chloride	50	50		1.41	71
p-Nitrobenzoyl chloride	9	91		1.6	71
p-Nitrobenzoyl chloride	27	73		1.74	71
p-Nitrobenzoyl chloride	50	50		1.80	71
Benzoyl chloride	15	85		1.6	40
Benzoyl chloride	20	80		1.5	68
Benzoyl chloride	43	67		1.9	68
Benzoyl chloride	40	60		1.7	69
2,4,6-Trimethylbenzoyl chloride	5	95		1.65	69
2,4,6-Trimethylbenzoyl chloride	6.7	93.3		1.5	68
Diphenylcarbamoyl chloride	100	0		1.1	72
N-Methyl, N′-acetylimidazolium ion	100	0	25	2.6	73
N-Methyl,N′-cinnamoyl-imidazolium ion	100	0	25	2.1	74
Acetylimidazolium ion	100	0	25	2.5	58
Acetylimidazole	100	0	25	2.7	58
Acetylpyridinium ion	100	0	25	5 ± 1	75, 67
N,O-Diacetylserinamide	100	0	100	1.0	76
Ethyldichloroacetate	100	0	25	*ca.* 5	77
Ethyldifluoracetate	100	0	25	2.1	77

Johnson[3] has listed possible transition states for reactions in which ^{18}O studies show that the formation of the tetrahedral intermediate is the rate-determining step. General base-catalysed process

H, δ+, O, H, H, O···C⋯O

XIX

H–O–CO⋯O δ−, L, H···O, H, H, δ+

XX

Transition state XIX involves the rate-determining formation of an anionic transition state and XX the rate-determining decomposition of the anionic transition state. General acid-catalysed process

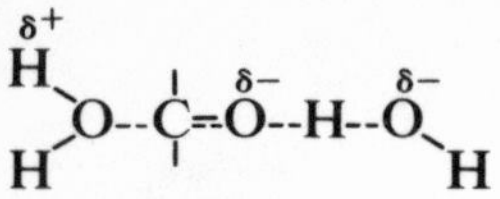

XXI

Possible cyclic processes

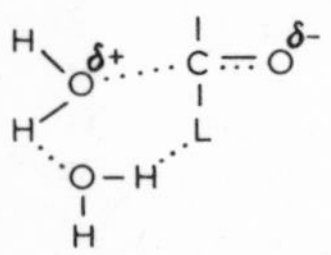

XXII 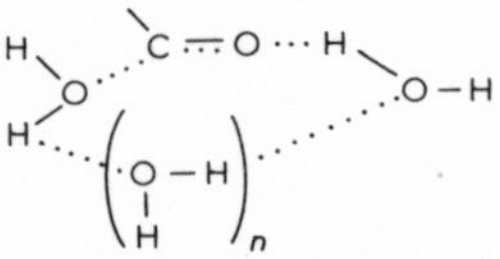 XXIII

The experimental evidence for ethyl trifluoracetate[38] hydrolysis is in accord with transition state XIX and either XIX or XX would be generally applied to other substrates, although transition states of type XXII or XXIII may be possible also. The general acid transition state XXI is unlikely because few intermolecular acid-catalysed ester hydrolyses are known.

Substrates with good leaving groups are often hydrolysed by mechanisms which involve rate-determining hydration of the carbonyl group. Mechanisms for this include specific acid–specific base pathways which involve protonation of the carbonyl group followed by rate-determining addition of hydroxide, *e.g.*

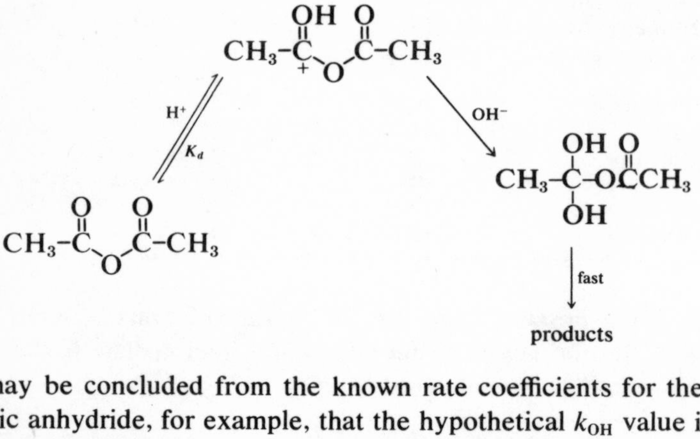

$$CH_3COOCOCH_3 \underset{K_d}{\overset{H^+}{\rightleftharpoons}} CH_3C(=OH^+)OCOCH_3 \xrightarrow{OH^-} CH_3C(OH)_2OCOCH_3 \xrightarrow{\text{fast}} \text{products} \qquad (13)$$

It may be concluded from the known rate coefficients for the hydrolysis of acetic anhydride, for example, that the hypothetical k_{OH} value is larger than a diffusion-controlled rate constant. If K_d is taken as about 10^7 for acetic anhydride, then from $k_{OH} = k_{H_2O}\, K_d/K_w$, where k_{H_2O} is the second-order water rate coefficient ($= 5.1 \times 10^{-5}$ l·mole^{-1}·sec^{-1}), $k_{OH} = 5.1 \times 10^{16}$ l·mole^{-1} sec^{-1}. The mechanism can therefore be ruled out, and the converse mechanism of rapid attack by the hydroxyl ion on the carbonyl group followed by a rate-determining protonation of the tetrahedral intermediate is also unlikely[3]. Reaction paths involving termolecular collisions involving hydrogen and hydroxyl are also unlikely since they would result in negative activation energies for hydrolyses[78].

1.6 THE UNIMOLECULAR MECHANISM OF SOLVOLYSIS

A unimolecular mechanism has been demonstrated for the hydrolysis and alcoholyses of certain carboxylic acid derivatives with good leaving groups, notably some acyl halides. There is a duality of mechanism, similar to that in solvolyses of alkyl halides, the ratio of the two paths depending upon the polarity of solvent, structure of the carboxylic acid derivative etc. For example, electron-withdrawing substituents increase the reactivity of acyl halides and esters as shown by the order of rates for alcoholysis[79a,79b], $Cl_3COCl > CHCl_2$-$COCl > CH_2ClCOCl > CH_3COCl$. Similarly, measurements of the rates of alkaline hydrolysis of esters and amides, of alcoholysis and aminolysis of substituted benzoyl chlorides (Table 5) show that *p*-substituted electron-withdrawing groups increase the reactivity.

In solvents of high polarity, the substituent effect is reversed, indicating that a change in mechanism has taken place from a bimolecular reaction to a rate-determining ionisation. In formic acid, the rate of reaction of *p*-nitrobenzoyl chloride is proportional to the concentration of water[80], showing the reaction to be second-order; in contrast, the rates of solvolysis of benzoyl chloride and toluoyl chloride are almost independent of the concentration of water. As might be expected, changes in the water concentration effect the rates of hydrolysis of the different halides in different ways. Crunden and Hudson[80] have shown that the rate of hydrolysis of *p*-methoxybenzoyl chloride is more sensitive to changes in water content than *p*-nitrobenzoyl chloride (Table 5), reflecting the more polar transition state in the former compound.

TABLE 5

RELATIVE RATES OF SOLVOLYSIS OF *p*-SUBSTITUTED BENZOYL CHLORIDES[80]

		Substituent				
Solvent	*Reactant*	NO_2	*Br*	*Cl*	CH_3	CH_3O
Formic acid	Water (1%)	0	0.086		6.6	Very fast
Acetone	Water (50%)	11.5	0.92	0.85	2.9	30
Acetone	Water (5%)	35	3.2		0.55	0.60
Ethanol	Ethanol	21.6	2.1	1.9	0.78	0.81
Ether	Ethanol (40%)	32.5	2.5	2.9	0.47	0.25
Benzene	Aniline	8.5		1.5	0.56	

References pp. 287–293

The duality of mechanisms is shown with clarity in the work of Gold *et al.*[81] in a study of the effect of added amine on the rate of reaction of benzoyl chloride in aqueous acetone. A comparison of the rate coefficients k_1 and k_0 for the reaction of benzoyl chloride, with and without 4-chloraniline, with the product compositions showed that, in 50% aqueous acetone, the ratio of anilide to acid was considerably greater than the rate ratio, whereas agreement with the rate ratio was found in 80% acetone–20% water as solvent. At the higher water concentration, approximately 40% of the anilide was produced in a non-rate-determining stage, supporting the mechanism

$$
\begin{aligned}
&\mathrm{PhCOCl} \rightleftharpoons \mathrm{PhCO^+ + Cl^-} \\
&\mathrm{PhCO^+} \xrightarrow{\mathrm{PhNH_2}} \mathrm{PhCONHPh} \\
&\mathrm{PhCO^+} \xrightarrow{\mathrm{H_2O}} \mathrm{PhCOOH}
\end{aligned}
\tag{14}
$$

Crunden and Hudson[80], by comparison of solvolysis rates in formic acid and aqueous acetone (although the same authors have stated that this may be coincidental), obtained a value of 40% S_N1 participation in aqueous acetone (65%); they also estimated the contribution of the S_N1 route to the solvolysis of substituted benzoyl halides (Table 6). It must, however, be emphasized that the nature of the intermediate trapped in the work of Gold *et al.*[81] could either be the acylium ion RCO^+ or the hydrated acylium ion $[RCO{\cdot}H_2O]^+$.

Bunton *et al.*[24] have shown that ^{18}O exchange into benzoyl chloride in 5% (v/v) aqueous dioxan as solvent is negligible and hence any pathway involving rate-determining bond breaking in the tetrahedral intermediate is ruled out under these conditions. However, in 25% (v/v) aqueous dioxan, the value of k_h/k_e for ^{18}O exchange is 25. This represents a degree of exchange which is by no means negligible in a solvent of increased polarity and hence one more likely to favour an S_N1-type mechanism. It is of interest to note that no trapping with acetic anhydride could be detected with 50% aqueous acetone.

The positive entropy of activation (+5,6 eu, see Table 2, p. 219)[57] for the hydrolysis of dimethylcarbamoyl chloride is an indication of a reaction

TABLE 6

% S_N1 REACTION IN THE SOLVOLYSIS OF *p*-SUBSTITUTED BENZOYL CHLORIDES[80]

Compound	k_1 (sec^{-1}) in		"% S_N1 in aq. acetone"
	65% aq. acetone	*Formic acid*	
p-NO_2–C_6H_4COCl	3.0×10^{-2}	1.0×10^{-6}	0
p-Br–C_6H_4COCl	2.2×10^{-3}	2.4×10^{-4}	10
C_6H_5COCl	8.9×10^{-3}	3.5×10^{-3}	40
p-$CH_3C_6H_4COCl$	28×10^{-3}	19×10^{-3}	70

which proceeds by an ionisation. Evidence for the existence of an intermediate by the isolation of dimethyl-3,3-tetramethylene urea on the addition of pyrrolidine to the aqueous reaction mixture. Although the rate of disappearance of dimethylcarbamoyl chloride is insensitive to the presence of strong nucleophiles such as hydroxide and azide[57,82], dimethylcarbamoyl azide can be isolated. Similarly, the rate of hydrolysis of diphenylcarbamoyl chloride is insensitive to added nucleophiles, although it reacts in a second-order process with amines[3] and the solvent isotope effect k_{H_2O}/k_{D_2O} of 1.1 is far lower than most other carboxylic acid derivatives, (see Table 4, p. 221). The reason for the S_N1 behaviour of the carbamoyl halides is the stability of the acylium ion, XXIV, closely related to the isocyanide structure, *viz.*

$$R_2\overset{+}{N}{=}C{=}O \longleftrightarrow R_2N{-}\overset{+}{C}{=}O^- \qquad (15)$$

XXIV

In view of the acylium ion mechanism for the hydrolysis of mesityl esters in strongly acid solution and the observation of a Van't Hoff *i* factor of 4 for mesitoic acid in concentrated sulphuric acid[83], it is not surprising that hydrolysis of mesityl chloride has been interpreted in terms of an acylium ion. This is supported by the observation of Bender and Chen[84] of the rate acceleration produced by electron-donating substituents in the 4-position of 4-substituted 2,6-dimethyl-benzoyl chlorides (in 1% water/99% acetonitrile) and also in the acid-catalysed hydrolysis of methyl esters of analogous esters. In addition, common ion effects are observed for the chloride, the ester rate is proportional to h_0[85], no ^{18}O exchange is observed for either the ester in acid or the chloride and the entropy of activation for the ester is positive[47]. However, at high values of pH, the observed rate expressions for both chloride and ester contain terms in hydroxide ion, both of which correlate with σ, giving $\rho + 1.2$, showing that the acylium ion path is not operative in these regions.

Bevan and Hudson[86] have reported that the hydrolysis of benzoyl fluoride is acid-catalysed and follows the $[H_3O^+]$, whereas other unhindered benzoyl halides do not. A parallel situation is found with the benzyl fluoride and chloride. Satchell[87] has likewise demonstrated the hydrolysis of acetyl fluoride and chloride to be acid-catalysed.

The existence of acylium ions in the reactions of acyl halides with Lewis acids has been amply demonstrated[88,89], but they have not so far been directly observed in hydrolysis reactions. To achieve direct observation of an acylium ion, it must, at best, be generated into a non-reactive environment if its rate of formation is slow, or be detectable in small concentrations if it is destroyed rapidly. The limiting factors are, therefore, the rate and mechanism of heterolysis and the stability and concentration of the acylium ion formed. In this context, few data are available on solvolyses of acyl halides in dipolar aprotic media in circumstances where the rates of nucleophilic attack on the acylium ion are reduced, or of acyl halides which would give readily detectable acylium ions even if only in small concentration.

2. Hydrolysis and alcoholysis of carboxylic acid halides

Investigations into the mechanism of hydrolysis and alcoholysis of acyl halides have been largely concerned with acyl chlorides and in particular with benzoyl chloride and the related aromatic acid chlorides. This was a result of the relatively slow rate of hydrolysis of benzoyl chloride compared with acetyl chloride (although their alcoholysis rates are easily measurable) and it is only comparatively recently[90] that stop-flow techniques have been used to measure the faster rate of hydrolysis. However, in spite of this limitation, considerable progress has been made towards elucidation of the mechanism or mechanisms of hydrolysis and alcoholysis of these halides.

What follows is, therefore, a summary, but not a completely comprehensive one, of the available data on the hydrolysis of acid chlorides. It is ordered under headings according to the structure of the halide although there is no intention to impute that this is the controlling factor in the solvolysis; indeed, as will be seen, changes of solvent can produce radical changes in mechanism. This topic has been the subject (in part) of four reviews, those of Bender[1], Hudson[2], Johnson[3], and Satchell[91] and particular aspects have been reviewed in the research papers of Kivinen[92] and Minato[93], to which the present author acknowledges a deep debt.

The detailed mechanism, or mechanisms, of the solvolysis of acid chlorides is still a matter of dispute. There are at least four possible mechanisms, (*a*)–(*d*) below, all of which have been proposed either to act separately or in various combinations, and there is a "unified" mechanism, that of Minato[93] which will be discussed later. The bimolecular mechanisms (*a*) and (*b*) differ in that (*a*) includes a tetrahedral intermediate whereas (*b*) does not. The former is commonly accepted as the most likely for the bimolecular mechanism and the arguments against (*b*) have been stated in the introduction. There is, however, good evidence for (*b*), at least in the case of the hydrolysis of chloracetyl chloride[94]. The "acylium ion" mechanism (*c*) and the hydrated carbonium ion mechanism (*d*) are both unimolecular mechanisms. Whereas the acylium ion XXVII has never been directly observed in hydrolysis or alcoholysis reactions, it is favoured as an intermediate by many workers, although it is kinetically indistinguishable from XXVIII.

$$(a)\quad R\text{-}\overset{\displaystyle O}{\overset{\|}{C}}\text{-}X + H_2O \rightleftharpoons \underset{\text{XXV}}{R\text{-}\overset{O^-}{\overset{|}{C}}\text{-}X \atop {}^{+}OH_2} \longrightarrow R\text{-}C{\overset{=O}{\underset{\diagdown O^+H_2}{}}} + X^- \qquad (16)$$

$$(b)\quad R\text{-}\overset{\displaystyle O}{\overset{\|}{C}}\text{-}X + H_2O \longrightarrow \underset{\text{XXVI}}{\left[H_2\overset{\delta+}{O}\cdots\overset{O}{\overset{\|}{C}}\cdots\overset{\delta-}{X} \atop R \right]} \longrightarrow H_2O^+\text{-}\overset{\displaystyle O}{\overset{\|}{C}}\text{-}R + X^- \qquad (17)$$

$$(c)\quad R\text{-}\overset{\displaystyle O}{\overset{\|}{C}}\text{-}X \rightleftharpoons \underset{\text{XVII}}{R\text{-}\overset{+}{C}{=}O + X^-} \begin{cases} \xrightarrow{H_2O} RCO_2H + H^+ + X^- \\ \xrightarrow[N^-]{} RCON + X^- \end{cases} \qquad (18)$$

$$(d)\quad R\text{-}\overset{\displaystyle O}{\overset{\|}{C}}\text{-}X + H_2O \underset{}{\overset{\text{fast}}{\rightleftharpoons}} R\text{-}\overset{OH}{\underset{X}{C}}\text{-}OH \overset{\text{slow}}{\rightleftharpoons} X^- + R\text{-}\overset{+}{C}{\overset{OH}{\underset{OH}{}}} \begin{cases} \xrightarrow{N^-} R\text{-}C{\overset{=O}{\underset{\diagdown N}{}}} + H_2O + X^- \\ \longrightarrow R\text{-}C{\overset{=O}{\underset{\diagdown OH}{}}} + X^- + H^+ \end{cases} \qquad (19)$$

XXVIII

The kinetic evidence which has been acquired in attempts to distinguish between these alternatives falls into the following classes.

(*i*) Effect of variation of X in RCOX.

(*ii*) Effect of variation of R in RCOX.

(*iii*) Effect of variation in solvent, both on individual halides and on a series of halides.

(*iv*) ^{18}O exchange into RCOX to detect a tetrahedral intermediate.

(*v*) Salt effects to detect nucleophilic catalysis and common ion effects to detect ionisation.

(*vi*) Solvent isotope effects.

(*vii*) Effect of added nucleophiles, *e.g.* amines, to trap electrophilic intermediates.

(*viii*) Acid and base catalysis.

For example, making certain assumptions, ^{18}O exchange studies should distinguish (*a*) and (*d*) from (*b*) and (*c*), since the latter should not give any ^{18}O

exchange. However, interpretation of ^{18}O exchange results for halides where (say) mechanisms (*a*) and (*c*) were operative would be difficult. A case such as benzoyl chloride[24] in 95% dioxan/water, where k_e is negligible, may be interpreted as a mixture of (*b*) and (*c*). It is then hard to see why k_e becomes appreciable, $k_h/k_e = 25$ for 25% (v/v) water in dioxan, if the mechanism does not change also. The explanation preferred is that mechanism (*a*) accounts for the bimolecular reaction and that very little exchange is seen because the rate of proton exchange in XXV is slow relative to the loss of X^-.

2.1 EFFECT OF VARIATION IN THE STRUCTURE OF RCOX

2.1.1 Variation in X

The carbon–halogen bond distances in acyl halides increase in the direction F < Cl < Br < I, and are similar, but slightly larger than, those of the alkyl halides (Table 7). Nuclear quadrupole resonance frequencies of halogen compounds suggest that the charge density on the chlorine atom of an acyl chloride is greater than that on an alkyl chloride (Table 8).

TABLE 7

BOND LENGTHS OF ALKYL AND ACYL HALIDES[96]

X	l_{C-X} in CH_3COX	l_{C-X} in CH_3X
F	1.37 ± 0.02	1.384
Cl	1.82 ± 0.02	1.779
Br	2.00 ± 0.04	1.936
I	2.21 ± 0.04	2.132

This evidence, together with the relatively low dissociation energies of acyl halides[95] suggests that the C–Cl bond is easily ionised. There is a marked rate increase in the acetyl series of F < Cl < Br < I. Acetyl fluoride hydrolyses only slowly in water at 30°C, in contrast to acetyl chloride which is readily hydrolysed even at −20°C. Kinetic data for acetyl chloride, bromide and iodide[100] (Table 9) show that the acetyl halides react more rapidly than benzoyl chloride. Comparison of data at −20°C gives the ratio Cl:Br:I = 1:350:7000 at $\gamma_{H_2O} = 0.333$ showing the order expected from the leaving group.

Later studies on acetyl chloride hydrolysis were those of Hudson and Moss[90], using a stopped-flow apparatus at 27°C and working with dioxan–water mixtures (19.6–75.6 v/v), and Cairns and Prousnitz[102] with acetone/water (15.35% v/v) at low temperatures. The value of the Grunwald–Winstein[103]

TABLE 8

^{35}Cl NUCLEAR QUADRUPOLE RESONANCE FREQUENCIES OF HALOGEN COMPOUNDS

Compound	*Eq. Q (Mc/s)*
Atomic ^{35}Cl	110.4[97]
CH_3Cl	75.13
$C_2H_5O{\cdot}CO{\cdot}Cl$	33.86[98]
$CCl_3{\cdot}COCl$	33.72[98]
$(CH_3)_2NCOCl$	31.8[98]
C_6H_5COCl	29.93[98]
$ClC(O)(CH_2)_4COCl$	29.98[98]
CH_3COCl	Predicted *ca.* 27.7[99]
NaCl	< 1[97]

parameter m was found to be 0.81 for dioxan/water[90] and 0.92 for acetone/water[100] similar to the values ($m \sim 1.0$) frequently obtained in S_N1 ionisations, favouring a mechanism of the type (*c*) or (*d*), that is a unimolecular mechanism similar to that postulated for benzoyl chloride in the more aqueous solvents. However, the rates of hydrolysis in dioxan/water are faster (23.5 times) than the rates in acetone/water for equal water concentrations, probably due to specific solvation by water molecules. Large rate increases on the addition of aniline show that bond formation is important and favour a mechanism of type (*d*).

TABLE 9

FIRST-ORDER RATE COEFFICIENTS FOR THE HYDROLYSIS OF ACETYL CHLORIDE, BROMIDE, AND IODIDE AND BENZOYL CHLORIDE IN ACETONE/WATER[100]

	k_{H_2O} (sec^{-1})			
γ_{H_2O}	CH_3COCl ($-20°C$)	CH_3COBr	CH_3COI ($-50°C$)	C_6H_5COCl ($+20°C$)
0.048			8.2×10^{-2}	
0.083	7.03×10^{-5}		1.23×10^{-1}	2.03×10^{-5}
0.167	2.09×10^{-4}		2.82×10^{-1}	4.33×10^{-5}
0.333		3.51×10^{-1} ($-20°C$)	~ 7 ($-20°C$)	
0.333		3.8×10^{-1} ($-50°C$)	9.1×10^{-1} ($-50°C$)	1.50×10^{-4}
0.500	6.9×10^{-3}	[a]		3.43×10^{-4}
0.750	1.32×10^{-1}			2.05×10^{-3}

[a] The rate coefficient for alcoholysis of acetyl bromide with ethylene chlorohydrin (2 mole$\cdot l^{-1}$ in dioxan) at 15°C was 1.17 sec^{-1}; that for acetyl chloride was 0.00507 sec^{-1} under the same conditions giving Cl : Br = 1 : 231, less than the ratio for hydrolysis (1 : 350)[101].

Swain and Scott[56] have measured the k_{RCl}/k_{RF} ratio in neutral or slightly acidic solutions for the hydrolysis of acetyl and benzoyl halides and compared them with those for triphenylmethyl halides. Values for the triphenylmethyl halides are ~ 10^4 greater than those for acyl halides (Table 10) reflecting the tendency for C–X bond breaking to be more complete than O–C formation at the transition state of trityl hydrolysis and the opposite tendency with the benzoyl halides. The C–F bond is harder to break but the carbon atom is made more electropositive.

TABLE 10

KINETIC DATA FOR SOLVOLYSIS OF ORGANIC CHLORIDES AND FLUORIDES (R-X) IN ACETONE/WATER[56,100]
(Temperature 25°C unless stated otherwise)

Compound R	*Compound* X	*H_2O in acetone solvent (% v/v)*	*k_1 (sec^{-1})*	*$\Delta S^{\ddagger}$ (eu)*	*E_a (kcal·mole^{-1})*	*k_{RCl}/k_{RF}*
Ph_3C	F		2.7×10^{-6}	−10	22.6	
	Cl		2.7	−17	12.5	1×10^{6}
CH_3CO	F	25	1.1×10^{-4}			
	Cl		8.6×10^{-1}	−14	13.9	7.8×10^{3}
C_6H_5CO	F	50	1.1×10^{-5} (0°C)	−7.1	18.8	39
	Cl		4.3×10^{-4} (0°C)			
C_6H_5CO	F	25	8.2×10^{-6}			88
	Cl		7.2×10^{-4}			

Combining the data of Swain and Scott[56] with that of Ugi and Beck[100], the approximate relative rates for RF, RCl, RBr, RI ($R = CH_3CO$) are $1.28\times10^{-4} < 1 < 350 < 7000$. (Note that the CH_3COF value is not corrected for solvent effects.)

As will be seen later, benzoyl fluoride exhibits Brönsted-acid catalysis in aqueous media[86] and was thought to be the only acyl derivative to do so although alkyl fluorides are known to behave in this way also. For benzyl fluoride, the acid-catalysed rate is reported to show dependence on h_0[104,105]. Satchell[87] studied the solvolysis of benzoyl and butyryl fluoride in 60:40 (v/v) dioxan–water mixtures (for which h_0 data are known)[106]. The solvolysis of each compound is catalysed by acid, but does not follow h_0. With excess of solvent the observed first-order loss of acyl derivative obeys the expression

$$-\mathrm{d}[\mathrm{RCOF}]/\mathrm{d}t = (k_o + k_c[\mathrm{H_3O^+}])[\mathrm{RCOF}] \qquad (20)$$

where the values for the second-order catalytic coefficients k_c are 2.3×10^{-3}, 1.2×10^{-3} and 1.7×10^{-4} l·mole^{-1}·sec^{-1} for acetyl, butyryl and benzoyl fluoride respectively, while k_0 values are 4.4×10^{-4}, 1.7×10^{-4} and $3.37\times$

10^{-5} sec^{-1}. The reactivity order is therefore acetyl > butyryl > benzoyl and the detailed mechanism is probably unlike that for benzoyl fluoride. Mechanisms involving protonation of the carbonyl oxygen would satisfy the dependence on $[H_3O^+]$ and the substituent effects which imply that nucleophilic approach of solvent is important, but only acyl fluorides exhibit this catalysis and catalysis *via* protonated oxygen might reasonably be expected for acyl chlorides and bromides. In later work, Satchell[87,107–109] showed that acid catalysis of acetyl chloride and bromide does occur, but he maintains the belief that this is different only in degree, not kind, from that seen in acyl fluorides since fluorine is known to form strong hydrogen bonds. The mechanism for fluorides is probably

$$\mathrm{RCOF + H_3O^+ \rightleftharpoons R\overset{\overset{\displaystyle O}{\|}}{C}FH^+ + H_2O \text{ fast}} \tag{21}$$

$$\mathrm{R\overset{\overset{\displaystyle O}{\|}}{C}FH^+ + H_2O \rightarrow R\text{-}\underset{\underset{\displaystyle OH_2^+}{|}}{\overset{\overset{\displaystyle O}{\|}}{C}} + HF \text{ slow}} \tag{22}$$

If this is the only acid-catalysed route to solvolysis little oxygen exchange will be seen.

Satchell[87] found that the acylation, in nitromethane, of β-naphthol by acetyl chloride to be catalysed by hydrogen chloride, and that added chloride ions depress the rate by a mass law effect. Both unionised and ionised acetyl chloride take part in the acylation and although the ionisation is kinetically independent of the amount of hydrogen chloride, the observed increase in rate by added hydrogen chloride can be explained most easily on the basis that the ionisation is dependent on the presence of HCl, the mechanism being

$$\mathrm{HCl \rightleftharpoons (H^+)_{solv} + Cl^-}$$

$$\mathrm{HCl + Cl^- \rightleftharpoons HCl_2^-}$$

$$\begin{array}{ccccc} \mathrm{CH_3COCl + HCl} & \rightleftharpoons & \mathrm{CH_3CO^+HCl_2^-} & \rightleftharpoons & \mathrm{CH_3CO^+ + HCl_2^-} \\ \text{ROH}\ \big\downarrow\ \text{slow, 1} & & \text{ROH}\ \big\downarrow\ \text{slow, 2} & & \text{ROH}\ \big\downarrow\ \text{slow, 3} \\ \text{Products} & & \text{Products} & & \text{Products} \end{array} \tag{23}$$

Contribution to the overall rate from steps 1 and 3 are approximately equal, and that from 2 is thought to be quite small and although an alternative mechanism such as *via* $(CH_3COClH)^+$ which fits the kinetics may be possible, and indeed may be present, it is probably not as prominent as that *via* $CH_3CO^+HCl_2^-$.

Satchell[107] found that the reaction of acetyl chloride with β-naphthol was accelerated by the addition of HBr and tetraethylammonium bromide, due to halogen exchange, *viz.*

$$\begin{aligned} AcCl + HBr &\rightleftharpoons AcBr + HCl \\ AcCl + Et_4NBr &\rightleftharpoons AcBr + Et_4NCl \end{aligned} \tag{24}$$

and that when acetyl halide and catalyst had a common halogen, an electrophilic catalysis was seen. This electrophilic catalysis is weaker than that found for the acyl fluorides.

Although in the reaction of acetyl chloride with β-naphthol (in nitromethane) unionised halide was thought to be implicated, Satchell[108] has shown that in the case of acetyl bromide (in acetonitrile) there is strong evidence that reaction proceeds *via* free and ion-paired acetylium ions, *viz.*

$$\begin{array}{ccccc} RCOBr & \rightleftharpoons & RCO^+Br^- + HBr & \rightleftharpoons & RCO^+ + HBr_2^- \\ & & \downarrow \scriptstyle{ArOH,\ slow} & & \downarrow \scriptstyle{ArOH,\ slow} \\ & & \text{Products} & & \text{Products} \end{array} \tag{25}$$

There is no detectable positive catalysis by added bromide ion, but there is a reduction in rate on adding chloride ions.

The reactions of β-naphthol and 4-methoxyphenol with acetyl, propionyl, butyryl, β-chloropropionyl and chloracetyl chlorides in acetonitrile produce some striking kinetic results[109]. The behaviour of acetyl, propionyl and *n*-butyryl chlorides fit reasonably well into the pattern for acetyl chloride in nitromethane and acetyl bromide in acetonitrile. However, with chloracetyl chloride the mechanism is essentially a synchronous displacement of covalently bound chlorine by the phenol and this process is powerfully catalysed by added salt with bond breaking being kinetically dominant. When no added salt is present the rate of hydrolysis of chloracetyl chloride is *ca.* 8000 times *slower* than that of acetyl chloride. Although, normally, in second-order acylation reactions, substituents with the greatest electron demand have been found to have the fastest rates, the reverse is true in this system. Satchell proposes that a route such as

$$ArOH + RCOCl + M^+Cl^- \longrightarrow \left[\begin{array}{c} Ar \diagdown \overset{\delta+}{} \quad R \quad \delta- \\ O..\overset{|}{\underset{||}{C}}..Cl..M^+Cl^-\ (\text{or } M^+) \\ H \diagup \quad\ O \end{array} \right]^{\ddagger} \tag{26}$$

transition state

$$\downarrow$$

$$Ester + HCl + M^+Cl^-$$

is the main catalytic route for both chloracetyl and β-chloropropionyl chlorides. Such assistance by Cl^- is unimportant for the reactions of acetyl chloride and bromide where reactants with an already fully ionised halide ion are always available, but, in acetonitrile, the reaction of chloracetyl chloride is solely dependent upon this very slow synchronous displacement. In media where structure is available to help the leaving group, bond formation will become dominant for chloracetyl chloride and this compound will react more rapidly than acetyl chloride, *e.g.* in hydroxylic media. Strong evidence against the presence of an ionisation route for chloracetyl chloride is the fact that added tetraethylammonium bromide has about the same effect on the rate as has the chloride. The reaction of chloracetyl chloride must be synchronous as in reaction (26), and not modifications of the carbonyl addition mechanism *via* a tetrahedral intermediate, *i.e.*

$$ArOH + RCOCl \underset{(B)}{\overset{(A)}{\rightleftharpoons}} R\text{-}\underset{\displaystyle OAr}{\overset{\displaystyle OH}{C}}\text{-}Cl \xrightarrow[(C)]{M^+Cl^-} RCOOAr + HCl \qquad (27)$$

This follows, since, if step C is slow, step B must also be slow and thus the addition intermediate would accumulate. Satchell found no spectrophotometric or kinetic evidence for this with either chloracetyl or β-chloropropionyl chlorides. This conclusion is obviously of wide potential significance in the hydrolyses of acyl derivatives and even the observation of oxygen exchange during solvolysis (one of the main supports of the tetrahedral intermediate postulate) does not rule reaction (26) out, since the acylation could remain synchronous in part.

Of the benzoyl halides, benzoyl chloride has been extensively investigated with respect to all the mechanistic criteria above and will be discussed at length in the following sections. Benzoyl fluoride has been investigated by Bevan and Hudson[80] and Satchell[87], who found the hydrolysis to be acid-catalysed, unlike that of benzoyl chloride, and the rate to increase rapidly with water concentration, (Table 11). Lithium chloride causes no increase in the rate and no salt effect can be detected. Swain and Scott[56] found that the Cl/F rate ratio for benzoyl chloride and fluoride increased with decreasing amount of water, and, inasmuch as comparisons between reactions carried out in different media are valid, that the value for the Cl/F ratio was found to be far smaller than for the alkyl and acetyl halides (Table 10, p. 230), indicating that bond breaking of the C–X bond in the transition state of the hydrolysis of benzoyl halides is far less important.

It will be seen in the next section that benzoyl chloride in 50% aqueous acetone reacts by both unimolecular and bimolecular mechanisms. The solvolysis of benzoyl fluoride, in addition to being acid-catalysed is also base-catalysed and, towards hydroxide ion, benzoyl fluoride actually reacts faster than ben-

TABLE 11

SOLVOLYTIC AND CATALYTIC RATE COEFFICIENTS FOR THE HYDROLYSIS OF BENZOYL FLUORIDE AT 30°C[86]

H_2O concn. (vol. %)	$10^4\ k_0$ (sec^{-1})	$10^4\ k_1$ (sec^{-1})	$10^4\ k_H$ ($l \cdot mole^{-1} \cdot sec^{-1}$)
9.1	0.00567	0.0455	0.792
15	0.0227	0.402	0.95
30	0.155		
37.5	0.297		
40	0.363	0.480	2.3
50	0.792		
70	3.63	3.88	5

zoyl chloride (~40%). This is in accord with the idea that the substitution of fluorine for chlorine makes the carbon of the acyl group more electropositive. Benzoyl bromide hydrolyses faster under neutral conditions than either benzoyl chloride or fluoride (Table 12), the relative rates in 50% acetone/water being Br:Cl:F::147:1:39.1. Hence, in the benzoyl series, the effect on the rate of hydrolysis of changing the halogen atom is considerably less than in the acetyl series, reflecting the smaller extent to which bond breaking is involved in the transition state for the hydrolysis of the benzoyl halides*. Hence it would be expected that the characteristics of unimolecular hydrolysis would be less marked in the benzoyl halides than in the acetyl halides. The faster rate of benzoyl and acetyl bromide may be used as a test for mechanism using added bromide ion.

TABLE 12

RATE OF HYDROLYSIS OF BENZOYL FLUORIDE, CHLORIDE AND BROMIDE IN 50% (v/v) ACETONE/WATER

X	Temp. (°C)	k_1
F	0.5	1.1×10^{-5}
Br	0.5	6.3×10^{-2}
Cl	0.7	4.3×10^{-4}

* Note, however, that the comparison is with measurements for the benzoyl halides in a more aqueous medium than the acetyl halides and that this would favour a greater contribution from the unimolecular mechanism in the benzoyl halides.

The hydrolyses of benzoyl and 4-chlorobenzoyl cyanides in water have recently been investigated by Hibbert and Satchell[111,112] and have proved to be particularly interesting. Acyl cyanides undergo two kinds of hydrolysis

$$RCOCN + H_2O = RCO_2H + HCN \tag{28}$$
$$RCOCN + 2H_2O = RCOCO_2H + NH_3 \tag{29}$$

Stoichiometry (28) is followed under neutral or in alkaline aqueous conditions and (29) in concentrated mineral acids. In acid solution reaction (28) is powerfully inhibited and in the absence of general acids or bases the rate of hydrolysis is a function of pH. At pH > 5.0 the reaction is first-order in OH^- but below this value there is a region where the rate of hydrolysis is largely independent of pH followed by a region where the rate *falls* as $[H_3O^+]$ increases. The kinetic data at various temperatures both with pure water and buffer solutions, the solvent isotope effect and the rate increase of the 4-chloro derivative ($\sim$ 2-fold) are compatible with the interpretation of the hydrolysis in terms of two mechanisms. These are a dominant bimolecular reaction between hydroxide ion and acyl cyanide at pH > 5.0 and a dominant water reaction at lower pH, the latter susceptible to general base catalysis and inhibition by acids. The data at pH < 5.0 can be rationalised by a carbonyl addition intermediate and are compatible with a two-step, but not one-step, cyclic mechanism for hydration. Benzoyl cyanide is more reactive towards water than benzoyl fluoride, but less reactive than benzoyl chloride and anhydride, an unexpected result since HCN has a smaller dissociation constant than HF or RCO_2H. There are no grounds, however, to suspect that an ionisation mechanism is involved.

These authors have also investigated the hydrolysis of acetyl and propionyl cyanides, the former reacting more rapidly than the latter. These two compounds undergo hydrolysis in neutral or alkaline conditions too rapidly to permit accurate measurements by the technique used (recording spectrophotometer). However, since the hydrolysis is inhibited by acids, rates were measured in perchloric acid concentrations of 1–2 *M* and above. The rate of hydrolysis passes through a minimum at $\sim$ 8 *M* perchloric acid concentration. A possible mechanism to explain such inhibition (which is a rare phenomenon) is

$$RCOCN + 2H_2O \underset{k_{-1}}{\overset{k_1}{\rightleftharpoons}} R\text{-}\overset{\displaystyle O^-}{\underset{\displaystyle OH}{C}}\text{-}CN + H_3O^+$$

$$RCOCN + H_2O + A^- \underset{k_{-3}}{\overset{k_3}{\rightleftharpoons}} R\text{-}\overset{\displaystyle O^-}{\underset{\displaystyle OH}{C}}\text{-}CN + HA$$

$$\text{R-}\underset{\text{OH}}{\overset{\text{O}^-}{\text{C}}}\text{-CN} \xrightarrow{k_2} \text{RCO}_2\text{H} + \text{CN}^- \quad (30)$$

This scheme predicts general base catalysis and general acid inhibition since the step controlled by k_1 is the base catalysis of one water molecule by another and there is no reason to suppose that bases other than water should not replace the catalytic water molecule in this step (*i.e.* k_3). From the principle of microscopic reversibility it follows that the general acid HA will inhibit the reaction as does H_3O^+ (*i.e.* k_{-3} and k_{-1}).

2.1.2 Variation in R

(a) R = aliphatic or substituted aliphatic

Early measurements established[79,80,113] that electron-withdrawing substituents increased the rate of hydrolysis, alcoholysis or aminolysis[114] of the acid chloride RCOCl, and that in the aliphatic series the rate decreased with

TABLE 13

RATES OF HYDROLYSIS AND ALCOHOLYSIS FOR RCOX

R	*Solvent, 10.9% v/v H_2O^{100} in acetone (−20°C) [RCOX] ≃ 4.1 × 10^{-2} mole·l^{-1}* k_R/k_{CH_3}	σ^*	*Alcohol/ ether[113] (25°C)*	*1 N H_2O in dioxan[115] (25°C)*
CH_3	1[a]	0.0	1[b]	1[c]
CH_3CH_2	0.693	−0.10		1.1
$CH_3(CH_2)_2$	0.537	−0.115		
$(CH_3)_3C$	0.0660	−0.165		0.247
$(CH_3)_2CH$				0.92
$CH_3(CH_2)_4$	0.496			
$CH_3(CH_2)_8$	0.404			
CH_2=CH-CH_2	0.406			
CH_3-CH=CH	0.36	+0.034		
$C_6H_5CH_2$	0.330	+0.21		
$ClCH_2$	18.7	+1.05	24	
Cl_2CH	2,850	+1.94	>50	
Cl_3C	> 10,000	+2.65	>50	
HO_2C	> 10,000			
Cl	927			
$C_5H_{10}N$	0.00019			
$((CH_3)_2CH)_2N$	0.0077			
CH_3O	0.00011	+1.45		

[a] $k_{CH_3COCl} = 10.9 \times 10^{-4}$ sec^{-1}.
[b] $k_{CH_3COCl} = 164 \times 10^{-4}$ sec^{-1}.
[c] $k_{CH_3COCl} = 8.9 \times 10^{-4}$ sec^{-1}.

chain length and steric bulk of R. Table 13 shows a compilation of data for this class of compounds. The substituents Cl_2CH, $ClCH_2$. CH_3, C_2H_5, and *n*-Pr are colinear on a log k_{rel} *versus* σ^* plot, of slope 1.25. We will return to this point later, but it seems unlikely that different mechanisms are in operation in the chloroacetyl and acetyl chlorides. This observation is in agreement with the later observation of Satchell[94] that, although the hydrolysis of chloroacetyl chloride, in acetonitrile, proceeds *via* a synchronous displacement at a far slower rate than acetyl chloride, in ethanolic solvents the rate would increase as the mechanism involved more ionisation. However, the very slow rates of the carbamic acid chlorides and methoxy acetic acid chlorides do indicate a different mechanism.

Acetyl chloride has been studied in some detail by Zimmerman and Yuan[116] using 5% water in acetone, Cairns and Prousnitz[102] using 15–35% v/v of water/acetone and Hudson and Moss[90] using 20–75% water/dioxan. Zimmerman and Yuan found a second-order dependence on water in the range studied and Archer and Hudson[117] indicated that the hydrolysis in dioxan also appeared to be third-order. The significance of the determination of "order" in solvolytic reactions has been investigated by Hudson *et al.* and these results will be discussed in a later section. The rate of hydrolysis of acetyl chloride is strongly increased by the addition of aniline: $k_2(PhNH_2)/k_2(H_2O) = 1{,}930$, *cf.* 96,000 for benzoyl chloride in 50% v/v aqueous dioxan). Swain and Scott[56] also found that the addition of hydroxide ion increased the rate of hydrolysis to 10^6 times that under neutral conditions. These same authors and Ugi and Beck[100] showed that the rate of hydrolysis substantially increased by the addition of amines (Table 14) for acetyl, pivaloyl and benzoyl chlorides, but that this has no effect on the rate of hydrolysis of diethyl carbamate, suggesting two types of mechanism with bond-making being more important than bond-breaking in the mechanism for acetyl chloride in hydroxylic solvents. Moreover, ^{18}O exchange of the carbonyl group increases with water in the 15–33%

TABLE 14

REACTION OF ACID CHLORIDES (4×10^{-2} mole·l^{-1}) WITH AMINES (0.1 *M*) IN WATER/ACETONE (10.9 vol % H_2O)[100]

R	*Temp.* (°C)	$k_{H_2O}\times(sec^{-1})$	$k_{amine}\times10^4\ (sec^{-1})$ *n-Butyl*	*Trimethyl*	*Triethyl*	*Pyridine*
CH_3	−40	0.975	$>10^5$	*ca.* 35,000	7000	$>10^5$
$(CH_3)_3C$	−40	0.027	$>10^5$	770	1.2	6600
$C_5H_{10}N$	+20	0.35	0.35			0.38
$[(CH_3)_2CH_2]_2N$	+20	13.0	13.0			12.0
C_6H_5	−40	0.0047	$>10^5$	$>10^5$	*ca.* 35,000	$>10^5$
2,4,6-$(CH_3)_3C_6H_2$	−40	1.22	14	1.2	1.2	1.79
2,4,6-$(CH_3)_3$-3,5$(CO_2)_2C_6$	+20	0.056	*ca.* 2000	25	7	93

water concentration range in dioxan[24] suggesting that ionisation follows initial hydration, *viz.* mechanism (*d*)[118] (see also p. 227)

$$\text{R-C}\begin{matrix}\nearrow\text{O}\\ \searrow\text{X}\end{matrix} + \text{H}_2\text{O} \xrightleftharpoons{\text{fast}} \text{R-}\overset{\text{OH}}{\underset{\text{X}}{\text{C}}}\text{-OH} \xrightleftharpoons{\text{slow}} \text{X}^- + \text{R-}\overset{+}{\text{C}}\begin{matrix}\nearrow\text{OH}\\ \searrow\text{OH}\end{matrix} \quad (31)$$

This kind of mechanism was first proposed by Berger and Olivier[119] and is open to criticism on the basis that in all ester and amide hydrolyses studied by ^{18}O exchange, the addition of water proceeds as the slow step followed by fast breakdown of the intermediate. However, in these compounds the intermediate cannot ionise as postulated in this mechanism and donation from O or N might well cause the addition of water to be the slow step. Mechanisms such as (*c*), p. 227, in which the rate-determining step is a direct rate-determining ionisation, have been postulated in view of the similar reactivities of benzoyl chloride in formic acid and 65% v/v aqueous acetone (solvents of equal ionising power). This observation would not support mechanism (*d*) in the case of acetyl chloride, but it has not yet been performed with acetyl chloride. Hudson and Moss[90] found that the Grunwald–Winstein equation held ($m = 0.81$) for a wide range of water concentration suggesting that a single mechanism is in operation with a transition state similar to that for benzoyl chloride in more aqueous solvents. Gold and Hilton[120], investigating the hydrolysis in pure water by the thermal maximum method of Bell and Clunie[121], proposed that their results were characteristic of a bimolecular mechanism. They arrived at this conclusion by means of some approximations and comparison with a second-order reaction of known mechanism, and were supported by the work of Cairns and Prousnitz[102] who found that the entropy of activation varied strongly with solvent dielectric constant, from −8.18 eu at $D = 26.3$ to −0.36 at $D = 46.4$. These later workers analysed their kinetic data in terms of the dielectric constant of the reaction medium and not in terms of log k_1 *versus* $[H_2O]$ plots. However, as pointed out by Hudson and Moss[90], the fact that rates of hydrolysis in dioxan are 3–5 times faster than in acetone ($D_{\text{dioxan}} = 2.21$, $D_{\text{acetone}} = 20.5$) means that specific solvation of the transition state by water molecules must be involved.

On balance, therefore, the hydrolysis of acetyl chloride proceeds in non-hydroxylic solvents by a mechanism which is a mixture of ionisation (*c*) and some direct displacement, possibly (*b*), and in hydroxylic solvents by an ionisation after the addition of water (*d*), possibly approximating to (*c*) in highly aqueous or hydroxylic media, although the apparent change in mechanism in the latter media may be a result of the change from specific to general solvation. This latter effect has been investigated in some detail by Hudson *et al.*[63,64].

(*b*) *Hydrolysis and alcoholysis of aromatic halides* ($R = Ar$)

The mechanism of hydrolysis of benzoyl chloride has been the subject of research for nearly half a century. In aqueous and partly aqueous media, the hydrolysis is not acid-catalysed[110,120,122–124] (indeed it may be slightly retarded at high acid concentrations[120]). However, the rate of hydrolysis in all solvents is increased by the addition of hydroxide ion *e.g.* in 50% acetone/water $k_2/k_1 = 20$. This observation is certainly not common to all benzoyl halides and there are wide variations in reactivity in differently substituted benzoyl halides and in different solvents.

The effect of substitution in the benzene ring on reactivity of the acyl chloride group was investigated by a series of workers, using hydrolysis and alcoholysis. Data from these workers are compiled in Table 15 for mono-substituted benzoyl chlorides.

The hydrolyses of substituted benzoyl chlorides are obviously effected by the nature of the substituents, but the wide variety of values of the relative rates, sometimes changing in relative magnitude, posed a difficult problem of interpretation to the early workers in the field. The data of Table 15, together with more data on the relative rates of polysubstituted benzoyl chlorides (*vide infra*) are now largely understood in terms of substituent effect by use of the Hammet σ values. Some of the earliest data, that of Berger and Olivier[119], is the most revealing, especially when data from the later work of Hudson and Wardhill[110] are added. Plotting log k_{rel} *versus* σ produces a smooth curve with a minimum at about $\sigma = 0.2$. 4-NO_2, 3-NO_2-3-Br, 3-Cl, mark the ascending curve, 4-CH_3O, 4-CH_3, 3-CH_3, H, the descending curve, with 4-Br and 4-Cl close to the minimum. This data, with 50% acetone/water solvent, clearly indicate a change of mechanism caused by the perturbation of the system by the substituent. 4-Methoxybenzoyl chloride reacts by a mechanism which is favoured by electron donation in the transition state, whereas the transition state for 4-nitrobenzoyl chloride is favoured by electron withdrawal. If these mechanisms are labelled ed and ew, the data can be used to classify the effects of the substituents in the different solvents depending on the slope of the line of the Hammett plot on which they lie (ed−)(ew+) (Table 16) *viz.* whether Hammett ρ is positive or negative.

Now, "S_N2" mechanisms such as (*a*) and (*b*), p. 227, would be favoured by electron withdrawal, making bond making more important and the attack of a nucleophile more likely. Unimolecular mechanisms will be favoured by electron donation and a very powerful electron-donating substituent should favour a rate-determining ionisation of type (*c*) rather than an initial pre-equilibrium of type (*d*). The shift of mechanism with increased polarity of the medium reflects the more polar transition state favoured by such a change.

Benzoyl chloride itself lies close to the minimum of the Hammett plot for 50% aqueous acetone, and is therefore of indeterminate mechanism. Ash-

TABLE 15

RELATIVE RATE COEFFICIENTS FOR THE HYDROLYSIS AND ALCOHOLYSIS OF SUBSTITUTED BENZOYL CHLORIDES

	R															
	2-NO₂	*4-NO₂*	*3-NO₂*	*2-Br*	*3-Br*	*2-Cl*	*3-Cl*	*2-I*	*3-I*	*4-I*	*4-Br*	*4-Cl*	*H*	*2-Me*	*3-Me*	*4-Me*
σ		0.778	0.71		0.391		0.373		0.35	0.28	0.23	0.23	0		−0.07	−0.17
H_2O/acetone[a,122b]	1.8	11.5	6.92	1.4	1.69	1.5	1.46				0.92	0.845	1		1.5	2.9
EtOH[b,79]	2.2	21.6	20.5	3.4	4.6	3.5	5.6	2.9	4.1	1.9	2.1	1.9	1	3.7	0.85	0.78
i-PrOH[c,80]	1.9	10.5								1.5	1.8	1.5	1			0.64
50% H_2O/acetone (v/v)[d,110]											0.91		1			3.28
5% H_2O/acetone (v/v)[e,126]		35									3.2		1			0.55
5% H_2O/dioxan (v/v)[f,126]													1			4.74*
65% H_2O/acetone (v/v)[g,80]		0.33									0.25		1			3.14
Formic acid[h,80]		0.0003									0.07		1			5.44
40% EtOH/ether[i,113]		19.0								1.9	0.91	1.9	1			0.60
10.9% H_2O/acetone[j,100]	0.51	56											1	1.54		
60% EtOH/benzene[k,128]		1.37			1.19								1		1.0	
EtOH[l,125]													1			

TABLE 15 (CONTINUED)

	R												
	2-OMe	*3-OMe*	*4-OMe*	*4-Ph*	*2-PhCO*	*3-PhCO*	*4-PhCO*	*3-CH_2Cl*	*4-CH_2Cl*	*3-$PhCH_2$*	*4-$PhCH_2$*	*2-Et*	*2-OEt*
σ		0.11	−0.27				0.459		0.184				
H_2O/acetone[a,122b]													
EtOH[b,79]	29	1.1	0.81										
i-PrOH[c,80]													
50% H_2O/acetone (v/v)[d,110]													
5% H_2O/acetone (v/v)[e,126]			0.6										
5% H_2O/dioxan (v/v)[f,126]			3.62										
65% H_2O/acetone (v/v)[g,80]													
Formic acid[h,80]													
40% EtOH/ether[i,113]			0.49										
10.9% H_2O/acetone[j,100]			0.35										
60% EtOH/benzene[k,128]			1.0										
EtOH[l,125]					v. fast	4.35	4.31	1.64	1.30	1.04	0.75	2.73	41.6

Temperatures: [a,b,d**,b] 0°C; [c,l,e,f] 25°C; [g,h] 9°C; [j] −20°C.
Kinetic method: [b,c,e,f,g,h,i,j,l] conductivity; [a] titration of chloride ions; [d] electrical; [k] both titration and conductivity – the same substituents were studied in benzene/ethanol, acetone/ethanol, and ether/ethanol at 0, 20, 40, 60 and 90 vol.%.

* Compare the results of Bunton and Lewis[127], who found $k_{4\text{-Me}}/k_H = 1.53$ with 66.7% water/dioxan (v/v) at 25°C.

** Temperature varied between 0.5 and 0.7°C.

TABLE 16

CLASSIFICATION OF SUBSTITUTED BENZOYL CHLORIDES ACCORDING TO THE PREDOMINANT MECHANISM OF HYDROLYSIS

<table>
<tr><th></th><th colspan="10">Substituents</th></tr>
<tr><th>Solvent</th><th>4-MeO</th><th>4-Me</th><th>3-Me</th><th>H</th><th>4-Cl</th><th>4-Br</th><th>3-Cl</th><th>3-Br</th><th>$3-NO_2$</th><th>$4-NO_2$</th></tr>
<tr><td>40%EtOH/60%Et_2O</td><td colspan="10">|—— ew+ ($\rho = 1.9$) ——|</td></tr>
<tr><td>EtOH</td><td>ed−</td><td colspan="9">|—— ew+ ——|</td></tr>
<tr><td>5%H_2O/95%Me_2CO</td><td>ed−</td><td colspan="9">|—— ew+ ——|</td></tr>
<tr><td>50%H_2O/50%Me_2CO</td><td colspan="4">|—— ed− ——|</td><td></td><td></td><td colspan="4">|—— ew+ ——|</td></tr>
<tr><td>1% water in formic acid</td><td colspan="10">|—— ed− ($\rho = -4.4$) ——|</td></tr>
</table>

down[129] measured the rate of alcoholysis of 4-nitrobenzoyl chloride in anhydrous ether with various alcohols and found the reaction to be apparent second-order in alcohol. These conditions favour a bimolecular mechanism for alcoholysis, and, as will be seen later, a value of 2 for the apparent order of the reaction with respect to solvent is a reasonable indication that the reaction is of the S_N2 type. (Values of up to 5 or 6 indicate a mechanism of the S_N1 type.) Gueskens *et al.*[130] have found polycyclic aromatic acid chlorides to react by the S_N2 mechanism in 50% (w/w) acetone/methanol.

Since benzoyl chloride reacts by a mechanism which is either a combination of S_N1/S_N2 (unimolecular/bimolecular) or some intermediate mechanism, a great deal of work has been directed towards elucidation of the factors which control its reactivity. In the context of the 3 and 4-substituted benzoyl halides, it is seen from the evidence of Hammett plots to react predominantly by a bimolecular mechanism in solvents of low polarity and by a unimolecular mechanism in solvents of high polarity.

Bohme and Schürhoff[131] postulated a change in mechanism on increasing the amount of water in dioxan, tetrahydrofuran or glycol dimethyl ether solvent, from breaks in the log k *versus* $[H_2O]$ plots. The increase in rate with increasing ionising power of the solvent clearly indicates a polar transition state. Peeling[132] showed that benzoyl halides exhibit a salt effect of comparable size to that of the benzyl halides. Hudson and Wardhill[110] showed that the rate of hydrolysis of benzoyl chloride in 5–30% water/acetone was proportional to the square of the water concentration (but at 30–75% the order is much higher, 4.0–~7.5)[128,133], and that the activation energy had a positive temperature coefficient. They expressed this behaviour in terms of the equation

$$dx/dt = k'[RCl][A][S]^m + k''[RCl][A]^n \quad (32)$$
$$k_3 = k'[S]^m + k''[A]^{n-1} \quad (33)$$

where S = inert solvent and A = substituting agent. n was found to equal 2;

assumption of $m = 1$ (*cf.* third-order) allows the linearity of k_3 with $[H_2O]$ at 0–30% water to be seen. In an inert solvent

$$\text{Rate} = k''[\text{RCl}][\text{A}]^2 \qquad (34)$$

Archer and Hudson[117] also deduced a change in mechanism from the marked increase in slope of the plot of the Kirkwood equation[134], a deduction later questioned by Gold *et al.*[135].

With 5–33.3 vol.% water/acetone mixtures, it is found[136] that common-ion salts have no effect on the rate of hydrolysis of benzoyl chloride whereas the rate in 15% (but not 33.3%) water is increased by the addition of neutral salts such as lithium bromide or potassium nitrate. The increase in ionic strength on the addition of neutral salts is not the major reason for the increase in rate and nucleophilic catalysis *via* the more easily hydrolysed benzoyl bromide was postulated.

Hydroxyl ions increase the rate of S_N2 reactions but have little or no effect on S_N1 reactions. The k_{OH^-}/k_{H_2O} ratio for benzoyl chloride in 50% water/acetone is about 600, as compared with values of 10^4–10^5 for S_N2 reactions and values of 10^2–10^3 for S_N1 reactions. In 25% aqueous acetone k_{OH^-}/k_{H_2O} is about 10^4 for both 4-nitro and 4-methoxybenzoyl chloride[133], but increasing the water content makes (k_{OH^-}/k_{H_2O})4-MeO $\ll$ (k_{OH^-}/k_{H_2O})4-NO_2, in agreement with the S_N1–S_N2 character for the mechanism with these two acyl halides. However, as pointed out by Bender and Chen[84], comparisons of this kind are valid only if the neutral and base-catalysed reactions are identical in mechanism. Indeed, the fact that k_{OH^-}/k_{H_2O} ratios have no fundamental significance was pointed out by Brown and Hudson[133]. However, they were used by Hudson and by others[127,137] to postulate mechanisms for the hydrolysis of polysubstituted benzoyl chlorides. Brown and Hudson found that hydroxide ion had no effect on the rate of hydrolysis of 2,4,6-trimethylbenzoyl chloride in 95% aqueous dioxan; a result contradicted by Bunton and Lewis[127] (using 5% aqueous dioxan) and Peeling[137] (using 80% alcoholic acetone), but which, however, is reasonable when the changes in solvent are considered, and generally in agreement with other mechanistic criteria in the work of Hudson. Bunton *et al.*[24] showed that there was negligible exchange of ^{18}O from the solvent to benzoyl chloride at 5% H_2O (v/v), but, at 25% H_2O (v/v) dioxan, k_h/k_e (p. 228) = 25. Also, these same authors found the rate of ^{36}Cl exchange between $Li^{36}Cl$ and benzoyl chloride in dry acetone to be much faster than with mesitoyl chloride. Exchange *via* an S_N1 mechanism would be expected to be faster than for an S_N2 mechanism, implying an S_N1 mechanism for benzoyl chloride in this exchange. In view of the solvent, this expectation must be wrong in the case of these two chlorides. For *p*-toluyl chloride no exchange was seen in either of these media and in 33% H_2O (v/v) $k_h/k_e = 51$. Bunton *et al.* therefore postulated a two-stage mechanism for the hydrolysis of benzoyl chloride and *p*-toluyl chloride in more aqueous media and for mesitoyl chloride in 5% water/acetone ($k_h/k_e = 31$).

Gold *et al.*[135] examined the influence of added aromatic amine on the hydrolysis of benzoyl chloride and found that in 20% (v/v) water/acetone the reaction was second-order but that in 50% water/acetone approximately half of the total reaction proceeds *via* a rate-determining process involving only benzoyl chloride. This was deduced on the basis of the excess reaction with amine at high water concentrations over that predicted from the behaviour at lower water concentrations. Kivinen[128] has studied and reviewed the ethanolysis of benzoyl chloride in a variety of solvents and found the tendency to proceed *via* an S_N2 mechanism to increase with substituents in the order, *p*-CH_3O < *m*-CH_3 ~ H ~ *m*-CH_3O < *m*-Br < *m*-NO_2 ~ *p*-NO_2, and the NO_2-substituted benzoyl chlorides to react *via* a pure S_N2 mechanism in pure alcohol. Ugi and Beck[100] showed that the rate of reaction of benzoyl chloride was accelerated by up to 10^8 times by amines, a far greater acceleration than that for mesitoyl chloride, and two years later Entelis and Nesterov[46] reported the detection of an intermediate in the reaction of benzoyl chloride with piperidine and piperazine in *n*-heptane (see p. 218). The rate coefficients for the reaction of benzoyl chloride with these nucleophiles are shown in Table 17.

TABLE 17

RATE COEFFICIENTS FOR THE REACTION $A+B \xrightarrow{k_1} AB \xrightarrow{k_2} C$ WHERE A = BENZOYL CHLORIDE, B = PIPERIDINE OR PIPERAZINE[46]

	k_2 (sec^{-1})	True second-order coefficient for step (1) ($l \cdot mole^{-1} \cdot sec^{-1}$)
Piperidine	6.8×10^{-2}	9.5×10^4
Piperazine	3.5×10^{-3}	5.6×10^4

In contrast to this demonstration of bimolecularity, Hall and Lueck[82] showed the possibility of acylium ion formation from benzoyl chloride by its reaction with mercuric perchlorate. In common with dimethylcarbamyl chloride, dimethylsulphamyl and tetramethyldiamidophosphochloridate, benzoyl chloride reacted readily to form the corresponding acylium ion; *n*-butyl chloroformate however was inert. Kivinen[138] studied the effect of mercuric chloride on the ethanolysis of 4-methoxybenzoyl chloride, benzoyl chloride and 4-nitrobenzoyl chloride and obtained the following approximate relative rates for the effect of mercuric chloride (0.30 *M*) in ethanol, 4-MeO, 2.91; 4-H, 1.00; 4-NO_2, 1.03, confirming the S_N2 character of the 4-nitrobenzoyl chloride ethanolysis. This interpretation was further reinforced by later work of Kivinen[139], who interpreted the activation volumes for the ethanolysis of anisoyl chloride and benzoyl chloride to be those for an S_N1 mechanism and an S_N2–S_N1 mixture respectively. Activation volumes for the solvolyses of some acyl and alkyl halides are given in Table 18.

TABLE 18

ACTIVATION VOLUMES (ΔV^*) FOR THE SOLVOLYSIS OF ALKYL AND ACYL HALIDES, RCl

R	*Solvent and temp. (°C)*	ΔV^* ($cm^3 \cdot mole^{-1}$)	*Ref.*
C_6H_5CO	EtOH, 0	−29.1	139
	THF† (2.2% H_2O/w), 20	−33.1	140
4-$MeOC_6H_4CO$	EtOH, 0	−20.1	139
	THF† (2.2% H_2O by wt), 20	−27.5	140
$(CH_3CO)_2O$‡	EtOH, 20	−20	141
C_2H_5	MeOH, 20	−22	142
t-Bu	H_2O/EtOH (80/20), 30	−22.2	143
$C_2H_5O{\cdot}CO$	H_2O, 0	−11.7	139

† THF represents tetrahydrofuran.
‡ For comparison.

Whereas the mechanism of benzoyl and the 3- and 4-substituted benzoyl chlorides appears to conform to a change from S_N1 to S_N2 or unimolecular to bimolecular mechanism, the effect of 2 substituents and of polysubstituents presents a continuing problem.

Relative rate coefficients for the solvolysis of polysubstituted benzoyl chlorides are collected in Table 19. Norris and Ware[125] found that 2:6 and 2:4:6 substituents affect the rate in two distinct ways. Halides with electron-donating groups (Me, MeO) reacted so fast in pure ethanol that the rate of ethanolysis could not be measured. Later measurements by Bender and Chen[84], Bunton and Lewis[127], Brown and Hudson[133] and Ugi and Beck[100] showed that this acceleration was greatest in acetonitrile (3,420 fold) and decreased as the solvent changed from aqueous acetone (10.9%) to aqueous dioxan (5%) for 2:4:6 trimethylbenzoyl chloride. On the other hand, electron-withdrawing groups decrease the rate considerably, in contrast to their opposite effect in the 4 position in low dielectric solvents. The differences in rate are far greater than those in the 3:4-substituted benzoyl chlorides and in the case of the 2:6 dimethyl and 2:4:6 trimethyl methoxy the faster rate of solvolysis suggests an ionisation mechanism.

Brown and Hudson[133] found that addition of hydroxyl ions, (0.0018 *M*) to 2:4:6 trimethylbenzoyl chloride in 45% aqueous acetone has no effect on the rate of hydrolysis. Bunton and Lewis[127], on the other hand, found that hydroxide

TABLE 19

RELATIVE RATE COEFFICIENTS FOR THE SOLVOLYSIS OF POLYSUBSTITUTED BENZOYL CHLORIDES

	k^*_{rel}					
Ref.	*Ref. 79*	*Ref. 125*[c]	*Ref. 84*[d]	*Ref. 127*[e]	*Ref. 133*[f]	*Ref. 100*[g]
2,4,6-tri-Me	F[a]		3.42×10^3	4.4	33.1	48
2,4,6-tri-Br	6.98×10^{-3}[b]	S				
2,4,6-tri-OMe				11.0		
2,4,6-tri-NO_2	S[b]				0.187	1.7
3,5-di-NO_2	S[b]					0.38
2,4,6-tri-Me						
2,4,6-tri-Et		F				
2,4,6-tri-Cl		S				
2,4-di-Me		6.8	3.2×10^2			
2,6-di-Me		F				31.4
2,6-di-MeO		F		4.36		23.2
2,4-di-Cl		7.7				
2,5-di-Cl		14.1				
3,4-di-Cl		7.2				
2,4-di-NO_2		31.8				
3,5-di-NO_2		F				145
2,6-di-NO_2		S				
4-MeO-2,6-di-Me			9.7×10^4			2.68×10^4
4-Br-2,6-di-Me			46			
4-NO_2-2,6-di-Me			7.4×10^{-2}			0.288

* $k_{C_6H_5COCl} = 1$. F = too fast to measure; S = too slow to measure.
[a] MeOH, 0°C.
[b] MeOH, 25°C.
[c] EtOH, 0°C.
[d] 99% acetonitrile/water.
[e] 95% dioxan/water.
[f] 10% water/acetone.
[g] 10.9% water/acetone.

ions (1 *M*) caused a thousand-fold increase in the rate of hydrolysis of 2:4:6 trimethylbenzoyl chloride in 45% dioxan/water. Peeling[137], using 80% acetone/ethanol, concluded from his own data that the hydrolysis of this compound is not an S_N1 reaction. The apparent contradiction can be reconciled by consideration of the three different solvents used.

2:4:6 Trimethylbenzoyl chloride hydrolyses in 45% dioxan/water by an S_N2-type mechanism. This is supported by the results of Bunton and Lewis[127] and also by previous data reported by Brown and Hudson[133], according to which the relative rate coefficients for hydrolysis of 4-methoxy and 4-methylbenzoyl chlorides are 0.362 and 0.472 ($k_{C_6H_5COCl} = 1$). The order of reaction rates 4-CH_3O < 4-CH_3 < H suggest an S_N2-type mechanism for 2:4:6 trimethylbenzoyl chloride.

In 95% acetone/water the relative rates of hydrolysis are 0.545 (4-MeO), 0.493 (4-Me) and 1.00 (H), indicating that 4-MeO benzoyl chloride is tending to react partly by an S_N1 mechanism. In 95% acetone/water, 2:4:6 trimethylbenzoyl chloride therefore reacts by an S_N1-type mechanism and would not be expected to show an effect with added hydroxide ion. In 80% (v/v) acetone/ethanol, the reaction is again S_N2-type.

In accord with this assignment is the observation of Hudson and Moss[144] that, with 95% aqueous acetone solvent, lithium perchlorate increased the rate of hydrolysis of 2,4,6-trimethylbenzoyl chloride, whereas chloride ions have no effect; this is in contrast to the behaviour of 4-nitrobenzoyl chloride whose rate of hydrolysis is decreased by lithium perchlorate and increased by chloride ions. The influence of solvent sorting[145] is probably small for S_N1 reactions in 90% aqueous acetone and neglecting ion pairing an observed salt effect reflects the effect of the ion atmosphere on the transition state. The same observation will not hold for either dioxan/water (because of the low dielectric constant of dioxan) and acetone/water of high water content (because of the extensive solvent sorting).

Bender and Chen[84] studied the hydrolysis of 2,4,6-trimethylbenzoyl chloride in 99% acetonitrile/water and found it to be subject to hydronium-ion catalysis and hydroxide-ion catalysis and to be retarded by added chloride ions, although other salts have little effect. Electron-donating 4-substituents (keeping the 2:6 methyl groups) accelerate the neutral and acid-catalysed hydrolyses but depress the base-catalysed hydrolysis. The Hammett ρ constants for the neutral, acid-catalysed and base-catalysed reactions are -3.85, -3.75 and $+1.20$. In neutral and acid-catalysed hydrolysis no ^{18}O exchange was found. This kind of evidence fits a mechanism in which there are two distinct mechanistic routes to hydrolysis, a unimolecular mechanism in neutral or acidic media and a bimolecular mechanism in basic media where a hydroxide ion adds to form a tetrahedral intermediate. The steric hindrance of the methyl groups could reduce the conjunction between the acetyl group and the aromatic ring. This would make the group more reactive in S_N2-type reactions, neglecting, for the present, the steric hindrance to tetrahedral intermediate formation, and more reactive also in S_N1-type reactions, since formation of the acylium ion would be favoured by steric release. In this context it is interesting to note that the sterically hindered acid chloride α-butyl-α-ethyl-γ-carbomethoxy-butyrylchloride is very sensitive to traces of moisture whereas the unhindered acid chloride γ-butyl-γ-ethyl-γ-carbomethoxy-butyryl chloride is not[146].

The ^{18}O exchange results of Bender and Chen[84] using 99% CH_3CN/H_2O contrast with those of Bunton *et al.*[24] who observed oxygen exchange in 95% dioxan/water for this compound and benzoyl and *p*-toluyl chloride. Acetonitrile/water (1%) is probably more similar to acetone/water (5%) than is dioxan/water (5%) or 80% v/v acetone/ethanol, a supposition supported by the

References pp. 287–293

fact that the ρ value for the hydrolysis of the 4-substituted 2,6-dimethylbenzoyl chlorides in 89.1% acetone/water studied by Ugi and Beck[100] (−3.88) is close to that for acetonitrile/water (1%) ($\rho = -3.85$).

Certain data, apart from that already discussed, argue against an S_N1-type mechanism for the hydrolysis of mesitoyl chloride, and by inference, benzoyl chloride. Benzoyl chloride is reported as reactive to diazomethane[147] but mesitoyl chloride is not. The formation of *t*-butyl mesitoate from mesitoyl chloride is known to be catalysed by pyridine[148], which it should not be if it proceeds *via* an acylium ion. Conductance measurements on mesitoyl chloride indicated no appreciable ionisation[149]. Minato[93] has used this evidence and the evidence of ^{18}O exchange and solvents effects to propose a general mechanism for the solvolysis of acyl chlorides, *via* a tetrahedral intermediate, *viz.*

$$\mathrm{R{-}\overset{\displaystyle O}{\overset{\|}{C}}{-}Cl} \underset{k_{-1}\ \text{fast}}{\overset{\text{slow}\ k_1[\mathrm{R'OH}]}{\rightleftharpoons}} \mathrm{R{-}\underset{O^-}{\underset{|}{\overset{\overset{H\ \ R'}{\backslash +/}}{\overset{\displaystyle O}{\overset{|}{C}}}}}{-}Cl} \underset{k_{-2}[\mathrm{R'OH}]}{\overset{k_2[\mathrm{R'OH}]}{\rightleftharpoons}} \mathrm{R{-}\underset{OH}{\underset{|}{\overset{OR'}{\overset{|}{C}}}}{-}Cl} \xrightarrow{k_3} \mathrm{R{-}\underset{OH}{\underset{|}{\overset{OR'}{\overset{|}{C^+}}}}+Cl^-} \tag{35}$$

whence

$$-\frac{\mathrm{d[RCOCl]}}{\mathrm{d}t} = \frac{k_1k_2k_3[\mathrm{RCOCl}][\mathrm{R'OH}]^2}{k_{-1}k_3 + k_{-1}k_{-2}[\mathrm{R'OH}] + k_2k_3[\mathrm{R'OH}]} \tag{36}$$

This mechanism explains much of the experimental evidence obtained from studies of the solvolysis of acyl chlorides, but it may not be in agreement (as was pointed out to Minato by a referee) with the linear relationship between σ^+ and $\log k/k_0$ for the hydrolysis of 4-substituted-2,6-dimethylbenzoyl chlorides[84]. Moreover, the common ion effect and electrophilic catalysis observed in the solvolysis of certain acid chlorides would possibly be explained by a simpler mechanism such as the S_N1 or hydration–ionisation mechanism. However, it is of interest to see how the mechanism applies to acetyl, benzoyl and mesitoyl chlorides. For acetyl chloride, k_1 and k_{-1} would be very large and the rate would approximate to

$$-\frac{\mathrm{d[RCOCl]}}{\mathrm{d}t} = \frac{k_1k_2}{k_{-1}}[\mathrm{RCOCl}][\mathrm{R'OH}]^2 \tag{37}$$

The *apparent* order in water is found to be 2 for solvent 1–5% water/acetone[116], 1.4 for 1–4% ethanol/carbon tetrachloride[144] and 2.0 for 1–40% ethanol/ether[144]. This, however, may be a result of specific solvation. As pointed out by Kivinen[92], Minato partly based his mechanism on different rates of solvolysis of acid chlorides in diethyl ether/ethanol and carbon tetrachloride/ethanol mixtures. This cannot be used as a criterion for detailed mechanism since, for example, Salomaa[150] has found that the apparent order with respect to ethanol of the rate of alcoholysis of methyl chloromethyl ether (an S_N1 reaction) in carbon tetrachloride/ethanol mixtures is 1.5–2.0, and that this apparent order is largely independent of mechanism in carbon tetrachloride/ethanol and also benzene/ethanol mixtures. Conversely, therefore, although using a mechanism such as Minato's leads to a prediction of the apparent order of reaction, this cannot be regarded as proving the mechanism.

The mechanism proposed by Minato is on no firmer ground when used to interpret the data for benzoyl and mesitoyl chlorides. The ^{18}O exchange for these compounds implies $k_3 \gg k_{-2}[\mathrm{R'OH}]$ since $k_h \gg k_e$. The kinetic equation thus reduces to

$$-\frac{\mathrm{d[RCOCl]}}{\mathrm{d}t} = \frac{k_1 k_2 [\mathrm{RCOCl}][\mathrm{R'OH}]^2}{k_{-1} + k_2[\mathrm{R'OH}]} \tag{38}$$

In a solvent of low polarity k_{-1} will be large and k_1 will be small when electron-donating substituents are present in the aromatic ring. When either or both of these conditions hold $k_{-1} \gg k_2[\mathrm{R'OH}]$, and

$$-\frac{\mathrm{d[RCOCl]}}{\mathrm{d}t} = \frac{k_1 k_2}{k_{-1}} [\mathrm{RCOCl}][\mathrm{R'OH}]^2 \tag{39}$$

Conversely, with media of high polarity and with electron-withdrawing substituents,

$$-\frac{\mathrm{d[RCOCl]}}{\mathrm{d}t} = k_1 [\mathrm{RCOCl}][\mathrm{R'OH}] \tag{40}$$

This, although correctly predicting the direction of the variation in apparent order in solvent for benzoyl chloride hydrolysis at 25°C in changing from 2.9 (2–5% water/dioxan) to 1.15 (5–15% water/acetone) does not do so unambiguously. The "unified" mechanism of Minato is, nevertheless, a very valuable contribution and may be preferred on the grounds that it uses the hydrated carbonium ion rather than the acylium ion, a species which although undoubtedly known, has never, to the knowledge of the present author, been detected in the media used for hydrolysis. However, there is still no evidence to decide between the mechanism of Minato and a combination of routes (*a*) or (*b*), and (*d*) (p. 227).

(*c*) *Acid halides with R=R'O or R'R"N*

The hydrolysis of alkyl chloroformates and carbamyl chlorides are two special cases of interest.

Alkyl chloroformates, ROC(=O)Cl, might be expected to hydrolyse rapidly by an $\mathrm{S_N1}$-type process, reminiscent of the hydrolysis of alkoxy ethers such as chloromethyl ether. Similarly, carbamates would be expected to react by a similar mechanism because of the possible stabilisation of the acylium ion formed, *viz.*

$$\mathrm{R{-}\overset{\underset{|}{R}}{\underset{+}{N}}{=}C{=}O \leftrightarrow R{-}\overset{\underset{|}{R}}{N}{-}\overset{+}{C}{=}O}$$

Bohme and Schürhoff[131] have measured the rate of hydrolysis of ethyl chloroformate and found it to be quite slow ($k_1 = 2.0 \times 10^{-4}\ \text{sec}^{-1}$ in water at 25°C). Hall[118], investigating the hydrolysis of ethyl chloroformate and dimethyl carbamyl chloride, found the rate sequence for neutral hydrolysis to be

$$CH_3COCl > C_6H_5COCl > C_2H_5OCOCl < (CH_3)_2NCOCl.$$

Rate parameters for the hydrolyses of some alkyl and acyl chlorides are compared in Table 20. The value of $\Delta S^{\ddagger}$ for dimethylcarbamyl chloride is close to that observed for the hydrolysis of *t*-butyl chloride in 90% water/10% dioxan[56b] (+8 eu).

TABLE 20

ACTIVATION ENERGIES AND ENTROPIES FOR THE HYDROLYSIS OF ACYL AND ALKYL CHLORIDES (RCl)

R	*Solvent*	E_a *(kcal·mole^{-1})*	$\Delta S^{\ddagger}$ *(eu)*	*Ref.*
CH_3	H_2O/Me_2CO 25/27	13.9	−14	56a
C_6H_5CO	H_2O/Me_2CO 50/50	18.8	−7.1	56a
EtOCO	H_2O	19.0	−12.4	56b
Me_2NCO	H_2O	21.6	+5.6	56b

Hydroxyl ions have no effect on the rate of reaction of dimethylcarbamyl chloride, which is, however, greatly reduced by chloride ions. Evidence for the existence of an intermediate was obtained by the isolation of dimethyl-3,3-tetramethylene on the addition of pyrrolidine to the aqueous reaction mixture. However, for chloroformates the analogy between acyl and alkyl chlorides breaks down, since acyl halides may be strongly conjugated in the ground state *as well* as in the transition state but alkyl halides (particularly α-chloromethyl-ethers) may not. The rate sequence found by Hall[118] may not only reflect the S_N2 reactivity for CH_3COCl and C_6H_5COCl and S_N1 reactivity for Me_2NCOCl, but also may be the rate sequence for S_N1 ionisation.

The primary products of the hydrolysis of ethyl chloroformate are hydrogen chloride, ethanol and carbon dioxide, the latter two compounds formed from the rapid decomposition of an intermediate monoethyl carbonate, *viz.*

$$C_2H_5OCOCl + H_2O \rightarrow C_2H_5OCOOH \rightarrow C_2H_5OH + CO_2 + HCl \qquad (41)$$

Hudson *et al.*[151,152] have concluded that the bimolecular solvolysis of ethyl chloroformate involves heterolysis of the carbon–chlorine bond and not heterolysis of the carbon–oxygen bond. Their data shows that the hydrolysis of ethyl chloroformate is a second-order reaction in water/acetone mixtures, methyl chloroformate reacting about 2.2 times as fast in 65% water/acetone at 50°C. Hydroxide ion accelerates the reaction (3.1×10^7 in 18% water/acetone and 3.4×10^6 in 85% water/acetone) and catalysis by hydroxide ion was observed with pure water as solvent by Hall[118]. There is some disagreement about the value for the hydrolysis rate coefficient for ethyl chloroformate in water and in other solvents (Table 21). To date, the data of Queen[153] (for pure water), Kivinen[92] (for ethanol) and Liemiu[101] (for methanol) must be considered the most accurate.

TABLE 21

REPORTED FIRST-ORDER RATE COEFFICIENTS FOR THE HYDROLYSIS OF ETHYL CHLOROFORMATE AT 25°C

Solvent	*k* *(sec⁻¹)*	*Ref.*
Water	2.1×10^{-4}	131
	3.87×10^{-4}	92
	3.671×10^{-4}	153
65% H_2O/35% Me_2CO	109×10^{-5}	152
	99×10^{-5}	92
15% Me_2CO/85% H_2O	208×10^{-6}[a]	151
	233×10^{-6}[b]	151
	(271[c]), 272×10^{-6}[d]	92

[a] Titrimetric.
[b] Volhard titration.
[c] Acid–base titration.
[d] Conductometric.

The rate sequence for methanolysis of alkyl chloroformates is $ClCH_2CH_2O$-COCl > MeOCOCl > EtOCOCl > *i*-PrOCOCl[101], and the opposite order is seen in formolysis[2] and a third sequence MeOCOCl > EtOCOCl < *i*-PrOCOCl for hydrolysis in aqueous acetone. The implied bimolecular, unimolecular, and mechanism-change process have not been firmly established. However, Green and Hudson[154] have indicated that, in the unimolecular reactions, alkyl–oxygen fission probably occurs after the rate-determining ionisation of the acyl–halogen bond. Under conditions of strong electrophilic catalysis, for example, the presence of silver ions, *iso*propyl chloroformate reacts by a rate-determining step to form a carboxonium ion[155,156], *viz.*

$$i\text{-PrOCOCl} + Ag^+ \xrightarrow{\text{slow}} i\text{-PrO}\overset{+}{C}{=}O \xrightarrow[NO_3^-]{\text{fast}} i\text{-PrONO}_2 + CO_2$$

TABLE 22

ACTIVATION PARAMETERS AT 25°C FOR THE HYDROLYSIS OF SOME ACID CHLORIDES[153]

	Chloroformate esters					*Dimethylcarbamylchloride*
	Ph	*Me*	*Et*	*n-Pr*	*i-Pr*	
$\Delta H^{\ddagger}$ (k·cal·mole^{-1})	14.110	16.199	17.070	17.131	24.102	20.263
$\Delta S^{\ddagger}$ (eu)	−19.79	−19.07	−17.00	−16.62	+10.12	3.50
$\Delta C_p^{\ddagger}$ (cal·deg^{-1}·mole^{-1})	−42.2	−36.0	−3.5	2.2	−9.1	−83.5

Queen[153] has recently measured the activation parameters $\Delta H^{\ddagger}$, $\Delta S^{\ddagger}$ and $\Delta C_p^{\ddagger}$ for the hydrolyses of Me, Et, *n*-Pr, *i*-Pr, and phenyl chloroformates and dimethyl carbamyl chloride in pure water (Table 22), and concluded that a change of mechanism (S_N2 to S_N1) takes place with increasing electron donation to the chlorocarbamyl group; the data, which include solvent isotope effects, are consistent with a unimolecular hydrolysis of *iso*propyl chloroformate and dimethyl carbamyl chloride and a bimolecular hydrolysis of the other four compounds.

Queen[157] has also measured the rates of the corresponding *s*-alkyl and aryl thiochloroformates and finds the series

$$\text{Ph} < \text{Me} < \text{Et} \approx n\text{-Pr} < i\text{-Pr} < t\text{-Bu}$$

which is understandable in terms of conjugative and hyperconjugative electron release to sulphur in the ground state. The reactions appear to be S_N1 process with no concomittant fragmentation.

Hudson *et al.*[158] have shown that N,N-dialkylcarbamates decompose in strongly acidic media to carbon dioxide, olefin, alkyl halide and alcohol, the rate of reaction of the secondary esters closely following h_0. This fact, together with the variation in the rate of hydrolysis of carbamates of cyclic alcohols with the ring size[154], shows that these reactions involve the intermediate formation of carbonium ions.

2.2 ROLE OF SOLVENT IN THE HYDROLYSIS OF ACID CHLORIDES

The preceding sections have shown the complexity of solvent effects in the solvolysis of acyl chlorides, and how ambiguities in the role of the solvent, particularly in its apparent reaction order, critically affect the assignment of detailed mechanism. It is the intention in this brief section to point to some of

TABLE 23

RATE COEFFICIENTS FOR ETHANOLYSIS OF BENZOYL CHLORIDE[159] AND THE EMPIRICAL SOLVENT PARAMETERS Z[160a] AND E_T[161]

Solvent	k*	Z (*kcal·mole*$^{-1}$)	E_T (*kcal·mole*$^{-1}$)
$(C_2H_5)_2O$	0.081		34.6
$CH_3COOC_2H_5$	0.16		38.1
Me_2CO	0.41	65.7	42.2
$C_6H_5CH_3$	0.49		33.9
$C_6H_5NO_2$	0.51		42.0
C_6H_6	0.61	54	34.5
C_6H_5Cl	0.62		37.5
CCl_4	0.73		32.5
C_2H_5OH	3.0	79.6	51.9
CH_3OH	18.9**	83.6	55.5

* k is a second-order rate coefficient, the concentration units being mole fractions and the time hours.

** Extrapolated from the data of Norris and Young[162].

the attempts to rationalise the effect of the solvent on the solvolysis of acid chlorides.

The effect of changing the solvent from non-polar to polar on the order and mechanism of the hydrolysis of 4-substituted benzoyl chlorides has been discussed earlier and is shown in Table 16 (p. 242). Probably the earliest systematic study was that of Norris and Haines[159] who determined the rate of ethanolysis of benzoyl chloride and found it to be very slow in ether, but faster in benzene and carbon tetrachloride. Their data are shown in Table 23, together with two solvent parameters of more recent invention. The first of these, the Kosower Z value[160a], based on the longest wavelength charge-transfer bond of 1-ethyl-4-carbomethoxypyridinium iodide, is more limited, particularly in more polar solvents, than the second, the E_T value of Dimroth *et al.*[161]. There is a good correlation between Z and E_T[160b]. A plot of log k for ethanolysis *versus* E_T is linear for a wide variety of solvents (Et_2O, CH_3CO_2-C_2H_5, Me_2CO, $C_6H_5NO_2$, C_2H_5OH and CH_3OH). However, the points for carbon tetrachloride, toluene, benzene and carbon tetrachloride all indicate that the rate of ethanolysis in these solvents is greater by a factor of ~ 10 than would be expected from the polarity of the solvent. The concentrations used by Norris and Haines (0.1 mole fraction benzoyl chloride and ethanol, 0.8 mole fraction solvent) do not exclude the possibility of significant differences in the real Z value of the reaction mixture from that of the pure solvent, but this clear separation of solvents into two groups, those capable of potentially strong hydrogen bonding and those not, and the linearity of the plot for the former, indicates that a common mechanism is operative in the group of solvents ether to methanol. The transition state in this mechanism is more influenced by the polarity of the solvent (approximately 5.8 times from the log k *versus* Z or E_T plots) than is that for the quaternisation of alkyl iodides[163], a reaction often referred to as a solvent-sensitive reaction.

The substituent effect data, discussed previously, shows that in pure ethanol the mechanism of ethanolysis of benzoyl chloride is mixed, but the reaction is subject to acceleration by electron-withdrawing substituents and is presumably largely bimolecular. The same is the case for 40% ethanol/ether and 5% water/acetone solvents. For 50% water/acetone solvent the point for hydrogen lies on the negative line as it does for formic acid; no E_T or Z values are available for formic acid (the betaines used to measure E_T values would be protonated). It would be of interest to know whether, for this series of solvents, a plot of log k *versus* Z or E_T reflects this change in predominant mechanism. Ugi and Beck[100] have shown that a plot of log k_{H_2O} *versus* Y, the Grunwald–Winstein parameter, is linear over a wide range of water concentrations in acetone (γH_2O 0.33–0.75) for the hydrolysis of benzoyl chlorides. There is a linear correlation between Z and Y for water/acetone over the range 60–100%[164], and if a plot of Z *versus* E_T for aqueous acetone is linear for the same range this would show that there is no large change of mechanism.

One approach to the problem of solvation has been to examine the effect of changes in the gross dielectric constant on the rate of reaction from both experimental and theoretical standpoints. Bohme and Schürhoff[131] employed this approach to the hydrolysis of benzoyl chloride by water in varying amounts of dioxan, tetrahydrofuran and glycoldimethylether. Plots of log k *versus* log (dielectric constant) for each solvent gave two lines showing a distinct break. When a linear relationship has been found to exist between the logarithm of the rate constant and the expression $(D-1)/(2D+1)$, or $1/D$ when $D > 7$, the slope of the plot assumes different values for different series of solvent mixtures[165]. Linear relationships of this kind have been established for the hydrolysis of acetyl chloride in dioxan/water mixtures[166] and for the hydrolysis of benzoyl chloride in acetone/water mixtures[117]. Bohme and Schürhoff[131] found that log k_{H_2O} varied linearly with $1/D$ with a number of binary solvent mixtures, and Kivinen[128] has found a linear relationship between log k_2 (the second-order rate coefficient) and log D for the ethanolysis of monosubstituted benzoyl chlorides in benzene/ethanol mixtures of 100–20% by weight. The plots of log k_2 *versus* $(D-1)/(2D+1)$ for this reaction have been found to be S-shaped, as predicted by Laidler[167] on theoretical grounds.

However, dielectric constant cannot be a decisive factor in determining the variation in solvolysis rate with solvent composition. If the amount of acetone in ethanol is increased from 0 to 50%, the dielectric constant changes from about 25 to 21 and remains practically constant at this value although the rate of ethanolysis of benzoyl chloride decreases as the 1.56 second power of the molar concentration of ethanol even in solvents, containing more than 50% acetone[168]. In the case of the hydrolysis of ethyl chloroformate in dimethyl sulphoxide (DMSO)/water mixtures the rate of hydrolysis increases as the concentration of DMSO increases, although this is decreasing the dielectric constant of the system. This conclusion was also drawn by Hudson[169], who stressed the importance of hydrogen bond formation between the solvent and the solute or between the solvent and the transition state which is disregarded when the theory of electrostatics is applied. The observation of the linearity of the plot of log k with E_T for potential hydrogen-bonding solvents (see p. 253) reinforces this point.

It has been suggested that the mean value of the slope, n, of the plots of log k *versus* $[H_2O]$ for a hydrolysis reaction gives the number of molecules participating in the rate-determining stage of the reaction, that is, the number of water molecules bound in the transition state[133,170,171]. Various values of n have been found, depending upon the substrate and the solvent. Those reported by Kivinen[128] for reactions of 4-substituted benzoyl chloride in benzene/ethanol, acetone/ethanol and diethyl ether/ethanol are all between 1 and 2, apart from that for 4-methoxybenzoyl chloride in 100–80% ethanol in diethyl ether which is close to 5. Hudson and Saville[63] have found that, for the ethanolysis of triphenylmethyl chloride, the slope increases from a value of about two at low alcohol concentrations to about six with the mixtures of high ethanol content. Hudson *et al.*[172,173] have proposed that the solvation effect on the reaction of acyl chlorides with alcohols in non-polar solvents is proportional to the concentration of self-associated alcohol over a wide concentration range. Thus, if the solution of alcohol in a non-polar solvent is represented as

$$\mathrm{ROH+ROH} \xrightleftharpoons{K_{1,1}} \mathrm{(ROH)_2},\ \mathrm{ROH+(ROH)_2} \xrightleftharpoons{K_{1,2}} \mathrm{(ROH)_3}\ \text{etc.}$$

$K_{1,1}$ and $K_{1,2}$ are equal in some cases[174] but usually $K_{1,1} < K_{1,2} < K_{1,3} < K_{1,4}$ etc.[175]. Hudson[173] assumes that the solvation of the transition state involves hydrogen-bonded species of the following kind

$$\mathrm{(ROH)}_i + \mathrm{R}^{\Delta+} \ldots \mathrm{X}^{\Delta-} \rightleftharpoons \mathrm{R}^{\Delta+} \ldots \mathrm{X}^{\Delta-} \ldots \mathrm{(ROH)}_i \tag{42}$$

and

$$k_\mu = \sum_{i=1} k_{i=1}[(\mathrm{ROH})_i] \tag{43}$$

where k_μ is the observed first-order rate coefficient. If the reactivity of the associated species $(\mathrm{ROH})_i$ is proportional to i, *viz.* $k_i = k \times i\,(i \neq 1)$, then

$$k_\mu = k_1[(\mathrm{ROH})_1] + k\sum_{i=2}^{i} i[(\mathrm{ROH})_i] \tag{44}$$

References pp. 287–293

When the reaction due to monomers can be neglected then

$$k_{\mu} = k \sum_{i=2}^{i} i[(ROH)_i] \tag{45}$$

or the rate is proportional to the amount of associated alcohol. The amount of associated alcohol can be determined spectroscopically in non-polar solvents and Hudson has shown that the second-order rate coefficients for the alcoholysis of acetyl and *p*-nitrobenzoyl chlorides in carbon tetrachloride solution increased rapidly with alcohol concentration and closely obeyed equation (45); this can, therefore, be ascribed to specific solvation (hydrogen bonding) of the transition state by alcohol molecules. As the total concentration of the alcohol decreases, its degree of dissociation increases and the reaction of the monomer may be detected, for example in the reaction between *n*-octanol (0.1–0.01 *M*) and *n*-caprylyl chloride in carbon tetrachloride at 40°C.

The treatment has been extended to dioxan/water and dioxan/alcohol mixtures, where the concentration of self-associated alcohol has to be calculated from activity coefficient data. It was found that alcoholysis of 4-nitrobenzoyl chloride in ether and dioxan can be accounted for solely on the grounds of specific solvation, but in the case of acetone some of the reaction proceeds by a mechanism without specific solvation, possibly due to dielectric solvation of the transition state. Table 24 shows the relative reactivities of associated alcohol in several solvents. Hudson *et al.*[172b] propose that in carbon tetrachloride the smallest associate is probably the trimer whereas in the ethers the corresponding associate has an open structure, *viz.*

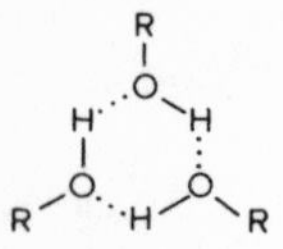

XXIX 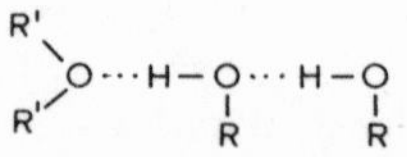 XXX

The reactivity of an associate may be related to the number of hydrogen bonds

TABLE 24

RELATIVE REACTIVITIES OF ASSOCIATED ALCOHOL WITH *p*-NITROBENZOYL CHLORIDE IN VARIOUS SOLVENTS AT 25°C[172b]

Solvent	CCl_4	Benzene	Dioxan	Ether	Acetone
$k_I/\Sigma i x_{Ai}$ (*sec*$^{-1}$)	0.0053	0.0033	0.0018	0.0155	0.0057

between the alcohol molecules. XXIX should therefore be three times more reactive than XXX. The dimer in benzene is found to be significantly less reactive (3.5 times) than the trimer in agreement with this interpretation. Moreover, this explains the greater reactivity of ethanol in benzene, carbon-tetrachloride, toluene and chlorobenzene than in solvents capable of hydrogen bonding as shown by a log $k_{\text{hydrolysis}}$ *versus* E_T plot.

2.3 KINETIC METHODS USED IN STUDIES OF THE SOLVOLYSIS OF ACID CHLORIDES

Most of the methods employed to follow the solvolysis of acid chlorides have been either conductometric or titrimetric. Conductometric methods were used by Branch and Nixon[113], Norris *et al.*[79,125,159,162], and Hudson and Wardhill[110]. However, in organic solvent mixtures, calibration is necessary since the amount of hydrogen chloride produced is not directly proportional to the measured conductance. Titrimetric methods usually employ a quenching technique using an inert solvent followed by titration. The hydrogen chloride is titrated with an amine which is inert to acid chlorides[128]. Spectrophotometric[178], potentiometric[117] and the concentration cell method have also been used to study the alcoholysis of acid chlorides. Table 25 shows the main methods used.

TABLE 25

KINETIC METHODS USED IN STUDIES OF THE SOLVOLYSIS OF ACID HALIDES

Method	*Ref.*
Titration	21, 56
Conductometric	24, 79, 80, 100, 113, 116, 125, 133, 136, 153, 159, 182
Concentration cell	110, 118
Potentiometric	117
Spectrophotometric	46, 85, 176
Stopped flow	96
Dilatometery	24, 135
Thermal maximum	120

3. Hydrolysis of amides

The hydrolyses of carboxylic acid amides, RCONR′R″, are catalysed by both acids and bases, and generally do not take place in neutral media. Ostwald[176] made the first quantitative measurements of the catalytic activity of various acids on hydrolyses and Reid[177] investigated the effect of substitution in the aromatic series. These workers were followed by several whose aim was to determine the effect on the rates of hydrolysis of changing the reaction medium and of structural effects in the substrate. A detailed bibliography of this period may be found in the paper by Bruylants[178]. During this period, it was established that substitution of the amides by halogens enhances the rate of alkaline hydrolysis[179], and that in the aliphatic series the length of the alkyl chain is an inverse function of the rate of hydrolysis[178].

Kinetic investigations of amide hydrolysis showed that the rate of hydrolysis in basic media is proportional to the concentration of amide and hydroxide ion. Similarly, early work[180–183] on the acid-catalysed hydrolysis of amides showed that the rate of acidic hydrolysis is, in general, proportional to the concentration of amide and hydroxonium ion. In several acidic hydrolyses, however, a maximum is observed in the pH–rate profile at 3–6 pH units, a phenomenon first reported by Benrath[184] and since supported by other workers[185–190]. This behaviour of amides is in contrast to the hydrolysis of nitriles whose rate constant increases continuously with the hydrogen ion concentration[191].

3.1 ACID CATALYSIS

Amides are reasonably strong bases, with pK_a's of the order of -1 to -3[189], and therefore are appreciably protonated in solutions of mineral acids[192–194]. The rate maxima are obtained in strongly acid conditions and the hydrolysis may be formulated

$$n\,H_2O + \text{amide} + H^+ \overset{K_a}{\rightleftharpoons} \text{amide.}H^{(+)}.(H_2O)_n \xrightarrow{k_r} \text{products}$$

where the value of n has been suggested to range from 4 to 5[195]. Protonation may take place at either the carboxyl oxygen atom or the nitrogen atom, *viz.*

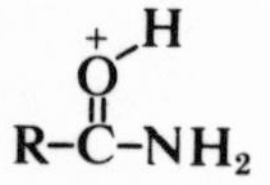

$$\underset{\text{XXXI}}{R{-}\overset{\overset{+}{O}{-}H}{\overset{\|}{C}}{-}NH_2} \qquad \underset{\text{XXXII}}{R{-}\overset{O}{\overset{\|}{C}}{-}\overset{+}{N}H_3}$$

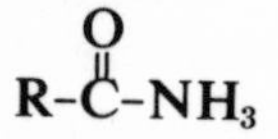

and there is some spectroscopic evidence from the work of Fraenkel and Niemann[196] that protonation takes place, at least in part, on the oxygen atom since it was observed that the methyl doublet of N,N-dimethylformamide is retained in solutions of high acidity. Infrared spectroscopic studies[197,198] on compounds of the type $RCON^+Et_3X^-$ and $R(COMe)N^+Et_2X^-$ and studies using ^{18}O-labelled amides support this view. Studies on *para*-substituted benzamides[193] indicate a correlation between pK_{BH^+}, and the σ value of the substituent, rather than the corresponding σ^+ value and indicate an absence of conjugation between the substituent and the protonated carboxyl group, which may be attributable to either non-coplanarity of the aryl and O-protonated amide or possibly to N-protonation.

A reasonably complete discussion of the role of acid in amide hydrolysis requires the inclusion of the Hammett acidity function (H_0)[199,200], and the controversial topic of the significance of this function in the determination of mechanism[201,202].

The hypothesis of a protonated amide intermediate is supported by the observation of the pH–rate maxima since once the amide is fully protonated, a further increase in acid concentration decreases the activity of water in the medium[188], the rate-determining step being the attack of a water molecule on the conjugate acid of the amide[203–205]. If the hydrolysis is of the form,

$$B + H_3O^+ \xrightleftharpoons{K_1} BH^+ + H_2O \xrightleftharpoons{K_2} M^+ \longrightarrow \text{hydrolysis products}$$

where K_1 and K_2 are equilibrium constants, M^+ is the transition state for the hydrolysis and k is a rate coefficient, then the Brönsted[206] equation for the rate of hydrolysis, v, is given by

$$v = k[M^+] = kK_2[B][H_3O^+]\, f_B f_{H_3O^+} (f_M + K_1) \qquad (46)$$

Therefore,

$$v = k_2[B][H_3O^+]$$

where

$$k_2 = \frac{K_2 k f_B f_{H_3O^+}}{K_1 f_{M^+}} \qquad (47)$$

and the experimental first-order rate coefficient, k_e, which gives the rate of the reaction in different concentrations of excess acid, may be shown to be

$$k_e = k_2 K[H_3O^+]/(K + h_0) \qquad (48)$$

For weak bases such as esters, $K \gg h_0$ and the observed[207] linear dependence of the experimental first-order coefficients on hydroxonium ion concentration is predicted. When $K \ll h_0$ then

$$k_e = k_2 K[H_3O^+]/h_0 \qquad (49)$$

and since h_0 increases with concentration much more rapidly than $[H_2O^+]$ for above 2 *M* mineral acid[207,208], then k_e will decrease in media of high acid concentration. It is, therefore, possible to calculate the pK_a of the amide from the acid concentration for maximal rate, using separately determined acidity functions. Edward and Meacock[189] performed these calculations for a number of amides and their data is shown in Table 26. Equation (48) also covers the variation in rate with acid concentration reasonably well, for the amides in this study, for the lower acid concentrations. However, at higher acid concentrations (7.6 *M*), the experimental values tend to be higher than predicted.

TABLE 26

RELATION BETWEEN p*K*'S OF AMIDES AND THE ACID CONCENTRATIONS (c_M IN MOLE·L^{-1}) IN WHICH THE RATES OF HYDROLYSIS ARE MAXIMAL AT 25°C[189]

Amide	*Acid*	C_M	*pK* *Predicted*	*pK* *Found*
$HCONH_2$[203]	HCl	6	−2.6	
	H_2SO_4	4.75	−2.75	
CH_2CONH_2[203]	HCl	3.25	−1.33	−0.5[210], −0.04[211], −1.40[194]
$CH_3CH_2CONH_2$[203]	HCl	3.20	−1.30	
	H_2SO_4	2.40	−0.130	
$C_6H_5CONH_2$	HCl	4.5	−0.90	−1.85 (18°±3°C)
	H_2SO_4	3.5	−2.01	
p-$MeOC_6H_4CONH_2$	H_2SO_4	3.0	−1.70	−1.60 (18°±3°C)
p-$NO_2C_6H_4CONH_2$	H_2SO_4	4.5	−2.60	−2.58 (18°±3°C)

The failure of the acidity function approach is shown in the work of Leisten[212], who studied the rates of hydrolysis of the thirteen amides listed in Table 27 in 5.9, 7.2 and 8.5 *M* perchloric acid. If the mechanism of hydrolysis is of the A-2 type[213], where a rapid reversible initial protonation of the amide is followed by the attack of water on to the conjugate acid, then the Zücker–Hammett hypothesis[214] would predict that the rates of hydrolysis in solutions of different acidity should be proportional to the acid concentration. Therefore, k_1 (the rate coefficient for the attack of water onto the conjugate acid) should be proportional to $[H_3O^+]/h_0$. In Table 27 the values of the first-order coefficients for hydrolysis at 5.86 and 7.19 *M* $HClO_4$ are given and compared with the calculated value.

The variations in rate are similar for different amides but the discrepancies between the observed and calculated values are large and in the same direction as those found by Edward and Meacock[189] for hydrolysis catalysed by sulphuric and hydrochloric acids. Similar failures are observed in the application of the Zücker–Hammett hypothesis to the hydrolysis of aliphatic esters[207]

TABLE 27

FIRST-ORDER RATE COEFFICIENTS FOR THE HYDROLYSES OF AMIDES IN AQUEOUS PERCHLORIC ACID SOLUTIONS AT 95°C WITH THE CALCULATED AND OBSERVED RATIOS FOR THE RATES AT TWO ACID CONCENTRATIONS[212]

	$10^4 \times k_1 (sec^{-1})$		k_1 in 5.86 M $HClO_4$/ k_1 in 7.19 M $HClO_4$	
Amide	5.86 M (H_0, −2.74)	7.19 M (H_0, −3.75)	*Observed*	*Calc.*
p-Nitrobenzamide		3.9		8.34
m-Nitrobenzamide	5.33	3.06	1.75	8.34
m-Bromobenzamide		1.6		8.34
p-Bromobenzamide		1.1		8.34
Benzamide	1.8	0.633	2.82	8.34
m-Toluamide		0.36		8.34
p-Toluamide	1.0	0.36	2.87	8.34
p-tert.-Butylbenzamide	0.92			8.34
Anisamide	0.59	0.20	2.93	8.34
o-Bromobenzamide		0.275		8.34
o-Nitrobenzamide		0.11		8.34
o-Toluamide		0.075		8.34
Acetamide	6.3	2.44	2.57	8.34

and ethylidene diacetate[215] and to amides in the work of Moodie *et al.*[216] and Bunton *et al.*[217].

Bunton *et al.*[217] measured the first-order rate coefficient, k, for the hydrolysis of benzamide, N-methylbenzamide and N,N-dimethylbenzamide in hydrochloric acid at 100°C and were unable to fit their results to any of the usual relationships. Plots of $\log k - \log [h_0/(K_{HA^+} + h_0)]$ against $\log a_w$ (a_w = the activity of water) are curved as are similar plots using h_A[218,219] (the acidity function derived from amides) and a plot of $\log k - \log [h_0/(K_{HA^+} + h_0)]$ against $H_0 + \log C_{H^+}$[220]. Bunton *et al.* therefore suggested that these relationships fail because the acid hydrolyses proceed *via* two distinct mechanistic paths, *viz.*

$$R'CONR_2 \xrightarrow[k_1]{H_3O^+} \left[H_2O..\overset{R}{\underset{HNR_2}{C}}{=}O \right]^+ (XXXIII) \longrightarrow \text{Products}$$

$$R'CONR_2 \underset{}{\overset{H^+}{\rightleftharpoons}} R'\overset{OH}{C}{-}NR_2 \xrightarrow[k_2]{H_2O} \left[H_2O..\overset{R'}{\underset{OH}{C}}..NR_2 \right]^+ \longrightarrow \text{Products} \qquad (50)$$

Formation of XXXIII could involve simultaneous attack of H^+ and H_2O[221]

or the addition of H^+ to give the N-protonated amide. The Brönsted–Bjerrum rate equation gives

$$\frac{k}{\alpha a_w} = \frac{k_1[H^+](1-\alpha)}{\alpha} + k_2\beta \qquad (51)$$

where k is the first-order rate coefficient and α is the amount of protonated amide. α can be calculated from the known pK_{HA} values of benzamide (−1.74), N-methylbenzamide (−1.7) and N,N-dimethylbenzamide (−1.4) and the H_A function[193,218–220,222]. Equation (51) fits the experimental data obtained by Bunton *et al.*[217] over a range of acid concentrations of 0.084–8.37 *M* hydrochloric acid, and they obtained the values in Table 28 for k_1 and k_2.

TABLE 28

RATE COEFFICIENTS FOR HYDROLYSIS AMIDES IN HYDROCHLORIC ACID[217]

	$10^4\ k_1$ ($l \cdot mole^{-1} \cdot sec^{-1}$)	$10^4\ k_2$ (sec^{-1})
Benzamide	3.75	16
N-Methylbenzamide	0.59	1.0
N,N-Dimethylbenzamide	0.95	2.8

The solvent deuterium isotope effects in 1 *M* and 6 *M* acid were small, having values of k_{H_2O}/k_{D_2O} of 1.15 and 0.90 at the two acid concentrations with benzamide. No oxygen exchange occurs between water and amides during acidic hydrolysis[223] in contrast to exchange observed in basic hydrolysis[224]. Bunton's mechanism, proposes that an alternative route to hydrolysis is *via* a concerted protonation of the amide nitrogen and attack of a water molecule on the carbon of the carboxyl group to give the N-protonated amide transition state, XXXIII which should be highly reactive.

Although there is some evidence for protonation of the amide at the carboxyl oxygen, there is no evidence which confines the protonation exclusively to this position and many of the kinetic results of earlier workers may be interpreted in terms of either kind of mechanism; in particular the results of Farber and Brieux[225] on the hydrolysis of ε-caprolactam in sulphuric acid fit equation (51) very well.

The hydrolysis of nicotinamide catalysed by hydrochloric acid, investigated by Jellinek and Gordon[226], does not show a maximum in the pH–rate profile. Since nicotinamide contains two basic groups, four equilibria have to be considered, *viz.*

$$RCONH_2 + H^+ \overset{K_1}{\rightleftharpoons} H^+RCONH_2$$

$$RCONH_2 + H^+ \overset{K_2}{\rightleftharpoons} RCONH_3^+$$

$$H^+RCONH_2 + H^+ \overset{K_3}{\rightleftharpoons} H^+RCO\overset{+}{N}H_3$$

$$RCO\overset{+}{N}H_3 + H^+ \overset{K_4}{\rightleftharpoons} H^+RCONH_3^+$$

where $H^+RCO\overset{+}{N}H_3$ is XXXIV

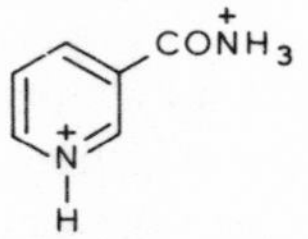

XXXIV

Jellinek and Gordon propose a mechanism in which a rapid protonation of the amide nitrogen is followed by a slow attack by a hydroxonium ion to give nicotinic acid, an ammonium ion and a hydrogen ion. Such a mechanism accords with their observation that the experimental rate coefficients increase more rapidly than would be the case if they were proportional to the first power of the hydrogen ion concentration.

If both $\overset{+}{H}RCO\overset{+}{N}H_3$ and $RCO\overset{+}{N}H_3$ can react with hydrogen ions, the rate of hydrolysis will be given by

$$v = k_1[RCO\overset{+}{N}H_3][H_3O^+] + k_2[H^+RCO\overset{+}{N}H_3][H_3O^+] \tag{52}$$

Values of the dissociation constant due to N in the ring of nicotinic acid and the amide were found to be 3.55×10^{-11} and 2.24×10^{-11} (from ultraviolet absorption spectra), whereas the value for the dissociation of amide groups is generally much smaller (acetamide, 3.1×10^{-15}). It is, therefore, reasonable to assume that K_1 and K_4 are large compared with K_2 and K_3. Using these assumptions, the observed rate, v, is given by

$$v = k_2 \frac{K_3[H^+][a][H_3O^+]}{1 + K_3[H^+]} \tag{53}$$

where $[a]$ = total amide concentration at time t. If hydrochloric acid is present in excess, a first-order reaction results, and at high concentrations of acid (since $K_3[H^+] \simeq (1 + K_3[H^+])$ a plot of k_{obs} *versus* $[H_3O^+]$ will be linear, as was observed. k_2 was found to be 7×10^{-5} (sec^{-1})/$[H_3O^+]$. A satisfactory fit to the experimental data was given by $K_3 = 0.67$. The proposed mechanism is

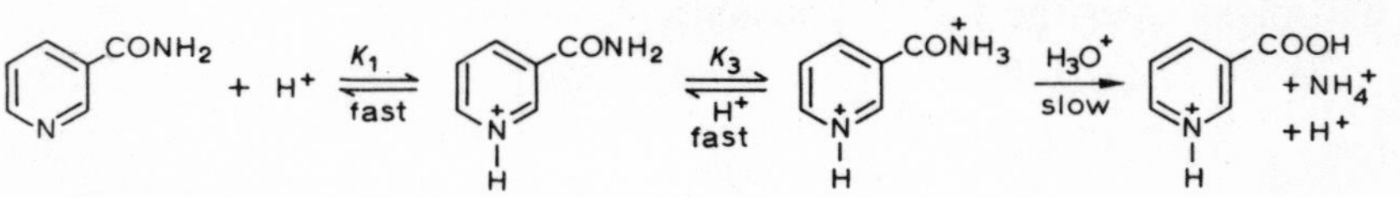

$$\text{(pyridine-3-CONH}_2) + H^+ \underset{\text{fast}}{\overset{K_1}{\rightleftharpoons}} \text{(N-H}^+\text{ pyridinium-3-CONH}_2) \underset{\text{fast}}{\overset{K_3}{\underset{H^+}{\rightleftharpoons}}} \text{(N-H}^+\text{ pyridinium-3-CO}\overset{+}{N}H_3) \xrightarrow[\text{slow}]{H_3O^+} \text{(N-H}^+\text{ pyridinium-3-COOH)} + NH_4^+ + H^+ \tag{54}$$

The acid-catalysed hydrolyses of nicotinamide and benzamide, involving a rapid protonation of the amide followed by slow reaction of the protonated amide with water, have similar activation energies[24] (18.8 and 21.7 kcal·mole^{-1}, respectively, compared with 20.5 kcal·mole^{-1} at 1.00 M HCl).

3.2 BASE CATALYSIS

The alkaline hydrolysis of carboxylic amides is similar in mechanism to that of esters, except that the concomitant oxygen exchange is generally faster than hydrolysis. There is some evidence that equilibration of the oxygens in the tetrahedral intermediate is not always faster than its breakdown[227–229], possibly due to the location of the negative charge on the nitrogen atom. Since the amide residue should be a poorer leaving group than either hydroxy or alkoxy groups, the ready return of the tetrahedral intermediate to reactants may be expected. The kinetics of alkaline saponification of esters has been extensively studied, but relatively little work has been done on the alkaline hydrolysis of amides and anilides. Meloche and Laidler[229] have studied *p*-substituted benzamides, Cason *et al.*[230,231], several branched-chain amides, Bruylants *et al.*[232,233] straight-chain aliphatic amides; Verkade *et al.*[234] have studied the alkaline alcoholysis of phenyl-substituted acetanilides, and Beichler and Taft[27] have investigated the alkaline hydrolysis of anilides.

The kinetic order of base-catalysed amide hydrolyses may vary considerably with the structure of the amide. Orders in hydroxide ion both smaller and larger than unity have been observed for a number of hydrolyses[28,29,33]. An order less than unity is observed when the amide itself is sufficiently acidic to be partially ionised, as in the hydrolysis of trifluoroacetanilide[27,28] and higher orders are observed in anilide hydrolyses in which it appears that the tetrahedral intermediate can be further ionised.

$$\underset{\text{XXXV}}{\text{Ar}'\text{–}\overset{\text{O}^-}{\underset{\text{OH}}{\text{C}}}\text{–NHAr}} \rightleftharpoons \underset{\text{XXXVI}}{\text{Ar}'\text{–}\overset{\text{O}^-}{\underset{\text{O}^-}{\text{C}}}\text{–NHAr}} \tag{55}$$

The hydroxide-catalysed hydrolysis of aliphatic amides is generally first-order in hydroxide ion. Bruylants and Kezdy[30] showed that, for a series of alkyl amides $RCONH_2$ ranging from $R = CH_3$ to $R = Cl_3C$, the rates of alkaline hydrolysis obey the Taft relationship

$$\log\left(\frac{k_i}{k_0}\right)_{OH^-} = \rho^* \cdot \sigma^* + E_s \tag{56}$$

with $\rho^* = +2.7$, and that in the case of trichloracetamide a fraction of the amide is ionised. The kinetic law then becomes

$$-\frac{d[\text{amide}]}{dt} = k_2[\text{amide}]\frac{[OH^-]}{1+K_i[OH]} \tag{57}$$

where K_i is the equilibrium constant for the process

$$RCONH_2 + OH^- \overset{K_i}{\rightleftharpoons} RCO\bar{N}H + H_2O$$

Dichloracetamide ($pK_i = 13.55$) also shows evidence of ionisation but monochloracetamide ($pK_i = 14.17$) does not; moreover Bruylants and Kezdy also found evidence for a termolecular mechanism in the hydrolysis of chloracetamide.

Similar evidence for the alkaline hydrolysis of 2,2,2-trifluoro-N-methylacetanilide(I) and other similar amides was obtained by Beichler and Taft[27], who showed that the rate of hydrolysis is given by the expression

$$-d[I]/dt = [I](k_2[HO^-] + k_3[HO^-]^2 \tag{58}$$

Table 29 gives the values of the second- and third-order rate coefficients for the hydrolysis of the N-methylacetanilides studied by Beichler and Taft.

TABLE 29

SECOND- AND THIRD-ORDER RATE COEFFICIENTS FOR THE HYDROLYSIS OF N-METHYLACETANILIDES, $RCO(NCH_3)C_6H_5$ IN WATER AT 25.5°C

Ionic strength 0.6 *M*

R	k_2 ($l \cdot mole^{-1} \cdot sec^{-1}$)	k_3 ($l^2 \cdot mole^{-2} \cdot sec^{-1}$)	k_3/k_2
CF_3	2.45	4.7×10^2	190
CHF_2	0.107	3.7	34
$C(CH_3)_2CN$	0.00062	0.00033	0.5
$CH_2OC_6H_5$	0.00025	0.00017	0.7
H	0.000233	0.00080	3.4
CH_2Cl	0.00086	0.00183	2.1
$CH_2\overset{+}{N}(CH_3)_2$	0.00012	0.00068	5.6

A similar but more complex expression has been given by Mader[28] for 2,2,2-trifluoroacetanilide. Schowen *et al.*[29,33] have investigated 2,2,2-trifluoro-N-methylacetanilide (XXXVIII) at pH 9.5–10[29] and 11.2–12.6[33], and shown that the hydrolysis in the lower pH range obeys the expression

$$-\frac{d[\text{XXXVIII}]}{dt} = [\text{XXXVIII}]k_0 + [\text{HO}^-]\left(k_1 + \sum_i k_i[\text{B}_i]\right) \tag{59}$$

corresponding to the superimposition of general base catalysis upon specific hydroxide ion catalysis. Their results are in accord with a rate-determining elimination of N-methylaniline from an intermediate adduct (Scheme A) or a less likely general base-catalysed attack by hydroxide ion in a concerted displacement (Scheme B).

Scheme A

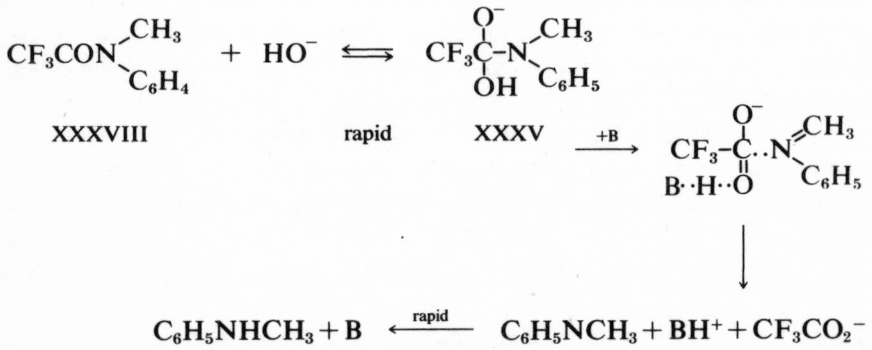

$$\underset{\text{XXXVIII}}{CF_3CON\begin{matrix}CH_3\\C_6H_4\end{matrix}} + HO^- \underset{\text{rapid}}{\rightleftharpoons} \underset{\text{XXXV}}{CF_3\overset{O^-}{\underset{OH}{C}}\text{-}N\begin{matrix}CH_3\\C_6H_5\end{matrix}} \xrightarrow{+B} CF_3\text{-}\overset{O^-}{\underset{B\cdot\cdot H\cdot\cdot O}{C}}\cdot\cdot N\begin{matrix}CH_3\\C_6H_5\end{matrix}$$

$$\downarrow$$

$$C_6H_5NHCH_3 + B \xleftarrow{\text{rapid}} C_6H_5NCH_3 + BH^+ + CF_3CO_2^-$$

Scheme B

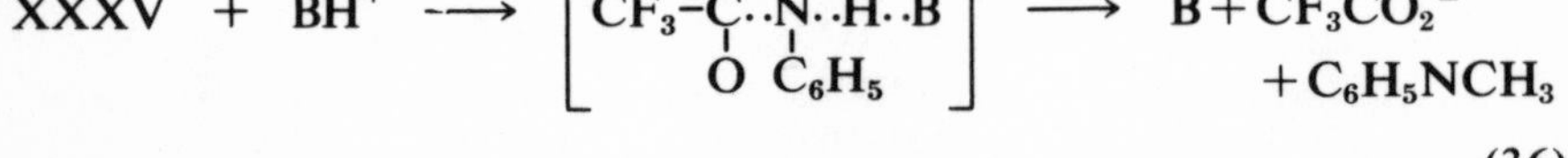

$$\text{XXXV} + B \rightleftharpoons \underset{\text{XXXVII}}{CF_3\text{-}\overset{O^-}{\underset{O^-}{C}}\text{-}N\begin{matrix}CH_3\\C_6H_5\end{matrix}} + BH^+$$

$$\text{XXXV} + BH^+ \longrightarrow \left[CF_3\text{-}\overset{O^-}{\underset{O}{C}}\cdot\cdot\overset{CH_3}{\underset{C_6H_5}{N}}\cdot\cdot H\cdot\cdot B\right] \longrightarrow B + CF_3CO_2^- + C_6H_5NCH_3 \tag{36}$$

Mechanisms A and B (above) are kinetically indistinguishable but may potentially be separated by the "reacting bond" and "solvation" rules of Swain *et al.*[235–237]. The Brönsted catalysis law holds approximately for the bases water, glycinate ion and hydroxide ion with $\beta \sim 0.3$, although it is of interest to note that glycine is above the line. Application of the reacting-bond rule of Swain and Thornton[235] leads to the prediction that amides which are not activated by electron-withdrawing groups on the carbonyl function will have both oxygen and nitrogen more basic in the transition state than acyl activated amides. The solvation rule[236,237] then predicts that the proton involved in general catalysis will be bound more tightly to the substrate moiety than in the cases where the amide has no electron-withdrawing substituent on the carboxyl group, and that β for the general catalysis will be small. The rate coefficient ratio, k_3/k_2, will decrease and the second-order term become difficult to detect experimentally. The conclusion is supported by the data of Beicher and Taft[27] for difluoro-N-methylacetanilide ($k_3/k_2 = 34$) and chloro-N-methylacetanilide ($k_3/k_2 = 2$). When $\beta = 0$ the mechanism will be one of specific hydroxide catalysis.

The hydrolysis of 2,2,2-trifluoro-N-methylacetanilide, XXXVIII, in aqueous solution at pH 11.2 to 12.6 obeys a more complicated expression than (59), *viz.*

$$-\frac{d[XXXVIII]}{dt} = k_{obs}[XXXVIII] = k_0[XXXVIII] + [HO^-]\frac{k_a(k_1 + k_2[HO^-])}{k_a + k_1 + k_2[HO^-]} \quad (60)$$

which is consistent with the intermediate formation of an adduct, and hydroxide ion and general base-catalysed conversion of this species to products, and not with a single conversion of (XXXVIII) to hydrolysis products. At low base concentrations (pH 9.5–10) general base catalysis is observed because the decomposition of the intermediate XXXV is rate-determining, but at higher base concentrations (pH 12) the reaction kinetics revert to simple first order in hydroxide ion because the hydroxide ion-catalysed decomposition of XXXVI to products then exceeds in rate the unimolecular reversion of XXXV to reactants and the addition step is then rate-determining. Elimination, rather than addition, is subject to general base catalysis, since in order for the reaction to go over from second order in base at lower base concentrations to first order in base at higher concentrations it must be the second of the two steps since the first involves two hydroxide ions. The observation of a change in rate-determining step with pH requires a multistep reaction pathway.

Schowen and Zuorick[29] have determined the solvent isotope effects for the hydrolysis to be $k_a^{H_2O}/k_a^{D_2O} = 1.0 \pm 0.2$ (for addition), $k_1^{H_2O}/k_1^{D_2O} = 3.3 \pm 0.7$ (for solvent-catalysed elimination) and $k_2^{H_2O}/k_2^{D_2O} = 2.2 \pm 0.3$ (for lyoxide ion-catalysed elimination). These values may be associated with the following activation processes.

$$HO^- + XXXVIII \longrightarrow \left[HO..\underset{CF_3}{\overset{N(CH_3)C_6H_5}{C}}{\cdots}O \right]^- \qquad k_a^{H_2O}/k_a^{D_2O} = 1.0 \pm 0.2$$

$$\left[\begin{array}{l} H \\ \quad O\cdot\cdot H\cdot\cdot O{\cdots}\overset{O}{\underset{CF_3}{C}}\cdot\cdot N \begin{array}{l} CH_3 \\ C_6H_5 \end{array} \\ H \end{array}\right]^-$$

$$HO^- + H_2O + XXXVIII \quad \text{or} \qquad k_1^{H_2O}/k_1^{D_2O} = 3.3 \pm 0.7$$

$$\left[O{\cdots}\overset{O}{\underset{F_3C}{C}}\cdot\cdot\overset{CH_3}{\underset{C_6H_5}{N}}\cdot\cdot H\cdot\cdot O \begin{array}{l} H \\ H \end{array}\right]^-$$

$$\left[H-O\cdot\cdot H\cdot\cdot O{\cdots}\overset{O}{\underset{CF_3}{C}}\cdot\cdot N \begin{array}{l} CH_3 \\ C_6H_5 \end{array}\right]^{2-}$$

$$2OH^- + XXXVIII \quad \text{or} \qquad k_2^{H_2O}/k_2^{D_2O} = 2.2 \pm 0.3$$

$$\left[O{\cdots}\overset{O}{\underset{F_3C}{C}}\cdot\cdot\overset{CH_3}{\underset{C_6H_5}{N}}\cdot\cdot H\cdot\cdot O-H \right]^{2-}$$

These results show that a rate-determining proton transfer is occurring in the overall process leading to products from the adduct of XXXVIII and hydroxide ion. A fully concerted process, strongly analogous to the mutarotation of glucose, is unlikely since in this case the weight of evidence[63] suggests that the transfer of the proton is not concerted with the rate-determining elimination of alkoxide ion from the carbonyl group. The authors[33] suggest that the carbon–nitrogen bond in the activated complex is not undergoing cleavage, and instead the rate-determining step is a single proton transfer to form XXXIX which then decomposes to give the stable products, trifluoroacetate ion and N-methylaniline.

$$CF_3-\overset{O^-}{\underset{O^-}{C}}\overset{+}{-}\overset{CH_3}{\underset{C_6H_5}{N}}-H$$

XXXIX

3.3 INTRAMOLECULAR CATALYSIS

Amide hydrolysis is not only subject to acid and base catalysis, but may be also brought about by a neighbouring group in the amide, the so-called "propinquity catalysis" effect. Wolfrom *et al.*[238] were among the first to suggest this possibility, in order to explain the observed hydrolysis of aldonamides, *viz.*

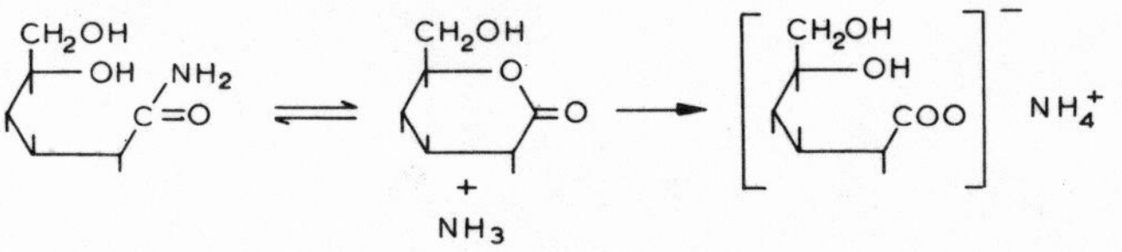

The rapid hydrolysis of hydroxylysylglycinamide in strong acid was postulated by Zahn and Zurn[239] to be a result of the intramolecular attack on a hydroxyl group, *viz.*

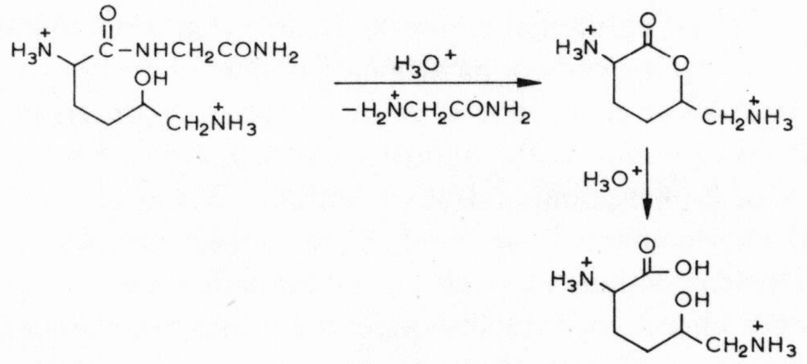

Similarly Zurn[240] found that the acidic hydrolysis of δ-hydroxyvaleramide and γ-hydroxybutyramide were much faster than the hydrolysis of the corresponding *n*-alkyl-carboamides.

Bruice and Hans-Marquardt[241] established the pH–rate profile for the hydrolysis of γ-hydroxybutyramide at 100°C in the neutral and alkaline region and found the observed rate to obey the equation

$$-\frac{\mathrm{d}[\text{amide}]}{\mathrm{d}t} = (k_{\text{OH}^-}[\text{OH}^-] + k_0)[\text{amide}] \tag{61}$$

The introduction of a hydroxyl group in the γ-position of butyramide increases the rates of alkaline and acid hydrolysis by factors of 15–20. In the neutral range, because of the spontaneous hydrolysis of γ-hydroxybutyramide, the introduction of a hydroxyl group has increased the rate of hydrolysis by at least 800. The mechanism suggested is

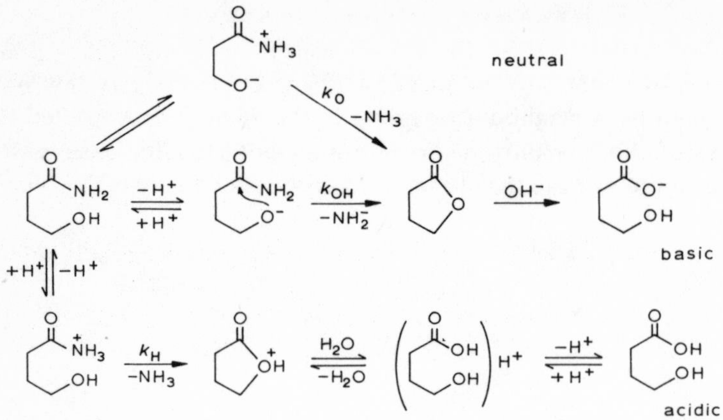

neutral

k_O $-NH_3$

$-H^+$ $+H^+$ k_{OH} $-NH_2^-$ OH^-

basic

$+H^+$ $-H^+$

k_H $-NH_3$ H_2O $-H_2O$ H^+ $-H^+$ $+H^+$

acidic

A more recent paper by Cunningham and Schmir[242] on the hydrolysis of 4-hydroxybutyranilide XL in neutral and alkaline solution suggests that intramolecular nucleophilic attack by the neighbouring hydroxyl group is followed by bifunctional catalysis by phosphate or bicarbonate buffers of the conversion of the tetrahedral intermediate to products. A quantitative comparison was made between the effects of buffer on the hydrolysis of 4-hydroxybutyranilide and on the hydrolysis of 2-phenyliminotetrahydrofuran, since both reactions proceed *via* identical intermediates. The mechanism suggested[243] states that the cyclisation of the hydroxyanilide ion yields an addition intermediate whose anionic form may either cleave to products or revert to reactant and whose neutral form invariably gives aniline and butyrolactone, *viz.*

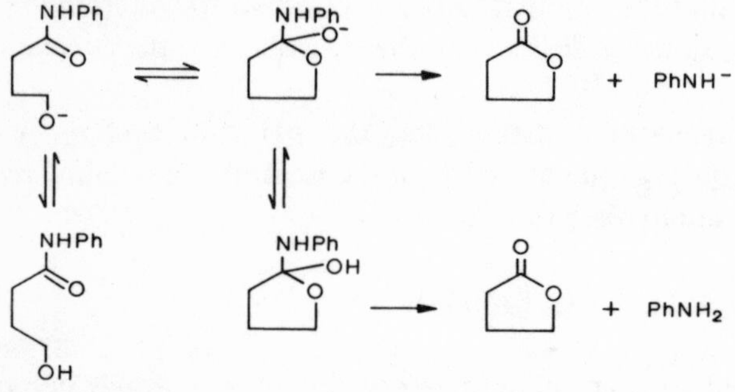

NHPh NHPh + $PhNH^-$

NHPh NHPh + $PhNH_2$

XL

At pH > 11, aniline production occurs *via* the anionic carbinolamine only and the overall reaction is first-order in hydroxide ion. At pH 8–10, the path *via* the carbinolanine gives a constant rate of hydrolysis and at pH 5.5–7 aniline arises solely through the neutral carbinolanine, present in its largest amounts. Eventually, acid catalysis leads to an increase in the rates of hydrolysis at lower pH. It is not necessary to involve a rate-determining attack by the neutral hydroxyl group to explain the pH-independent region. Moreover, when phosphate and carbonate are used as buffers, a non-linear dependence of the rate of hydrolysis upon total buffer concentration at fixed pH is seen. This non-linear variation suggests the existence of an addition intermediate whose formation is rate-determining at low buffer concentrations while at high buffer concentrations, the rate-limiting step is the formation of the intermediate. Phosphate buffer is far more effective than imidazole, in contrast to the reverse observation in the displacement at acyl carbon[244]. It is proposed that the complexes between the neutral intermediates and the acidic forms of the buffers decompose to yield aniline and are of the forms shown below, XXXXI and XXXXII, *viz.*

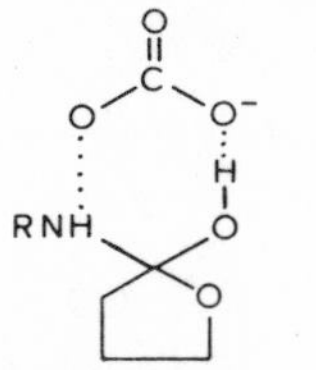

XLI 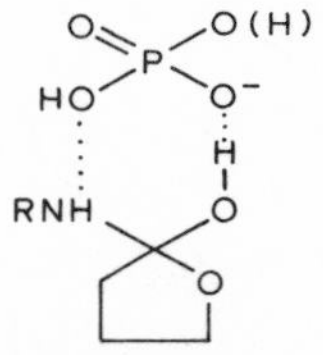 XLII

The specific acceleration of phosphoric acid and bicarbonate as catalysts has been attributed to bifunctional action in other reactions[245–248].

Bruice and Tanner[249] reported the first unequivocal instance of intramolecular general base catalysis of the hydrolysis of an amide bond by a neighbouring hydroxyl group. Using substituted *o*-hydroxybenzamides, including 5-nitrosalicylamide, salicylamide and *o*-methoxybenzamide, they measured their rates of hydrolysis in aqueous solution at 100°C and ionic strength 1.0 *M* with KCl. Benzamide and *o*-methoxybenzamide both give linear plots of log k_{obs} *versus* pH, whereas salicylamide and 5-nitrosalicylamide show plateaus over the regions 8–10 and 6–8 pH units respectively. If intramolecular general base catalysis is operative, then electron-withdrawal by the meta-nitro group should increase the susceptibility of the amide bond to nucleophilic attack, whereas the resultant large decrease in the basicity of the phenolate ion would result in a poorer general base catalyst.

The relative sizes of the Hammett ρ and Brönsted α constants will determine the relative rate of 5-nitrosalicylamide. If intramolecular base catalysis applies, then 5-nitrosalicylamide should hydrolyse more rapidly, since the nitro group will increase the susceptibility of the amide bond to attack by hydroxide ion and increase the efficiency of the phenolic hydroxyl as a general acid catalyst. The value of k_{obs} at the plateau region was found to be 18 times smaller for the 5-nitrosalicylamide than for salicylamide; a mechanism of intramolecular general base catalysis is, therefore, the preferred mechanism.

Amide hydrolysis may be assisted by a neighbouring carboxyl group. Leach and Lindley[250] found that, although asparagine or asparaginyl glycine hydrolyse *via* a normal acid hydrolysis to give ammonia and aspartic acid or aspartyl-glycins, the hydrolysis of glycyl-L-asparagine and L-leucyl-L-asparagine XLIII is dependent upon the mole fraction of substrate in the undissociated form. A mechanism involving intramolecular participation of the undissociated carboxyl group was suggested for XLIII.

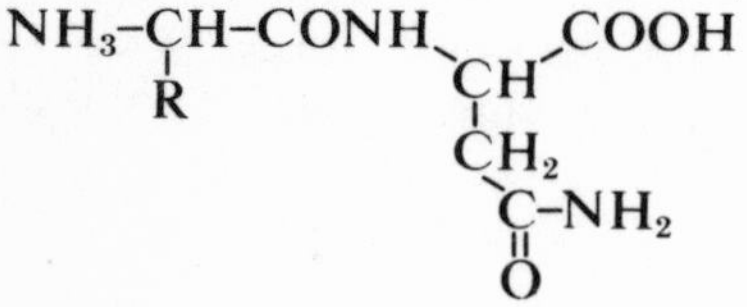

```
NH3-CH-CONH     COOH
    |       \  /
    R        CH
             |
             CH2
              \
               C-NH2
               ||
               O
```

XLIII

The work of Bender *et al.*[251,252] on the hydrolysis of the mono-amide and mono-methylester of phthalic acid indicated that the amide hydrolysis proceeded *via* the anhydride. However, it has been reported that the undissociated carboxyl group is the reactive species in the hydrolysis of such esters as methylhydrogen 2,3-di-(*t*-butyl)succinate and methylhydrogen-3,6-dimethylphthalate[253].

The hydrolysis of *o*-carboxyphthalimide, investigated by Zerner and Bender[254] shows evidence of intramolecular catalysis by the carboxyl group. Direct perpendicular attack of the carboxyl group on the imide carbonyl group is not sterically feasible and intramolecular general base or general acid catalysis is preferred. The pH rate profile for the hydrolysis of *o*-carboxy-pthalimide shows a maximum at pH 2.9, whereas that of phthalimide only a minimum of ~ 2.3 pH units, as expected for a continuation of the hydrogen ion- and hydroxide ion-catalysed reactions. The observed deuterium solvent kinetic isotope effect ($k_{obs}^{H_2O}/k_{obs}^{D_2O} = 2.6$ at pH 3) for the hydrolysis of *o*-carboxy-phthalimide, does not allow a differentiation to be made between a mechanism of general acid or of general base catalysis. The two mechanisms may be abbreviated as shown below.

General base catalysed

$$BH^+ \underset{K'_{a_1}}{\overset{\text{fast}}{\rightleftharpoons}} B + H^+; \; B + H_2O \underset{k_2}{\overset{k_1}{\rightleftharpoons}} \text{XLIV} \tag{62}$$

$$\text{XLIV} + H^+ \underset{K'_{a_2}}{\overset{\text{fast}}{\rightleftharpoons}} (\text{XLIV})H^+; \; (\text{XLIV})H^+ \xrightarrow{k_3} \text{products} \tag{63}$$

General acid catalysed

$$B + H^+ \underset{\text{fast}}{\overset{1/K'_{a_1}}{\rightleftharpoons}} BH^+; BH^+ + H_2O \underset{k'_2}{\overset{k'_1}{\rightleftharpoons}} (XLIV)H^+ \tag{64}$$

$$(XLIV)H^+ \underset{\text{fast}}{\overset{K'_{a_2}}{\rightleftharpoons}} H^+ + XLIV \xrightarrow{k'_3} \text{product} \tag{65}$$

where

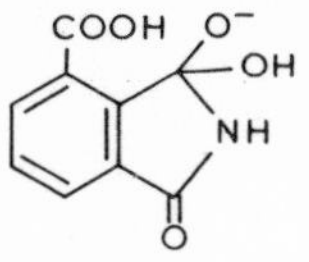

XLIV

Both mechanisms give expressions for k_{obs} which are consistent with the observed pH–rate profiles. Morawetz *et al.*[255,256] have reported schemes in which both a carboxyl group and a carboxylate ion participate in hydrolysis, and a similar mechanism of concerted catalysis has been proposed for the hydrolysis of *o*-carboxyphthalanilic acid[256] which exhibits a bell-shaped pH–rate profile.

Hoagland and Fox[257] investigated the rates of hydrolysis of some phthaloyl and nitrophthaloyl-ω-amino acids, *viz.*

at near neutrality, and showed that at pH 8.5 the length of the alkyl chain has little effect on the rate of hydrolysis, but that for both the 3 and 4-nitrophthaloyl series of acids the kinetic data show that N-methylcarboxylate and N-ethylcarboxylate at pH 7 are hydrolysed at an appreciably greater rate than derivatives having a longer alkyl chain between carboxylate and the imide nitrogen atom. pH–rate profiles also show that all phthaloyl derivatives of glycine or β-alanine are hydrolysed at greater rates than would be expected by extrapolation of the linear region above pH 8 to below pH 8. Carboxylate catalytic constants, $10^4\,k_{COO^-}$(sec^{-1}), are 3.2 ($n = 1$), 3.1 ($n = 2$), 1.2 ($n = 3$) and 0.15 ($n = 5$) for the 3-nitrophthaloyl amino acids. Two possible mechanisms for catalysis by the carboxylate group which are indistinguishable from the data are the direct addition of the carboxylate group to the carbonyl group as a nucleophilic catalyst or general base catalysis of the addition of a molecule of water, *viz.*

(*a*) *Nucleophilic catalysis*

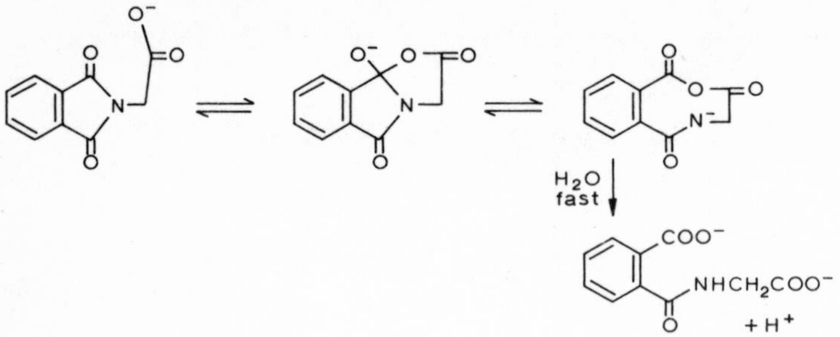

(*b*) *General base catalysis*

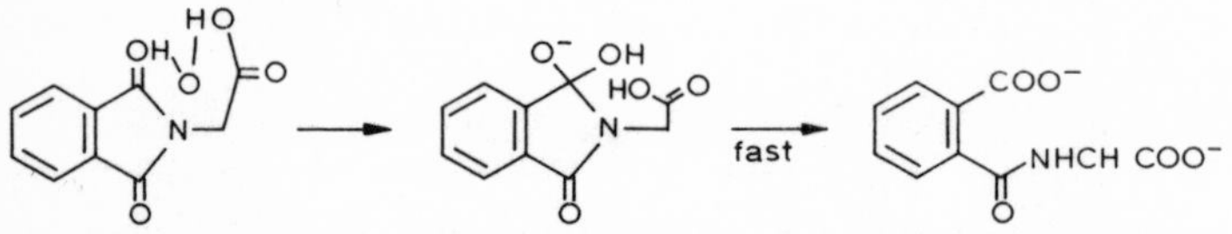

(66)

3.4 HYDROLYSIS OF DIAMIDES, LACTAMS AND N-ACYL-AMMONIUM IONS

The hydrolysis of diamides has been studied by Bruylants *et al.*[178,258,259], who found that, for complete hydrolysis, the reaction may be looked at as a sequence of two consecutive–competitive reactions, *viz.*

$$\underset{\text{(diamide)}}{A} \xrightarrow[+B]{k_I} E + \underset{\text{(monoamide)}}{C} \xrightarrow[+B]{k_{II}} \underset{\text{(diacid)}}{D} + E \tag{67}$$

$$E = NH_4^+ \text{ or } NH_3$$

Rate coefficients for the hydrolyses may then be determined by concentration–time or concentration–concentration relationships, in general using an excess of acid or base to reduce the kinetics to a simpler form.

Table 30 shows the activation parameters and relative rates of hydrolysis in both acid and base hydrolysis for the diamides investigated by Bruylants *et al.* The ratio $k_I/2k_0$, which gives the relative rate (corrected statistically) of the diamide to monoamide, is always greater than unity for alkaline hydrolysis and increases with decreasing distance between the groups, although one amide group will enhance the reactivity of a second one by virtue of its electron-withdrawing ability. In particular, for the alkaline hydrolysis of maleamide k_I ($= 3.8 \times 10^{-1}$ l·mole^{-1}·sec^{-1}) is greater than the rate coefficient for acetamide by two orders of magnitude, whereas that of fumaramide is only slightly greater. In acid-catalysed hydrolysis, the presence of a neighbouring amide

TABLE 30

KINETIC DATA FOR THE ACID AND BASE HYDROLYSIS OF SOME DIAMIDES
(k_0 refers to the hydrolysis of acetamide).

Amide		Alkaline ($k_0 = 6.34 \times 10^{-4}$ $l \cdot mole^{-1} \cdot sec^{-1}$ at 65°C)				Acid ($k_0 = 2.63 \times 10^{-4}$ $l \cdot mole^{-1} \cdot sec^{-1}$ at 65°C)			
		$\Delta H^{\ddagger}$ (kcal·mole⁻¹)	$\Delta S^{\ddagger}$ (eu)	$k_I/2k_0$ at 65°C	k_{II}/k_0 at 65°C	$\Delta H^{\ddagger}$ (kcal·mole⁻¹)	$\Delta S^{\ddagger}$ (eu)	$k_I/2k_0$ at 65°C	k_{II}/k_0 at 65°C
$H_2NOC\text{-}CONH_2$	I	9.25	−19.2	99		20	−6.1	0.24	
	II	11.55	−25.1		3.2	20	−4.6		0.95
$H_2NOC\text{-}CH_2\text{-}CONH_2$	I	16.0	−7.8	12.7		18.4	−6.1	0.22	
	II	11.4	−25.0		0.5	16.5	−15.3		0.74
$H_2NOC\text{-}(CH_2)_2\text{-}CONH_2$	I	16.5	−8.3	2.22		18.6	−6.8	0.93	
	II	13.9	−20.8		0.8	19.4	−5.8		0.73
$H_2NOC\text{-}CH{=}CH\text{-}CONH_2$	I	15.5	−11.3	6.64				0.19	
(*trans*)	II	10.3	−29.7		1.9				0.38
cis	I	14.8	−4.9	299		21.7	−2.7	0.22	
	II	18.0	−10.4		0.34	18.7	−7.0		3.2

group reduces the reactivity and the effect increases with the decreasing chain length between the groups. However, the rate ratio for the hydrolysis of the second amide group varies in the opposite direction, the hydrolysis of maleamic acid being considerably more rapid than the hydrolysis of a maleamide. Moreover, in the acid-catalysed hydrolysis of succinamide, maleamide and phthalamic acid a monomolecular mechanism of hydrolysis is observed. With these amides there is propinquity catalysis of some kind, the reaction proceeding *via* the imide, whereas in the reaction of *o*-benzamido-N,N-dicyclohexylbenzamide (XLV) participation is by the carbonyl oxygen[260]. This was shown by the isolation of benzoylanthranil in 80% yield from a reaction carried out in a solution of dry dioxan saturated with hydrogen chloride, the mechanism being

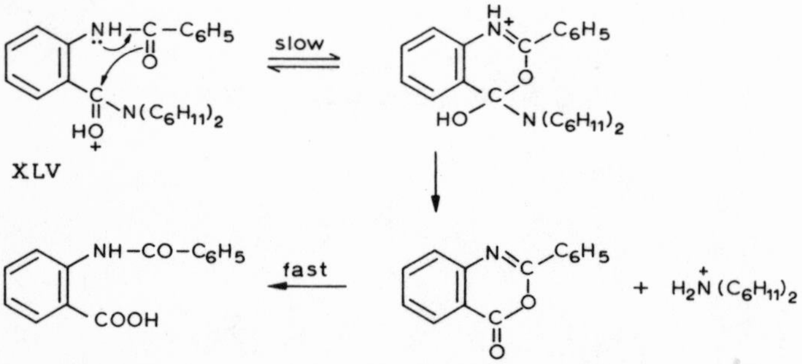

XLV

slow

fast

$+ \; H_2\overset{+}{N}(C_6H_{11})_2$

The hydrolysis of lactams parallel those of lactones, the 6-membered ring lactams being hydrolysed in basic media faster than either the 5- or 7-membered lactams[261] (Table 31). Further reduction in ring size greatly decreases the hydrolytic stability as seen in the α and β-lactams and β-lactones. Penicillin being the most obvious example of the β-lactam case.

The hydrolysis of N-acylammonium ions must also be considered as being

TABLE 31

LABILITY OF LACTAMS AND LACTONES TO BASE-CATALYSED HYDROLYSIS

Ring size	*Lactam*[261] (k_{OH^-})*rel.**	*Lactone*[262] $10^4\,k_{OH^-}$** $(l \cdot mole^{-1} \cdot sec^{-1})$
0	1.0	
5	0.94	1,480
6	4.53	55,000
7	0.44	2,500

* Temp., 75°C; solvent, H_2O.
** Temp., 0°C; solvent, 60/40 dioxan/water (vol. %).

related to the hydrolysis of amides, although they are postulated in the direct nucleophilic catalysis of the hydrolysis of carboxylic esters, acid chlorides and anhydrides. For example, pyridine in the absence of an acetate buffer catalyses the hydrolysis of acetic anhydride, *viz.*

$$(CH_3CO)_2O + C_5H_5N \underset{k_{-1}}{\overset{k_1}{\rightleftharpoons}} CH_3{-}\overset{O}{\overset{\|}{C}}{-}\overset{+}{N}C_5H_5 + CH_3COO^-$$

$$\xrightarrow[H_2O]{k_2} CH_3COOH + NC_5H_5 + H^+ \qquad (68)$$

and the catalytic coefficient for pyridine is about 30,000 times greater than that for acetate ions in 50% aqueous acetone at 25°C[263]. However, in aqueous acetate buffer, the catalytic coefficient for pyridine (k_p) is inversely proportional to the acetate concentration[264] and shows a solvent isotope effect (k_{H_2O}/k_{D_2O}) of 5 ± 1. k_p can be shown to be given by

$$k_p = k_1 k_2 [H_2O]/(k_{-1}[OAc^-] + k_2[H_2O]) \qquad (69)$$

Therefore, if $k_{-1}[OAc^-] \gg k_2[H_2O]$ then an inverse relationship between k_p and $[OAc^-]$ and a large solvent isotope effect are reasonable, since the isotope effect arises from the conversion of H_2O to $H_2O{-}R^+$ in the transition state with the corresponding lowering of the zero point energy of the water molecule. Johnson[265] has shown that no catalysis is experienced by acetic anhydride, in the presence of a buffer 1.06 *M* in acetate and 0.54 *M* in acetic acid, due to free pyridine (0.063 *M*) in 40% dioxan–60% water (v/v) at 35°C. Catalysis by free pyridine is observed with 100% water[264], the explanation being that the overall rate coefficient is given by

$$k_{obs} = k_w + k_{Ac}[OAc^-] + k_p[C_5H_5N] \qquad (70)$$

where k_w is the rate coefficient for the reaction with water and k_{Ac} is the catalytic coefficient for acetate ion. Thus, on going to a lower dielectric solvent, k_p is reduced relative to $k_w + k_{Ac}[OAc^-]$, because the value of k_1 is reduced and particularly because the value of k_{-1} (ion recombination) is raised more rapidly than k_w and k_{Ac} are lowered ($k_p \propto 1/k_{-1}[OAc^-]$). Bunton *et al.*[266] have obtained evidence that acetylpyridinium ion undergoes general base-catalysed hydrolysis, as is also suggested by the large solvent kinetic isotope effect of 5.5[267,268].

N-acylimidazoles may be prepared[269–271], and because of their relevance to enzyme chemistry (*via* histidine[272]) they have been the subject of much quantitative investigation[269]. For the 4(5)-substituted imidazoles or imidazole itself, the N-acyl imidazole intermediates are hydrolytically unstable. The stability increases in the order N-acylpyridines < acylimidazoles ≪ amides. However, N-acetylbenzimidazole has a half-life of 21 h at room temperature[273] as compared with a half-life of 40 min for N-acetylimidazole under the same conditions[273], and with N-benzoylimidazole which is stable for days in cold carbonate solution[274].

Jencks and Carriuolio[275] have shown that the pH–rate profile for the hydrolysis of N-acetylimidazole (25°C, solvent H_2O) is given by

$$k_{obs} = 2.8\,[\mathrm{AcIMH^+}] + 0.005\,[\mathrm{AcIM}] + 19{,}000\,[\mathrm{AcIM}][\mathrm{OH^-}] \qquad (71)$$

where $\mathrm{AcIMH^+}$, AcIM are given in equation 72 and where pK'_a for the ionisation

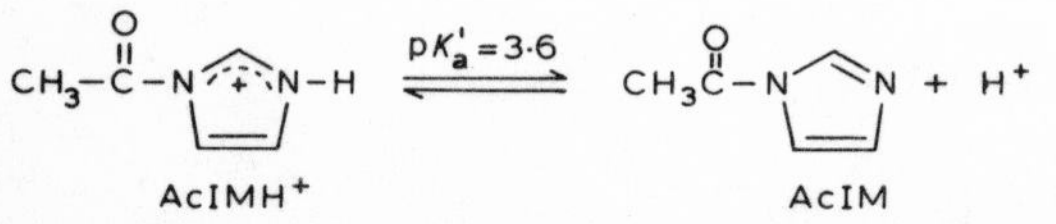

$$\underset{\mathrm{AcIMH^+}}{\mathrm{CH_3{-}\overset{O}{\overset{\|}{C}}{-}N\overset{+}{\bigcirc}N{-}H}} \; \underset{}{\overset{pK'_a = 3.6}{\rightleftharpoons}} \; \underset{\mathrm{AcIM}}{\mathrm{CH_3\overset{O}{\overset{\|}{C}}{-}N\bigcirc N}} + \mathrm{H^+} \qquad (72)$$

is 3.6. The spontaneous hydrolysis of $\mathrm{AcIMH^+}$, shown by the plateau in the pH–rate profile below 2 pH units is probably a bimolecular reaction with water rather than a unimolecular decomposition to the acylium ion since the reaction is strongly depressed by concentrated salt or acid (*cf.* amides in strong acids), and values for k_{H_2O}/k_{D_2O} of 2.5 and $\Delta S^{\ddagger} = -30.2$ eu are reported. Since the neutral imidazole is a good leaving group, very little oxygen exchange would be expected in the hydrolysis of $\mathrm{AcIMH^+}$, an expectation confirmed by Bunton[276]. The neutral hydrolysis of AcIM

$$\mathrm{AcIM + H_2O = AcOH + IMH}$$

is probably a general acid–base-catalysed attack involving

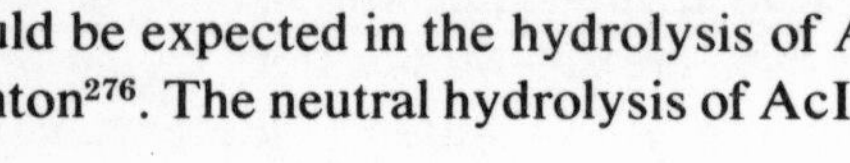

$$\mathrm{H_2O \cdots H \cdots \overset{H}{\overset{|}{O}} \cdots \overset{O}{\overset{\|}{\underset{CH_3}{\underset{|}{C}}}}{-}N\bigcirc N \cdots H \cdots OH}$$

A class of acylpyridinium salts that has been investigated quantitatively is the carbamoyl pyridinium compounds. Diphenylcarbamoyl-pyridinium chloride was synthesised in 1907[277]. 1-(N,N-Dimethylcarbamoyl)-pyridinium chloride (XLVI) was found by Johnson and Rumon[278] to decompose to its component parts in non-hydroxylic solvents, dielectric constants from 5 to 95, *viz.*

$$(CH_3)_2N-\overset{O}{\overset{\|}{C}}-\overset{+}{N}C_5H_5\ Cl^- \underset{k_2}{\overset{k_1}{\rightleftharpoons}} C_5H_5N + (CH_3)_2N-\overset{O}{\overset{\|}{C}}-Cl \quad (73)$$

XLVI

but not in hydroxylic solvents with dielectric constants from 11 to 80. This was attributed to the variation of relative nucleophilicity of pyridine and chloride in the two types of solvent. In water (XLVI) reacts with nucleophiles by direct nucleophilic displacement and only hydroxide and water terms were found in the rate coefficients (k_0) which were independent of buffer concentration. No catalysis by H_3O^+ was seen up to 1 *M* HCl[232]. A linear free energy correlation of the reactivity of XLVI compared with that of *p*-nitrophenylacetate indicates that anionic nucleophiles do not show any special reactivity to the cationic substrate. It is interesting to note that Johnson and Ramon showed that the neutral hydrolysis term was not due to an S_N1-type of mechanism to give the dimethylcarbamoyl cation[188], *viz.*

$$(CH_3)_2N-\overset{O}{\overset{\|}{C}}-\overset{+}{N}C_5H_5 \underset{k_{-1}}{\overset{k_1}{\rightleftharpoons}} (CH_3)_2N\overset{+}{\cdots}C\cdots O + C_5H_5N$$

$$\downarrow k_2$$

$$(CH_3)_2N-CO_2H \quad (74)$$

$$\downarrow \text{fast}$$

$$\text{products}$$

whence $k_{H_2O} = k_1k_2/(k_{-1}[C_5H_5N]+k_2)$, since there was a lack of a mass law effect of added pyridine.

However, k_{obs} *versus* $[C_5H_5N]$ is linear in the pH region used (5.2–5.3) in which the water reaction would be expected to be most important, and also $\Delta S^\ddagger = -27$ eu for the water reaction, whereas S_N1 ester hydrolyses characteristically have values of $\Delta S^\ddagger$ of 0–10 eu.

3.5 EXPERIMENTAL METHODS USED IN KINETIC STUDIES OF AMIDE HYDROLYSIS

A wide variety of analytical methods has been utilised to follow the rates of amide hydrolysis. The paper of Bruylants[178] contains a valuable comparison of the more widely used methods. These are

(*a*) The titration of ammonia. The amount of ammonia in the reaction mixture can, for example, be determined colorimetrically after Nesslerisation. This method was successfully applied to the hydrolysis of numerous aliphatic amides in both acid and base solutions.

(*b*) The amount of organic acid may be determined after removal of alkali, etc. by a cation exchange. Bruylants found that this method worked well for succinamide.

TABLE 32

ALKALINE HYDROLYSIS OF ACETAMIDE AT 25°C

[*NaOH*]	$k_2 \times 10^5$ ($l \cdot mole^{-1} \cdot sec^{-1}$)	*Method*
0.050	3.77	Nessler reagent
0.200	3.73	Ion exchange
0.500	3.75	Gas volumetric
1.00	3.73	Gas volumetric
1.00	3.78	*UV* spectrophotometric

(*c*) Unreacted amide may be determined spectrophotometrically (*UV*, 200–230 mμ). Table 32 shows the reproducibility achieved between the results of different authors using these methods[183,279,280].

Benrath[184] used a dilatometric method to follow the acid-catalysed hydrolysis of acetamide, whereas Leisten[212] and Rabinovitch and Winkler[187] both analysed for the liberated ammonia by the "formol" method[281].

Edwards and Meacock[189], following amide hydrolyses in strong acid solution, used a UV spectrophotometric method. Bunton *et al.*[224] followed the hydrolysis of NN-dimethylbenzamide by adding an alcoholic solution of 1-fluoro-2-4-dinitrobenzene and alkali to aliquots of reaction mixture and determining the N,N-dimethyl-2,4-dinitroaniline spectrophotometrically at 381 mμ.

Biechler and Taft[27], also used a UV spectrophotometric method for anilides, including trifluoroacetanilide, as did Johnson[265] for the hydrolysis of dimethylcarbamoylpyridinium chloride. The latter compound was also studied by the use of NMR and by conductivity measurements by the same author and it was found that these techniques give complementary results. Hoagland and Fox[257] used a pH-Stat method to follow the hydrolyses of phthaloyl and nitrophthaloyl-amino acid, whereas Jellinek and Gordon[226] used a polarographic method to follow the hydrolysis of nicotinamide.

4. Hydrolysis of carboxylic anhydrides

The hydrolysis of carboxylic anhydrides

$$RCO \cdot O \cdot OCR + H_2O = 2RCOOH \tag{75}$$

has been studied quantitatively since the beginning of the century, particularly with regard to the effects of various catalysts, subject as the hydrolyses are to acid, base, and nucleophilic catalysis. Early work such as that of Ostwald[282] and Lunden[283] has been reviewed in the papers of Verkade[284] and Skrabal[285]. Szabo[286] reported that the hydrolysis of acetic anhydride was catalysed by hydrogen ions and acetate ions. Orton and Jones[287] studied the hydrolysis of acetic anhydride in acetic acid, in acetic anhydride and acetone, and in various mixtures of these solvents, and found that non-ionised acids are effective catalysts in a solvent containing over 90% acetic acid. The reactions were also studied by Verkade[284] by conductivity methods and by Benrath[288] by a densitometric method. Kilpatrick[289], using a dilatometric method, measured the velocity of hydration of acetic anhydride at 0°C and found $k = 0.00045$ sec^{-1} in pure water. Added sodium chloride depressed the rate, and added acid increased it ($k_{H_3O} = 0.00052$ sec^{-1}) but acetate ion was 20% more effective ($k_{Ac^-} = 0.00064$). Formate ion was found to be a remarkably good catalyst ($k_{HCOO^-} = 0.66$). In sodium propionate, propionic acid and sodium butyrate/butyric acid the reaction was found to be slower than in water alone. Moreover, the reaction under these conditions no longer exactly follows the first-order rate law. The kinetic data in these media indicate the intermediate formation of mixed anhydrides, *viz.*

$$\begin{aligned} &RCOO{\cdot}COR + R'COO^- = RCOO{\cdot}OCR' + RCOO^- \\ &RCOO{\cdot}COR' + H_2O = RCOOH + R'COOH \end{aligned} \qquad (76)$$

Kilpatrick's results[289–291] clearly indicate a reactivity order for the aliphatic anhydrides of formic > acetic > propionic > butyric, since verified experimentally. The observed catalysis by acetate ion must be a case of general base catalysis.

The acid catalysis of the hydrolysis of acetic anhydride was later investigated by Gold and Hilton[292] using aqueous solutions of hydrochloric, perchloric, sulphuric and phosphoric acids at 0°C. The first-order rate coefficient for hydrolysis is approximately proportional to h_0 for all of these acids over a wide range of concentrations and this correlation is far better than with log $[H^+]$. The mechanism proposed by these authors is

$$(CH_3CO)_2O + H^+ \rightleftharpoons [(CH_3CO)_2OH]^+ \qquad \text{equilibrium} \qquad (77)$$

$$\text{A-1}\begin{cases} [(CH_3CO)_2OH]^+ \rightarrow [\overset{*}{M}{}^+] \rightarrow CH_3CO^+ + CH_3COOH & \text{slow} \qquad (78) \\ [CH_3CO^+] + H_2O \rightarrow CH_3COOH + H^+ & \text{fast} \qquad (79) \end{cases}$$

where $[\overset{*}{M}{}^+]$ represents the transition state for the rate-determining step. Gold and Hilton[292] established that acetyl chloride was not an intermediate when hydrochloric acid is used as a catalyst, but it is an intermediate in acetylations by acetic anhydride in acetic acid containing hydrogen chloride[293].

Bunton and Perry[294] measured the rates of hydrolysis of acetic, benzoic, mesitoic, acetic–benzoic and acetic–mesitoic anhydrides in aqueous dioxan. They found that the hydrolyses of acetic, acetic–benzoic and benzoic anhydrides in this medium, catalysed by perchloric acid, proceeded by A-2 mechanisms (see below), and those of acetic–mesitoic and mesitoic anhydrides predominantly by the A-1 mechanism. Since the hydrolysis of acetic anhydride in aqueous media was thought to go by the A-1 mechanism, there appeared to be a change in mechanism for this anhydride with changing solvent. Three mechanistic criteria were applied; the Zücker–Hammett hypothesis[295], the values of the entropies of activation, and the position of bond fission in the hydrolysis of the unsymmetrical anhydride. For a mixed anhydride, RCO-OOCR′, and $H_2{}^{18}O$ the A-2 and A-1 mechanisms are shown in their simplest form below.

$$\begin{matrix} R'CO \\ \quad O \\ RCO \end{matrix} + H^+ \underset{}{\overset{\text{fast}}{\rightleftharpoons}} \left[\begin{matrix} R'CO \\ \quad O.H \\ RCO \end{matrix}\right]^+ \tag{80}$$

$$\text{A-1} \quad \left[\begin{matrix} R'CO \\ \quad O.H \\ RCO \end{matrix}\right]^+ \underset{\text{fast}}{\overset{\text{slow}}{\rightleftharpoons}} RCO_2H + R'\overset{+}{C}O\,H_2{}^{18}O \longrightarrow R'C\begin{matrix} =O \\ -{}^{18}OH \end{matrix} + H^+ \tag{81*}$$

$$\text{A-2} \quad \left[\begin{matrix} R'CO \\ \quad O.H \\ RCO \end{matrix}\right]^+ + H_2{}^{18}O \longrightarrow R'CO_2H + RC\begin{matrix} =O \\ -{}^{18}OH \end{matrix} + H^+ \tag{82}$$

The hydrolysis of acetic anhydride in water was thought to be an A-1 reaction, being both h_0-dependent and having $\Delta S^{\ddagger} = +2.2$ eu. However, for the hydrolysis in aqueous dioxan, Bunton and Perry[294] found values for $\Delta S^{\ddagger}$ which suggest that the mechanism is not of the A-1 type (see Table 33), and found that k_{D_2O}/k_{H_2O} is 1.45 (in 60/40 dioxan/water). This value for the solvent isotope effect is in the range normally associated with A-2 reactions; A-1 reactions are usually found to have $k_{D_2O}/k_{H_2O} > 2$. They conclude that the mechanism in aqueous dioxan is bimolecular where the activity coefficient ratio (f_S/f_{X^*}) varies with the acidity as does the ratio f_B/f_{BH^+}, where S represents the initial and X* the transition state (containing S, H^+ and H_2O) and B is a Hammett base.

Acetic–mesitoic anhydride hydrolysis was found to be dependent upon h_0, to have $\Delta S^{\ddagger} = -3.8$ eu and bond fission was found to be predominantly mesitoyl–oxygen. This evidence strongly suggests the A-1 mechanism in-

* Rapid equilibration of the ^{18}O will take place between the carbonyl and hydroxyl oxygens.

TABLE 33

KINETIC DATA FOR SOLVOLYSIS OF ANHYDRIDES IN PERCHLORIC ACID AT 0°C[294]

Anhydride	*Solvent*	*Slope of Hammett plot*	$\Delta H^{\ddagger}$ *(kcal·mole⁻¹)*	$\Delta S^{\ddagger}$ *(eu)*	*Bond fission*	*Mechanism*
Acetic	Dioxan/ water					
	60/40 (v/v)	0.7	16.5	−16.1		A-2(?)
	40/60 (v/v)	0.7	15.4	−19.3		A-2(?)
	H_2O	1	21.4	+2.2		(A-1)*
Benzoic	Dioxan/ water					
	60/40 (v/v)	$k_1 \propto [HClO_4]$	18.8	−18.9		A-2
Mesitoic			21.8	−8.6		A-1
Acetic–benzoic		*ca.* 0.6	15.7	−22.4	CH_3CO–O	A-2
Acetic–mesitoic		1.1	18.1	−3.8	ArCO–O	A-1

* See later in this section.

volving the formation of a mesitoylium ion. Acetic–benzoic anhydride hydrolysis, on the other hand, was found to satisfy the criteria for an A-2 mechanism. The acetyl–oxygen bond is broken both in acidic and neutral hydrolysis, the rate is not dependent upon h_0 and $\Delta S^{\ddagger} = -22.4$ eu. The rate of hydrolysis is one third that of acetic acid under comparable conditions. Benzoic anhydride (which is not very soluble in water) gave results suggesting an A-2 mechanism with a $\Delta S^{\ddagger}$ value of -18.9 eu and a non-linear dependence of k_1 upon h_0 in dioxan/water.

A re-investigation of the mechanism of the hydrolysis of acetic anhydrides in water was made by Bunton and Fendler[4]. They found that the hydrolysis of trimethylacetic anhydride follows an A-2 mechanism both in water and in aqueous dioxan and is slower than that of acetic anhydride. If acetic anhydride follows an A-1 mechanism in water, both anhydrides should have the same mechanism and both should have similar reactivities in aqueous acids. If the A-2 mechanism is followed, steric effects should make trimethylacetic anhydride the less reactive compound. The entropy of activation of the hydrolysis catalysed by perchloric acid was re-estimated, taking into account the electrolyte effect of the perchlorate ions upon the rate of spontaneous hydrolysis, and a new value for $\Delta S^{\ddagger}$ of -35 eu (for the spontaneous hydrolysis) and -10 eu at 2 M $HClO_4$ for the acid-catalysed hydrolysis was obtained, compared with the value of $+2.2$ eu obtained previously. The new value is in the order of magnitude of $\Delta S^{\ddagger}$ for A-2 reactions but smaller than that observed for trimethylacetic anhydride under similar acidic conditions (-26 eu). Plots of log k against log $[H^+]$ or H_0 are curved for both anhydrides and no evidence was found for an A-1 mechanism for the hydrolysis of trimethylacetic anhydride.

Nucleophilic and general base catalysis of the hydrolysis of acetic anhydrides are closely related. The hydrolysis of acetic anhydride is catalysed by formate and nitrite ions by nucleophilic catalysts[296]. Evidence for the postulated intermediate in the latter case, acetyl nitrite, has been obtained by the addition of α-naphthylamine to the system, trapping the intermediate to form 4-amino-1,1-azonaphthalene, *viz.*

$$CH_3COOOCCH_3 + NO_2^- \xrightarrow{\text{slow}} CH_3COO^- + CH_3\overset{\displaystyle O}{\overset{\|}{C}}\text{-}ONO \tag{83}$$

$$CH_3\overset{\displaystyle O}{\overset{\|}{C}}\text{-}ONO \xrightarrow[H_2O]{\text{fast}} CH_3COO^- + ONO^- (+2H^+)$$

$$CH_3\overset{\displaystyle O}{\overset{\|}{C}}\text{-}ONO \xrightarrow[\text{fast}]{\alpha\text{-naphthylamine}} \text{4-amino-1,1-azonaphthalenes}$$

Bunton and Fendler[297] have shown that fluoride ion will catalyse the hydrolysis of acetic and succinic anhydrides in water and aqueous dioxan. In water, the rate of loss of acetic anhydride is greater than the rate of formation of acetic acid, showing the build up of acetyl fluoride. The hydrolysis of succinic anhydride is also catalysed by fluoride ion but no build up of succinyl fluoride is seen. The catalysis of anhydride hydrolysis by pyridine has been discussed in the previous section, and it is perhaps sufficient to mention here that the catalytic coefficient for pyridine is about 30,000 times greater than that for acetate ion in 50% aqueous acetone at 25°C.

4.1 COMPARISON OF THE MECHANISMS OF ANHYDRIDE HYDROLYSIS

Although, so far in this section, we have discussed the kinetic evidence in terms of the A-1 and A-2 mechanisms favoured by Bunton *et al.* (p. 224), the same points which give rise to differences in detailed interpretation of mechanism for hydrolysis of acyl halides apply with equal weight to the hydrolysis of anhydrides. Butler and Gold[67,68,298,299] have studied the spontaneous hydrolysis of acetic anhydride and presented the following scheme.

$$CH_3\overset{O}{\overset{\|}{C}}OAc + H_2O \rightleftharpoons CH_3\underset{OH}{\overset{OH}{C}}\text{-}OAc \qquad \text{rapid} \tag{84}$$

$$CH_3\underset{OH}{\overset{OH}{C}}\text{-}OAc + H_2O \rightleftharpoons CH_3\underset{OH}{\overset{O^-}{C}}\text{-}OAc + H_3O^+ \qquad \text{rapid} \tag{85}$$

$$CH_3\underset{OH}{\overset{O^-}{C}}\text{-}OAc + HX \longrightarrow CH_3\underset{OH}{\overset{O^- \quad H^+}{C\text{-}O\text{-}Ac}} + X^- \qquad \text{slow} \tag{86}$$

$$CH_3\underset{OH}{\overset{O^- \quad H^+}{C\text{-}O\text{-}Ac}} \longrightarrow 2\,CH_3C\begin{matrix}\nearrow O \\ \searrow OH\end{matrix} \tag{87}$$

Two successive pre-equilibria lead to an anionic intermediate followed by a rate-determining proton transfer from the conjugate acid of the catalyst to an oxygen atom of the intermediate. In the case of catalysis by acetate ion, these authors have proposed that their experimental data require that the transition state contain one acetic anhydride, one acetate and at least one water molecule.

Their mechanism has been criticised by Johnson[300] and by Koskikallio[301] on several grounds. Hipkin and Satchell[302] have proposed a direct synchronous displacement mechanism

$$CH_3\overset{O}{\overset{\|}{C}}OAc + 2H_2O \xrightarrow{\text{slow}} CH_3\underset{\underset{H\ \ OH_2}{H-O}}{\overset{O}{\overset{\|}{C}}}\text{-}O\text{-}Ac \xrightarrow{\text{fast}} 2\,CH_3C\begin{matrix}\nearrow O \\ \searrow OH\end{matrix} + H_2O \tag{88}$$

of a piece with their proposals for the hydrolysis of acyl halides. They suggest that, although bond-breaking is more important than bond-making in these reactions, the reaction is still essentially bimolecular and not unimolecular. Whether this is still the case in more polar solvents is open to dispute[303]. Points against the concerted mechanism are the observed large decrease in the activation energy (*i.e.* a large negative value of $\Delta C_p^{\ddagger}$) with rising temperature and the marked increase in the rate of hydrolysis with decreasing water content that are seen in the hydrolysis of open-chain simple anhydrides and which point to the existence of a highly polar intermediate in the reaction.

The mechanism proposed by Johnson[300] for the general base-catalysed hydrolysis of acetic anhydride, *viz.*

$$\begin{array}{ccccc} CH_3\overset{O}{\overset{\|}{C}}{-}O{-}Ac + B + H_2O & \xrightarrow[\text{slow}]{k_1} & CH_3\overset{O^{\delta-}}{\overset{\|}{C}}{-}O{-}Ac \\ & & \delta^-\,\underset{H}{O}\cdot\cdot H\cdot\cdot B\delta^- & \longrightarrow & CH_3{-}\overset{O^-}{\underset{OH}{C}}{-}O{-}Ac \\ & & & & \downarrow \text{fast} \\ CH_3\overset{O}{\overset{\|}{C}}{-}O{-}Ac + B + H_2O & \xrightarrow[\text{slow}]{k_1} & CH_3\overset{O^{\delta-}}{\overset{\|}{C}}{-}O{-}Ac \\ & & \delta^-\,\underset{H}{O}\cdot\cdot H\cdot\cdot B\delta^- & \xrightarrow{\text{fast}} & \text{products} \end{array} \quad (89)$$

is in general agreement with the mechanism proposed later by Kivinen[303] for the neutral hydrolysis, *viz.*

$$CH_3{-}\overset{O}{\overset{\|}{C}}{-}O{-}Ac + 2\,H_2O \rightleftharpoons \left[CH_3{-}\overset{O}{\overset{\|}{C}}{-}O{-}Ac \atop \underset{H}{O}\cdot H\cdot\cdot OH_2 \right] \rightleftharpoons CH_3{-}\overset{O^-\;H_3O^+}{\underset{\underset{H}{O}}{C}}{-}O{-}Ac$$

$$\Downarrow\Uparrow \quad (90)$$

$$CH_3{-}\overset{O}{\overset{\|}{\underset{\underset{H}{O}}{C}}} + \underset{H}{O}{-}Ac + H_2O \rightleftharpoons \left[CH_3{-}\overset{O}{\overset{\|}{C}}{-}O{-}Ac \atop \underset{H}{O}\;H{-}OH_2 \right]$$

Kivinen proposes that the neutral hydrolysis of acetic anhydride is promoted by water acting as a weak base. The solvent isotope effect, $k_{H_2O}/k_{D_2O} \simeq 3$, is suggestive of general base catalysis.

In summary, therefore, the detailed mechanism of the hydrolysis of carboxylic anhydrides is still in doubt and we must hope for further experimental evidence to clarify the position. As for the hydrolysis of the other carboxylic acid derivatives dealt with in this chapter, none of the mechanistic criteria, that have been used to interpret the kinetic data, gives an unambiguous interpretation, resulting in a situation where details of mechanism are open to argument. This is particularly the case for solvolysis reactions where uncertainty as to the structure and effect of the solvent preclude a firm assignment of transition state structures. This is not to say that the mechanisms are not understood, and many valuable extrapolations as to structural and solvent effects may be made on the basis of these mechanisms, based as they are on extensive experimental data.

REFERENCES

1 M. L. BENDER, *Chem. Rev.*, **60** (1960) 53.
2 R. F. HUDSON, *Chimia (Aaran)*, **15** (1961) 394.
3 S. L. JOHNSON, *Advan. Phys. Org. Chem.*, **5** (1967) 237.
4 T. C. BRUICE AND S. J. BENKOVIC, *Bioorganic Mechanisms*, Wiley, London and New York, 1966.
5 C. K. INGOLD, *Structure and Mechanism in Organic Chemistry*, Cornell University Press, Ithaca, New York, 1953.
6 F. KLAGES, *Chem. Ber.*, **82** (1949) 358.
7. J. L. FRANKLIN, *Ind. Eng. Chem.*, **41** (1949) 1070.
8 G. W. WHELAND, *Resonance in Organic Chemistry*, Wiley, New York, 1955.
9 F. LYNEN, E. REICHERT AND L. RUEFF, *Annalen.*, **14** (1951).
10 L. PAULING AND R. B. COREY, *J. Am. Chem. Soc.*, **74** (1952) 3964.
11 M. ST. C. FLETT, *Trans. Faraday Soc.*, **44** (1948) 767.
12 M. ST. C. FLETT, *J. Chem. Soc.*, (1951) 962.
13 N. FUSON, M. L. JOSIEN AND E. M. SHELTON, *J. Am. Chem. Soc.*, **76** (1954) 2526.
14 R. N. JONES, W. F. FORBES AND W. A. MUELLER, *Can. J. Chem.*, **35** (1957) 504.
15 E. J. COREY, T. H. TOPIE AND W. A. WOZNIAK, *J. Am. Chem. Soc.*, **77** (1955) 5415.
16 E. J. COREY, H. J. BURKE, *J. Am. Chem. Soc.*, **77** (1955) 5415.
17 C. S. FOOTE, *J. Am. Chem. Soc.*, **86** (1964) 1853.
18 M. J. S. DEWAR, *Electronic Theory of Organic Chemistry*, Oxford University Press, London, 1949.
19 R. L. SCHOWEN, *J. Am. Chem. Soc.* **88** (1966) 1215.
20 M. M. DAVIES, *Trans. Faraday Soc.*, **36** (1940) 333.
21 A. L. HENNE, M. S. NEWMAN, L. L. QUILL AND R. A. STAINFORTH, *J. Am. Chem. Soc.*, **69** (1947) 1819.
22 A. LAPWORTH AND R. H. F. MANSKE, *J. Chem. Soc.*, (1928) 2533.
23 M. L. BENDER, *J. Am. Chem. Soc.*, **73** (1951) 1626.
24 C. A. BUNTON, T. A. LEWIS AND D. R. LLEWELLYN, *Chem. Ind. (London)*, (1954) 1154.
25 M. E. ZAUGG, V. PAPENDICK AND R. J. MICHAELS, *J. Am. Chem. Soc.*, **86** (1964) 1399.
26 M. L. BENDER AND R. J. THOMAS, *J. Am. Chem. Soc.*, **83** (1961) 4183.
27 S. S. BEICHLER AND R. W. TAFT JR., *J. Am. Chem. Soc.*, **79** (1957) 4927.
28 P. M. MADER, *J. Am. Chem. Soc.*, **87** (1965) 3191.
29 R. L. SCHOWEN AND G. W. ZUORICK, *J. Am. Chem. Soc.*, **88** (1966) 1223.
30 A. BRUYLANTS AND F. KEZDY, *Record Chem. Progr.*, **21** (1960) 213.
31 M. T. BEHUNE AND E. M. CORDES, *J. Org. Chem.*, **29** (1964) 1255.
32 K. R. LYNN, *J. Phys. Chem.*, **69** (1965) 687.
33 R. L. SCHOWEN, H. JAYARAMAN AND L. KERSHNER, *J. Am. Chem. Soc.*, **88** (1966) 3373.
34 B. ZERNER AND M. L. BENDER, *J. Am. Chem. Soc.*, **83** (1961) 2267.
35 W. P. JENCKS AND M. GILCHRIST, *J. Am. Chem. Soc.*, **86** (1964) 5616.
36 E. S. HAND AND W. P. JENCKS, *J. Am. Chem. Soc.*, **84** (1962) 3505.
37 M. L. BENDER AND H. D'A. HECK, (1966) unpublished work.
38 L. R. FEDOR AND T. C. BRUICE, *J. Am. Chem. Soc.*, **87** (1965) 4138.
39 M. L. BENDER AND R. D. GINGER, *J. Am. Chem. Soc.*, **77** (1957) 348.
40 C. A. BUNTON, N. A. FULLER, S. G. PERRY AND V. J. SHINER, *J. Chem. Soc.*, (1963) 2918.
41 C. A. BUNTON AND J. H. FENDLER, *J. Org. Chem.*, **30** (1965) 136.
42 M. L. BENDER, *J. Am. Chem. Soc.*, **75** (1953) 5986.

43 W. D. PHILLIPS, *J. Chem. Phys.*, **23** (1955) 1363.
44 H. S. GUTOWSKY AND C. H. HOLM, *J. Chem. Phys.*, **25** (1956) 1228.
45 R. M. HAMMAKER AND B. A. GUGLER, *J. Mol. Spectr.*, **17** (1965) 356.
46 S. G. ENTELIS AND O. V. NESTEROV, *Proc. Acad. Sci., U.S.S.R., Chem. Sect. (Engl. Transl.)*, **148** (1963) 174.
47 M. L. BENDER AND R. J. THOMAS, *J. Am. Chem. Soc.*, **83** (1961) 4189.
48 G. H. HAGGIS, J. B. HASTED AND T. J. BUCHANAN, *J. Chem. Phys.*, **20** (1952) 1452.
49 R. E. ROBERTSON, *Progr. Phys. Org. Chem.*, **4** (1968) 213.
50 M. EIGEN, *Discussions Faraday Soc.*, **39** (1965) 1.
51 F. FRANKS, E. M. ARNETT, W. G. BENTRUDE, J. J. BURKE AND P. McC. DUGGLEBY, *J. Am. Chem. Soc.*, **85** (1963) 1350.
52 E. M. ARNETT, W. G. BENTRUDE, J. J. BURKE AND P. McC. DUGGLEBY, *J. Am. Chem. Soc.*, **87** (1965) 1541.
53 A. MOFFAT AND H. HUNT, *J. Am. Chem. Soc.*, **81** (1959) 2082.
54 J. KOSKIKALLIO, D. POULI AND E. WHALLEY, *Can. J. Chem.*, **37** (1959) 1360.
55 L. EBERSON, *Acta Chem. Scand.*, **18** (1964) 534.
56a C. G. SWAIN AND C. B. SCOTT, *J. Am. Chem. Soc.*, **75** (1953) 246.
56b C. G. SWAIN, R. CARDINAUD AND A. D. KETLEY, *J. Am. Chem. Soc.*, **77** (1955) 934.
57 H. K. HALL, *J. Am. Chem. Soc.*, **77** (1955) 5993.
58 W. P. JENCKS AND J. CARRIUOLO, *J. Biol. Chem.*, **234** (1959) 1280.
59 S. L. JOHNSON AND K. A. RUMON, *J. Am. Chem. Soc.*, **87** (1965) 4782.
60 J. B. HYNE, *J. Am. Chem. Soc.*, **82** (1960) 5129.
61 J. MARTIN AND R. E. ROBERTSON, *paper presented at Spring Meeting of the American Chemical Society, Pittsburgh, March 31st, 1966.*
62 B. E. CONWAY, *Can. J. Chem.*, **37** (1959) 613.
63 R. F. HUDSON AND B. SAVILLE, *J. Chem. Soc.*, (1955) 4114.
64 R. F. HUDSON AND I. STELZER, *Trans. Faraday Soc.*, **54** (1958) 213.
65 G. ASKNES, *Acta Chem. Scand.*, **14** (1960) 1475, 1526.
66 A. R. BUTLER AND V. GOLD, *J. Chem. Soc.*, (1961) 2305.
67 A. R. BUTLER AND V. GOLD, *Proc. Chem. Soc.*, (1960) 15.
68 A. R. BUTLER AND V. GOLD, *J. Chem. Soc.*, (1962) 2212.
69 C. A. BUNTON, N. A. FULLER AND S. G. PERRY, *Chem. Ind. (London)*, (1960) 1130.
70 T. C. BRUICE AND W. C. BRADBURY, *J. Am. Chem. Soc.*, **87** (1965) 4838.
71 S. L. JOHNSON, *Thesis*, Massachusetts Institute of Technology, 1959.
72 S. L. JOHNSON, H. M. GIRON AND G. L. TAN, (1966) unpublished work, cited in ref. 3.
73 R. WOLFENDEN AND W. P. JENCKS, *J. Am. Chem. Soc.*, **83** (1961) 4390.
74 S. L. JOHNSON AND G. L. TAN, (1966) unpublished work, cited in ref. 3.
75 A. R. BUTLER AND V. GOLD, *J. Chem. Soc.*, (1961) 4362.
76 B. M. ANDERSON, E. H. CORDES AND W. P. JENCKS, *J. Biol. Chem.*, **236** (1961) 455.
77 W. P. JENCKS AND J. CARRIUOLO, *J. Am. Chem. Soc.*, **83** (1968) 1743.
78 V. GOLD, *Trans. Faraday Soc.*, **44** (1948) 506.
79a J. F. NORRIS, E. V. FASCE AND C. J. STAUD, *J. Am. Chem. Soc.*, **57** (1935) 1415.
79b J. F. NORRIS AND D. V. GREGORY, *J. Am. Chem. Soc.*, **50** (1928) 1813.
80 E. W. CRUNDEN AND R. F. HUDSON, *J. Chem. Soc.*, (1956) 501.
81 V. GOLD, J. HILTON AND E. G. JEFFERSON, *J. Chem. Soc.*, (1954) 2756.
82 M. K. HALL AND C. H. LUECK, *J. Org. Chem.*, **28** (1963) 2818.
83 H. P. TREFFERS AND L. P. HAMMETT, *J. Am. Chem. Soc.*, **59** (1937) 1708.
84 M. L. BENDER AND M. C. CHEN, *J. Am. Chem. Soc.*, **85** (1963) 30, 37.
85 C. T. CHMIEL AND F. A. LONG, *J. Am. Chem. Soc.*, **78** (1956) 3326.
86 C. W. L. BEVAN AND R. F. HUDSON, *J. Chem. Soc.*, (1953) 2187.
87 D. P. N. SATCHELL, *J. Chem. Soc.*, (1963) 555, 558.
88 B. P. SUSZ AND D. CASSIMATIS, *Helv. Chim. Acta*, **44** (1961) 395.
89 G. A. OLAH, S. J. KUHN, W. S. TOLGEYSI AND E. B. BAKER, *J. Am. Chem. Soc.*, **84**, (1962) 2733.
90 R. F. Hudson and G. E. Moss, *J. Chem. Soc.*, (1962) 5157.

91 D. P. N. Satchell, *Quart. Rev.*, **17** (1963) 160.
92 A. Kivinen, *Acta Chem. Scand.*, **19** (1965) 845.
93 H. Minato, *Bull. Chem. Soc. Japan*, **37** (1964) 316.
94 J. M. Briody and D. P. N. Satchell, *J. Chem. Soc.*, (1965) 168.
95 A. S. Carson and H. A. Skinner, *J. Chem. Soc.*, (1949) 936; (1950) 656.
96 P. W. Allen and L. E. Sutton, *Trans. Faraday Soc.*, **47** (1951) 236.
97 C. H. Townes and B. P. Dailey, *J. Chem. Phys.*, **17** (1949) 782.
98 P. J. Bray, *J. Chem. Phys.*, **23** (1955) 703.
99 P. J. Bray, *J. Chem. Phys.*, **22** (1954) 1787.
100 I. Ugi and F. Beck, *Chem. Ber.*, **94** (1961) 1839.
101 R. Leimu, *Chem. Ber.*, **70** (1937) 1040.
102 E. J. Cairns and J. M. Prousnitz, *J. Chem. Phys.*, **32** (1960) 169.
103 E. Grunwald and S. Winstein, *J. Am. Chem. Soc.*, **70** (1948) 846.
104 C. G. Swain and E. T. Spalding, *J. Am. Chem. Soc.*, **82** (1960) 6104.
105 A. K. Coverdale and G. Kohnstam, *J. Chem. Soc.*, (1960) 3806.
106 C. A. Bunton, J. B. Ley, A. J. Rhind-Tutt and C. A. Vernon, *J. Chem. Soc.*, (1957) 2327.
107 D. P. N. Satchell, *J. Chem. Soc.*, (1963) 564.
108 D. P. N. Satchell, *J. Chem. Soc.*, (1964) 3724.
109 D. P. N. Satchell, *J. Chem. Soc.*, (1965) 168.
110 R. F. Hudson and J. E. Wardhill, *J. Chem. Soc.*, (1950) 1729.
111 F. Hibbert and D. P. N. Satchell, *J. Chem. Soc.*, (1967) 653.
112 F. Hibbert and D. P. N. Satchell, *J. Chem. Soc.*, (1967) 755.
113 G. E. K. Branch and A. C. Nixon, *J. Am. Chem. Soc.*, **58** (1936) 2499.
114 E. G. Williams and C. N. Hinshelwood, *J. Chem. Soc.*, (1934) 1079.
115 D. Weller, *Ph.D Thesis*, London, 1955.
116 G. Zimmerman and C. Yuan, *J. Am. Chem. Soc.*, **77** (1955) 332.
117 B. L. Archer and R. F. Hudson, *J. Chem. Soc.*, (1950) 3259.
118 H. K. Hall, *J. Amer. Chem. Soc.*, **77** (1955) 5993.
119 G. Berger and S. C. J. Olivier, *Rec. Trav. Chim.*, **46** (1927) 516, 861.
120 V. Gold and J. Hilton, *J. Chem. Soc.*, (1955) 838.
121 R. P. Bell and J. C. Clunie, *Proc. Roy. Soc. (London), Spr. A*, **212** (1952) 16.
122 S. C. J. Olivier and G. Berger, (a) *Rec. Trav. Chim.*, **45** (1926) 452; (b) **46** (1927) 516; (c) **46** (1927) 609.
123 H. Bohme, *Chem. Ber.*, **74** (1941) 248.
124 R. Leiumu and P. Salomaa, *Acta Chem. Scand.*, **1** (1947) 353.
125 J. F. Norris and V. W. Ware, *J. Am. Chem. Soc.*, **61** (1939) 1418.
126 D. A. Brown and R. F. Hudson, *J. Chem. Soc.* (1953) 883.
127 C. A. Bunton and T. A. Lewis, *Chem. Ind (London)*, (1956) 180.
128 A. Kivinen, *Ann. Acad. Sci. Fennicae, Ser. A* (1961) 108.
129 A. A. Ashdown, *J. Am. Chem. Soc.* **52** (1930) 268.
130 G. Gueskens, M. Planchon, J. Nasielski and R. Martin, *Helv. Chim. Acta*, **42** (1959) 522.
131 H. Bohme and W. Schürhoff, *Chem. Ber.*, **28** (1951) 84.
132 E. R. A. Peeling, *Ph.D. Thesis*, London, 1946.
133 D. A. Brown and R. F. Hudson, *J. Chem. Soc.*, (1963) 3352.
134 J. G. Kirkwood, *J. Chem. Phys.*, **2** (1934) 351.
135 V. Gold, J. Hilton and E. G. Jefferson, *J. Chem. Soc.*, (1954) 2756.
136 B. L. Archer, R. F. Hudson and J. E. Wardhill, *J. Chem. Soc.*, **182** (1953) 888.
137 E. R. A. Peeling, *J. Chem. Soc.*, (1959) 2307.
138 A. Kivinen, *Suomen Kemistilehti, B*, **36** (1963) 163.
139 A. Kivinen, *Suomen Kemistilehti, B*, **40** (1967) 19.
140 H. Heydtmann and H. Stieger, *Chem. Ber.*, **70** (1966) 1095.
141 M. W. Perrin, *Trans. Faraday Soc.*, **34** (1938) 144.
142 S. D. Hamman and D. R. Teplitzky, *Discussions Faraday Soc.*, **22** (1965) 70.

143 J. B. HYNE, H. S. GOLENKIN AND W. G. LAIDLAW, *J. Am. Chem. Soc.*, **88** (1966) 2104.
144 R. F. HUDSON AND G. MOSS, *J. Chem. Soc.*, (1964) 2982.
145 E. GRUNWALD AND A. F. BUTLER, *J. Am. Chem. Soc.* **82** (1960) 5647.
146 J. CASON AND K. W. KRAUS, *J. Org. Chem.*, **26** (1961) 2624.
147 M. S. NEWMAN in *Steric Effects in Organic Chemistry*, M. S. Newman (Ed.), Wiley, New York, 1956, p. 227.
148 S. G. COHEN AND A. SCHMEIDER, *J. Am. Chem. Soc.*, **63** (1941) 3382.
149 See ref. 147, p. 226.
150 P. SALOMAA, *Ann. Univ. Turku. Ser. A.*, **14** (1953) No. 1.
151 R. F. HUDSON AND M. GREEN, *J. Chem. Soc.*, (1962) 1055, 1076.
152 E. W. GRUNDEN AND R. F. HUDSON, *J. Chem. Soc.*, (1961) 3748.
153 A. QUEEN. *Can. J. Chem.*, **45** (1967) 1619.
154 M. GREEN AND R. F. HUDSON, *J. Chem. Soc.*, (1962) 1076.
155 D. M. KEVILL AND G. H. JOHNSON, *Chem. Commun.* (1966) 235; *J. Am. Chem. Soc.*, **87** (1965) 928.
156 R. BOSCHAN, *J. Am. Chem. Soc.*, **81** (1959) 3341.
157 A. QUEEN, Private Communication.
158 R. F. HUDSON, R. J. G. SEARLE AND A. MANCUSO, *Helv. Chim. Acta*, **50** (1967) 997.
159 S. F. NORRIS AND E. C. HAINES, *J. Am. Chem. Soc.*, **57** (1935) 1425.
160 E. M. KOSOWER, *An Introduction to Physical Organic Chemistry*, Wiley, New York, 1968, (a) p. 301, (b) p. 303.
161 K. DIMROTH, C. REICHARDT, T. SIEPMANN AND F. BOHLMANN, *Annalen*, **661** (1963) 1.
162 J. F. NORRIS AND H. H. YOUNG, *J. Am. Chem. Soc.*, **57** (1935) 1420.
163 E. M. KOSOWER, *J. Am. Chem. Soc.*, **80** (1958) 3267.
164 E. M. KOSOWER, *J. Am. Chem. Soc.*, **80** (1958) 3253.
165 K. WIBERG, *Physical Organic Chemistry*, Wiley, New York, 1964.
166 V. GOLD AND J. HILTON, *J. Chem. Soc.*, (1955) 3303.
167 K. J. LAIDLER, *Suomen Kemistilehti, A*, **33** (1960) 44.
168 A. KIVINEN, unpublished results.
169 R. F. HUDSON, *Z. Electrochem.*, **68** (1954) 215.
170 E. TOMMILA, M. TIILIKAINEN AND A. VOIPPO (a) *Ann. Acad. Sci. Fennicae, Ser. A II*, (1955) 65; (b) *Acta Chem. Scand.*, **9** (1955) 975.
171 E. TOMMILA AND A. HELLA, *Ann. Acad. Sci. Fennicae, Ser. AII* (1954) 53.
172a R. F. HUDSON AND G. W. LOVEDAY, *J. Chem. Soc. B*, (1966) 766.
172b R. F. HUDSON, G. W. LOVEDAY, S. FLISZAR AND G. SALVADORI, *J. Chem. Soc. B*, (1966) 769.
173 R. F. HUDSON, *Chem. Ber.*, **68** (1964) 215.
174 H. KEMPTE AND R. MECKE, *Z. Physik. Chem. Ser. B*, **46** (1940) 229.
175 W. C. COBURN AND E. GRUNWALD, *J. Am. Chem. Soc.*, **80** (1958) 1318.
176 W. OSTWALD, *J. Prakt. Chem.*, **27** (1883) 1.
177 (a) E. E. REID, *Am. Chem. J.*, **21** (1899) 284; (b) **24** (1900) 397; (c) **45** (1911) 327.
178 A. BRUYLANTS, *Bull. Classe. Sci. Acad. Roy. Belg.*, (5) **27** (1941) 189.
179 E. CALVERT, *J. Chim. Phys.*, **30** (1933) 144.
180 J. C. CROCKER, *J. Chem. Soc.*, (1907) 91, 593.
181 S. KILPI, *Z. Physik. Chem. (Frankfurt)*, **80** (2) (1912) 165.
182 H. VON EULER AND E. RUDBERG, *Z. Anorg. Allgem. Chem.*, **127** (3) (1923) 244.
183 N. VON PESKOFF, J. MEYER, *Z. Physik. Chem. (Frankfurt)*, **82** (1913) 129.
184 A. BENRATH, *Z. Anorg. Chem.*, **151** (1926) 53.
185 H. VON EULER AND A. E. ÖLANDER, *Z. Physik. Chem. (Leipzig)*, **131** (1927) 107.
186 T. W. J. TAYLOR, *J. Chem. Soc.*, (1930) 2741.
187 B. S. RABINOVITCH AND C. A. WINKLER, *Can. J. Res.*, **20** B (1942) 73.
188 V. K. KRIEBLE AND K. A. HOLST, *J. Am. Chem. Soc.*, **60** (1938) 2976.
189 J. T. EDWARD AND S. C. R. MEACOCK, *J. Chem. Soc.*, (1957) 2000.
190 J. T. EDWARD, H. P. HUTCHINSON AND S. C. R. MEACOCK, *J. Chem. Soc.*, (1955) 2520.
191 V. K. KRIEBLE AND C. I. NOLL, *J. Am. Chem. Soc.*, **61** (1939) 560.

192 A. R. Katrizky, A. J. Waring and K. Yates, *Tetrahedron*, **19** (1963) 465.
193 J. T. Edward, H. S. Chang, K. Yates and R. Stewart, *Can. J. Chem.*, **38** (1960) 1518.
194 A. R. Goldfarb, A. Mele and N. Gustein, *J. Am. Chem. Soc.*, **77** (1955) 6194.
195 R. B. Horner and R. B. Moodie, *J. Chem. Soc.*, (1963) 4377.
196 G. Fraenkel and C. Niemann, *Proc. Nat. Acad. Sci. US*, **44** (1958) 688.
197 R. Gompper and P. Altreuther, *Z. Anal. Chem.*, **170** (1959) 205.
198 R. Stewart and L. J. Muenster, *Can. J. Chem.*, **39** (1961) 401.
199 M. A. Paul and F. A. Long, *Chem. Rev.*, **57** (1957) 1.
200 F. A. Long and M. A. Paul, *Chem. Rev.*, **57** (1957) 935.
201 J. F. Bunnett, *J. Am. Chem. Soc.*, **83** (1961) 4956, 4968, 4973.
202 R. B. Martin, *J. Am. Chem. Soc.*, **84** (1962) 4130.
203 V. K. Krieble and K. A. Holst, *J. Am. Chem. Soc.*, **60** (1938) 2971.
204 A. Hantzsch, *Chem. Ber.*, **64** (1931) 661.
205 T. W. J. Taylor and W. Baker, *Organic Chemistry of Nitrogen*, N. V. Sidgwick (Ed.), Clarendon Press, Oxford, 1945, p. 144.
206 J. N. Brönsted, *Z. Physik. Chem. (Leipzig)*, **102** (1922) 169.
207 R. P. Bell, A. L. Dowding and J. A. Noble, *J. Chem. Soc.*, (1955) 3106.
208 L. P. Hammett and A. J. Deyrup, *J. Am. Chem. Soc.*, **54** (1931) 2721.
209 L. P. Hammett and M. A. Paul, *J. Am. Chem. Soc.*, **56** (1934) 827.
210 N. F. Hall, *J. Am. Chem. Soc.*, **52** (1930) 5115.
211 H. Lemaire and H. J. Lucas, *J. Am. Chem. Soc.*, **73** (1951) 5198.
212 J. A. Leisten, *J. Chem. Soc.*, (1959) 765.
213 C. K. Ingold, *Structure and Mechanism in Organic Chemistry*, Bell, London, 1953, p. 785.
214 L. Zucker and L. P. Hammett, *J. Am. Chem. Soc.*, **61** (1939) 2779.
215 R. P. Bell and B. Lukianeko, *J. Chem. Soc.*, (1957) 1686.
216 R. B. Moodie, P. D. Wale and T. J. Whaite, *J. Chem. Soc.*, (1963) 4273.
217 C. A. Bunton, C. O'Connor and T. A. Turney, *Chem. Ind. (London)*, (1967) 1835.
218 K. Yates, J. B. Stevens, *Can. J. Chem.*, **43** (1965) 529.
219 K. Yates, J. C. Riordan, *Can. J. Chem.*, **43** (1965) 2328.
220 J. F. Bunnett and F. P. Olsen, *Can. J. Chem.*, **44** (1966) 1899, 1917.
221 H. Zimmerman and J. Randolph, *Angew. Chem. Intern. Ed. Engl.*, **4** (1965) 40.
222 K. Yates, J. B. Stevens and A. R. Katrizky, *Can. J. Chem.*, **42** (1957) 64.
223 M. L. Bender, R. D. Ginger and K. C. Kemp, *J. Am. Chem. Soc.*, **76** (1954) 3350.
224 C. A. Bunton, B. Nayak and C. O'Connor, *J. Org. Chem.* **33** (1961) 572, 1968.
225 S. J. Farber and J. A. Brieux, *Chem. Ind. (London)*, (1966) 599.
226 H. H. G. Jellinek and A. Gordon, *J. Phys. Colloid. Chem.*, **53** (1949) 996.
227 M. L. Bender, R. Ginger and J. P. Unik, *J. Am. Chem. Soc.*, **80** (1958) 1044.
228 C. A. Bunton and D. N. Spatcher, *J. Chem. Soc.*, (1956) 1079.
229 I. Meloche and K. J. Laidler, *J. Am. Chem. Soc.*, **73** (1951) 1712.
230 J. Cason, C. Gastaldo, D. L. Glusker, J. Allinger and L. B. Ash, *J. Org. Chem.*, **18** (1953) 1129.
231 J. Cason and H. Wolfhagen, *J. Org. Chem.*, **14** (1949) 155.
232 A. W. De Roo and A. Bruylants, *Bull. Soc. Chim. Belges*, **63** (1954) 140.
233 M. Willems and A. Bruylants, *Bull. Soc. Chim. Belges*, **60** (1951) 191.
234 B. M. Wepster and P. E. Verkade, *Rec. Trav. Chim.*, **67** (1948) 411, 425; **68** (1949) 77; **69** (1950) 1393; H. J. Biekart, H. B. Dessens, P. E. Verkade and B. M. Wepster, *Rec. Trav. Chim.*, **71** (1952) 1245.
235 C. G. Swain and E. R. Thornton, *J. Am. Chem. Soc.*, **84** (1962) 817.
236 C. G. Swain, D. A. Kuhn and R. L. Schowen, *J. Am. Chem. Soc.*, **87** (1965) 1553.
237 C. G. Swain and J. Worosz, *Tetrahedron Letters*, (1965) 3199.
238 M. L. Wolfrom, R. B. Bennett and J. D. Cram, *J. Am. Chem. Soc.*, **80** (1958) 944.
239 H. Zahn and L. Zurn, *Annalen*, **613** (1958) 76.
240 L. Zurn, *Annalen*, **613** (1958) 76.
241 T. C. Bruice and F. Hans-Marquardt, *J. Am. Chem. Soc.*, **84** (1962) 365.
242 B. A. Cunningham and G. L. Schmir, *J. Am. Chem. Soc.*, **89** (1967) 917.

243 B. A. Cunningham and G. L. Schmir, *J. Am. Chem. Soc.*, **88** (1966) 551.
244 T. C. Bruice and R. Lapinski, *J. Am. Chem. Soc.*, **80** (1958) 2265.
245 A. J. Hubert, R. Buyle and B. Margitay, *Helv. Chim. Acta*, **46** (1963) 1429.
246 M. Brenner and W. Hoffer, *Helv. Chim. Acta*, **44** (1961) 1794.
247 P. B. Hamilton, *J. Biol. Chem.*, **158** (1945) 375.
248 B. Glutz and H. Zollinger, *Angew. Chem. Intern. Ed. Engl.*, **4** (1965) 440.
249 T. C. Bruice and D. W. Tanner, *J. Org. Chem.*, **30** (1965) 1668.
250 S. J. Leach and H. Lindley, *Trans. Faraday Soc.*, **49** (1953) 915.
251 M. L. Bender, Y. L. Chow and F. Chloupek, *J. Am. Chem. Soc.*, **80** (1958) 5380.
252 M. L. Bender, F. Chloupek and M. C. Nevea, *J. Am. Chem. Soc.*, **80** (1958) 5384.
253 L. Eberson, *Acta Chem. Scand.*, **16** (1962) 2245.
254 B. Zerner and M. L. Bender, *J. Am. Chem. Soc.*, **83** (1961) 2267.
255 H. Morawetz and I. Oreskes, *J. Am. Chem. Soc.*, **80** (1958) 2591.
256 H. Morawetz and J. Shafer, *J. Am. Chem. Soc.*, **84** (1962) 3783.
257 P. D. Hoagland and S. W. Fox, *J. Am. Chem. Soc.*, **89** (1967) 1389.
258 M. Vigneron, P. Crooy, F. Kezdy and A. Bruylants, *Bull. Soc. Chim. Belges.*, **69** (1960) 616.
259 P. Crooy, *Ph.D. Thesis*, Catholic University, Louvain, 1961.
260 T. Cohen and J. Lipowitz, *J. Am. Chem. Soc.*, **83** (1961) 4866.
261 M. Gordon, *Ph.D. Thesis*, Manchester University, 1950.
262 R. Huisgen and H. Ott, *Tetrahedron*, **6** (1959) 253.
263 V. Gold and S. L. Bafna, *J. Chem. Soc.*, (1953) 1406.
264 V. Gold and A. R. Butler, *J. Chem. Soc.*, (1961) 2305.
265 S. L. Johnson, *J. Phys. Chem.*, **67** (1963) 495.
266 C. A. Bunton, N. A. Fuller, S. G. Perry and V. J. Shiner Jr., *Tetrahedron Letters*, **14** (1961) 458.
267 V. Gold and A. R. Butler, *J. Chem. Soc.*, (1961) 4362.
268 C. A. Bunton and V. J. Shiner Jr., *J. Am. Chem. Soc.*, **83** (1961) 3207.
269 E. R. Stadtmann, in *The Mechanism of Enzymology*, W. D. McElroy and B. Gloss (Eds.), John Hopkins Press, Baltimore, 1954, p. 581.
270 T. C. Bruice, in *Methods in Enzymology*, S. P. Colowick and N. O. Kaplan (Eds.), Academic Press, New York, 1963, Vol. VI, p. 606.
271 H. A. Staab, *Angew. Chem. Intern. Ed. Engl.*, **1** (1962) 351.
272 M. Bergman and L. Zervas, *Z. Physiol. Chem.*, **175** (1928) 145.
273 H. A. Staab, *Chem. Ber.*, **90** (1957) 1320.
274 O. Gerngross, *Chem. Ber.*, **46** (1913) 1908.
275 W. P. Jencks and J. Carriuolio, *J. Biol. Chem.*, **234** (1959) 1272.
276 C. A. Bunton, *J. Chem. Soc.*, (1963) 6045.
277 J. Herzoy, *Chem. Ber.*, **40** (1907) 1831.
278 S. L. Johnson and K. A. Rumon, *J. Phys. Chem.*, **68** (1964) 3149.
279 O. Reitz, *Z. Electrochem.*, **44** (1938) 693.
280 S. Widequist, *Arkiv. Kemi*, **4** (1952) 429.
281 K. Myrback, *Ann. Rev. Biochem.*, **8** (1934) 59.
282 W. Ostwald, *J. Prakt. Chem.*, **135** (1883) 1.
283 H. Lunden, *Medd. Kgb. Vetensk. Nobelinst.*, **2** (1911) 1.
284 P. E. Verkade, *Rec. Trav. Chim.*, **35** (1915) 79, 299.
285 A. Skrabal, *Monatsh.*, **43** (1922) 493.
286 R. Szabo, *Z. Physik. Chem.* (*Leipzig*), **122** (1926) 405.
287 K. J. P. Orton and M. Jones, *J. Chem. Soc.*, **101** (1912) 1708.
288 A. Benrath, *Z. Physik. Chem.* (*Leipzig*), **67** (1909) 501.
289 M. Kilpatrick, *J. Am. Chem. Soc.*, **50** (1928) 2891.
290 M. Kilpatrick, *J. Am. Chem. Soc.*, **52** (1930) 1410.
291 M. Kilpatrick and M. L. Kilpatrick, *J. Am. Chem. Soc.*, **52** (1930) 1418.
292 V. Gold and J. Hilton, *J. Chem. Soc.*, (1955) 843.
293 D. P. N. Satchell, *Chem. Ind.* (*London*), (1958) 1442.

294 C. A. Bunton and S. G. Perry, *J. Chem. Soc.*, (1960) 3071.
295 L. Zucker and L. P. Hammett, *J. Am. Chem. Soc.*, **61** (1939) 2785, 2791.
296 E. B. Lees and B. Saville, *J. Chem. Soc.*, (1962) 2262.
297 C. A. Bunton and J. H. Fendler, *J. Org. Chem.*, **32** (1967) 1547.
298 A. R. Butler and V. Gold, *J. Chem. Soc.*, (1961) 2305.
299 A. R. Butler and V. Gold, *J. Chem. Soc.*, (1962) 976.
300 S. L. Johnson, *J. Am. Chem. Soc.*, **84** (1962) 1729.
301 J. Koskikallio, *Acta Chem. Scand.*, **17** (1963) 1417.
302 J. Hipkin and D. P. N. Satchell, *J. Chem. Soc.*, (1965) 1057.
303 A. Kivinen, *Suomen Kemistilehti, B*, **38** (1965) 159.

INDEX

B

C

D

F

Q

R

U